# Introduction to 3D Spatial Visualization: An Active Approach

Workbook by: Sheryl A. Sorby
Software by: Anne F. Wysocki
Based on work by: Beverly Baartmans and Shery Sorby

THOMSON
DELMAR LEARNING™

Austalia Canada Mexico Singapore Spain United Kingdom United States

THOMSON

DELMAR LEARNING

Introduction to 3D Spatial Visualization: An Active Approach
by Sheryl Sorby
Software by Anne F. Wysocki

**Business Unit Director:**
Alar Elken

**Executive Editor:**
Sandy Clark

**Acquisitions Editor:**
James Devoe

**Development Editor:**
John Fisher

**Executive Marketing Manager:**
Maura Theriault

**Channel Manager:**
Mary Johnson

**Marketing Coordinator:**
Sarena Douglass

**Executive Production Manager:**
Mary Ellen Black

**Production Manager:**
Larry Main

**Production Editor:**
Stacy Masucci

**Technology Project Manager:**
David Porush

**Technology Project Specialist:**
Kevin Smith

**Editorial Assistant:**
Mary Ellen Martino



Printed in the United States of America
1 2 3 4 5 XX 06 05 04 03 02

For more information contact Delmar Learning, Executive Woods 5 Maxwell Drive, PO Box 8007, Clifton Park, NY 12065-8007
Or find us on the World Wide Web at: http://www.delmar.com



Library of Congress Catalog Card Number: 1-4018-1389-5

# Preface

The ability to visualize in three dimensions has been shown to be an important skill for people who intend to study in scientific and technical fields. Well-developed spatial skills have been linked to success in engineering, computer science, chemistry, medicine, mathematics, and architecture to name just a few. Design is central to engineering and well-developed spatial skills have been shown to be critical to a person's ability to develop creative design solutions to problems. Well-developed spatial skills have also been linked to a person's ability to interact with a computer in performing database manipulations and to a person's ability to understand various aspects of structural chemistry. Doctors, who must learn to use modern-day laparoscopy tools, require well-developed 3-D spatial skills. Architects must often visualize how a new structure will look as well as how it interacts with its surroundings when designing a new building.

In educational psychology research, the distinction is often made between "spatial ability" and "spatial skills." The difference between the two is described briefly in the following. Spatial ability is defined as the innate ability to visualize that a person has before any formal training has occurred, i.e., a person is born with ability. However, spatial skills are learned or are acquired through training. As with any other type of skill (writing, mathematics, etc.) some people may have a higher degree of innate ability than others, however, most people can eventually acquire the skill through patience and practice. The materials in this text will assist you in developing your 3-D spatial visualization skills.

These materials contain nine separate modules to help you develop your 3-D spatial visualization skills. For each module, there is a software as well as a workbook component. To maximize your skill development, we suggest that you first work through the appropriate software module on the CD-ROM. After you complete the software module, work through the pages for that module in the workbook. Because sketching with pencil and paper have been shown to be particularly helpful in the development of 3-D spatial skills, you will find considerable benefit in completing the workbook pages, many of which require hand sketching. You should probably work through the modules in the order they are presented on the CD-ROM and in this workbook, but you can do them in any order if you like. One exception to this is that you should complete and understand Module 4 (Rotation of Objects about a Single Axis) before you attempt Module 5 (Rotation of Objects about Two or More Axes) since these two modules are meant to be completed sequentially.

The multimedia software found on the CD-ROM works on either a PC or a MacIntosh platform and requires no additional software to run.

Good luck and have fun!

To begin using the software from a PC:
1) Insert the CD-ROM into the CD drive of your PC.
2) Click *Start* and then *Run* from the main Windows screen.
3) In the dialogue box that appears, click *Browse* to locate your CD drive (typically Z:).
4) Double-click the *PC* folder, double-click the filename "0menu.exe," then click *OK*.
5) Wait momentarily for the software to load.
6) Click the name of the module you would like to access.

To begin using the software from a MacIntosh computer:
1) Insert the CD-ROM into the CD drive of your computer. You should see the CD-ROM icon on your desktop.
2) Double-click the CD-ROM icon to open the CD-ROM folder.
3) Double-click the "mac" folder.
4) Double-click the "0menu.exe" file. This will open a menu, from which you can choose which modules to explore.

Thanks especially to Michigan Tech students David W. Shafer and Robert D. Sims III for their invaluable assistance in creating and editing the figures for this workbook. Thanks also to Robert E. Landsperger (rel) for his technical support in the preparation of this work.

You guys are the best!

This material is based upon work supported by the National Science Foundation under Grant No. DUE-9752660. Any opinions, findings, and conclusions or recommendations expressed in this material are those of the authors and do not necessarily reflect the views of the National Science Foundation.

Table of Contents

# Isometric Drawings & Coded Plans

Isometric views are useful for showing a 3-dimensional object on a 2-dimensional sheet of paper. An isometric view is defined as if you were looking down a diagonal of a cube on the object:

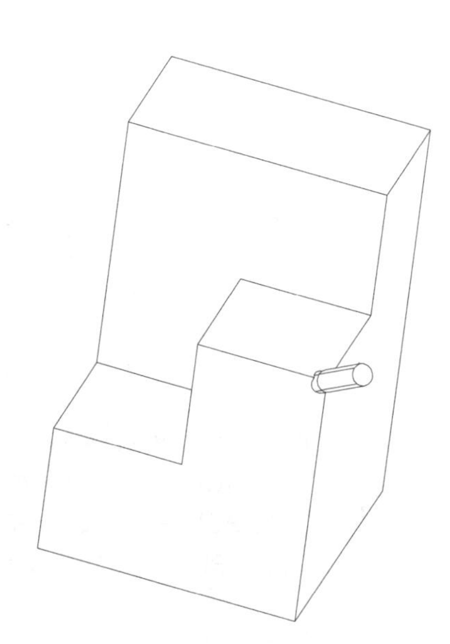

3-D Object

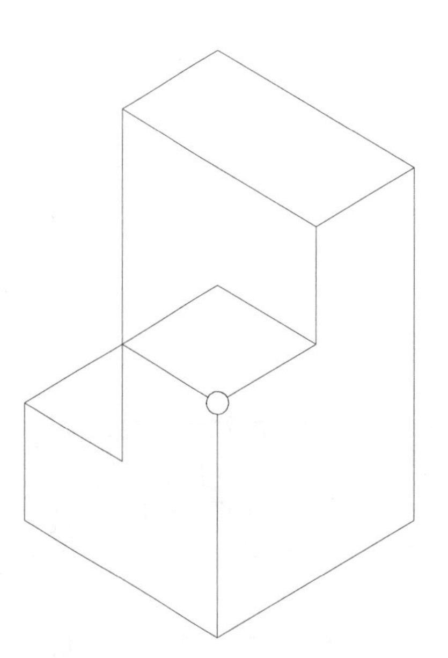

Isometric View of Object

Isometric Dot Paper is used as an aid in making isometric sketches. It consists of a grid of dots that are arranged equidistant from one another. The lines connecting the dots meet at an angle of 120 degrees with respect to one another. Isometric Grid Paper is similar to Isometric Dot Paper except that the dots are connected to form a grid on the page.

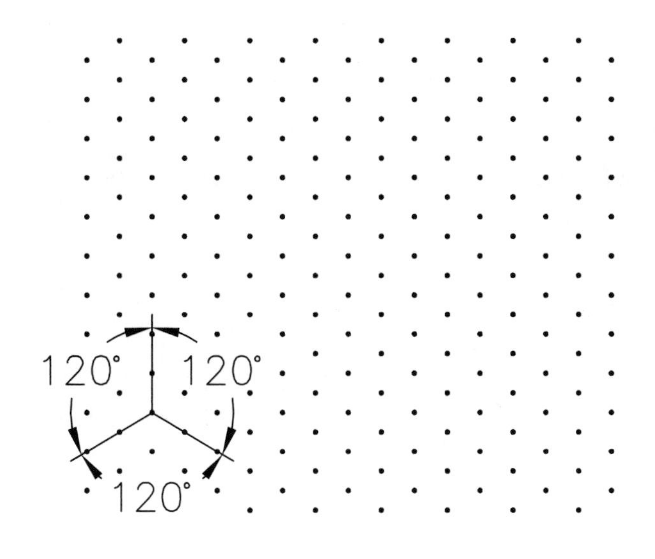


Isometric Dot Paper

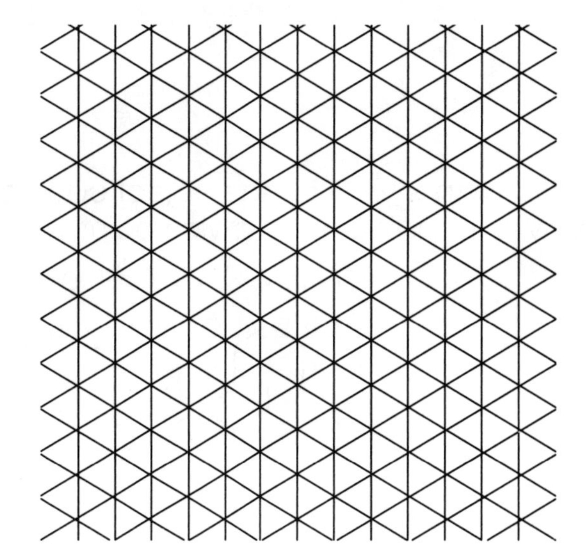

Isometric Grid Paper

When making an isometric drawing of an object created from blocks, do not show each individual cube on the object. Show only the visible surfaces and edges. An edge exists where two surfaces intersect.

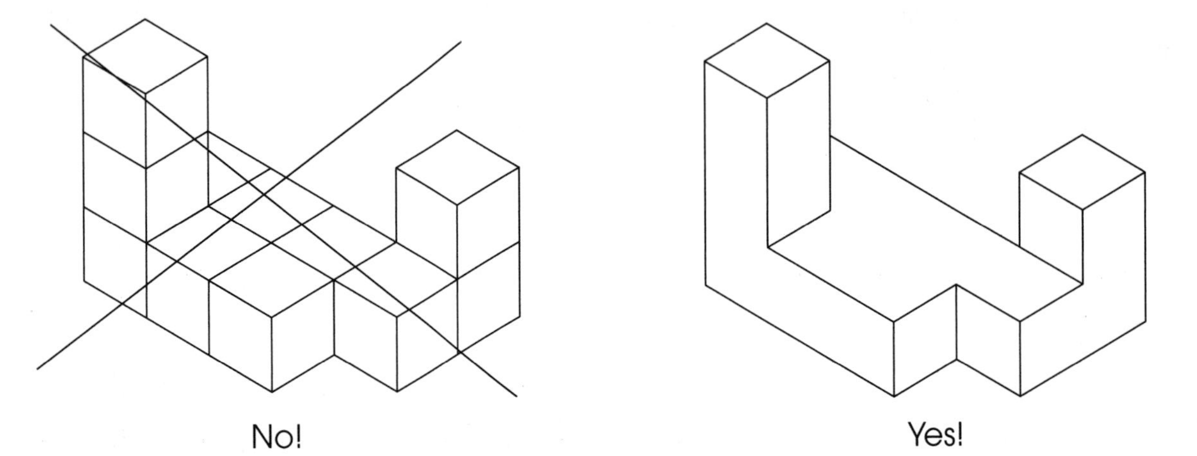


Coded Plans are used to represent a 3-dimensional object on a 2-dimensional sheet of paper. Each number on the coded plan represents the height of the stack of cubes at that location. Coded Plans are built up from the page to the heights specified. Corners on the Coded Plans are marked to indicate the viewpoint of the observer with respect to the object.

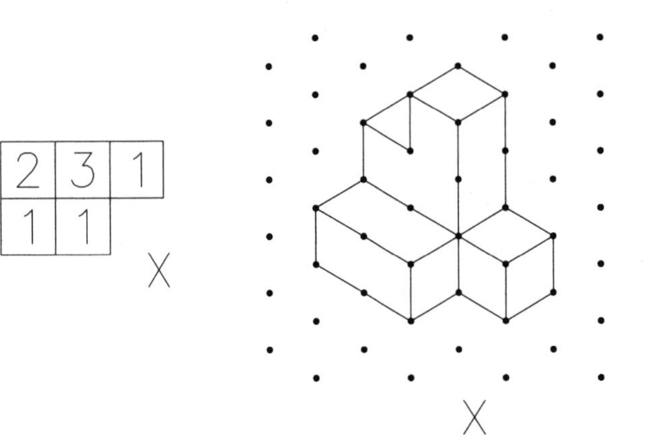


Different corner views of the object can be drawn. Notice that the object appears differently depending on your viewpoint.

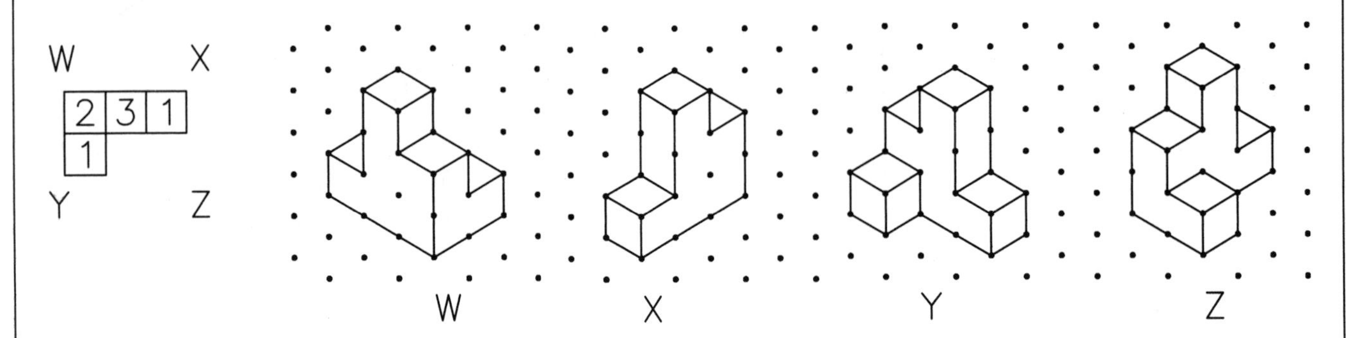

When sketching a corner view of an object, sketch one surface at a time until the view is complete.

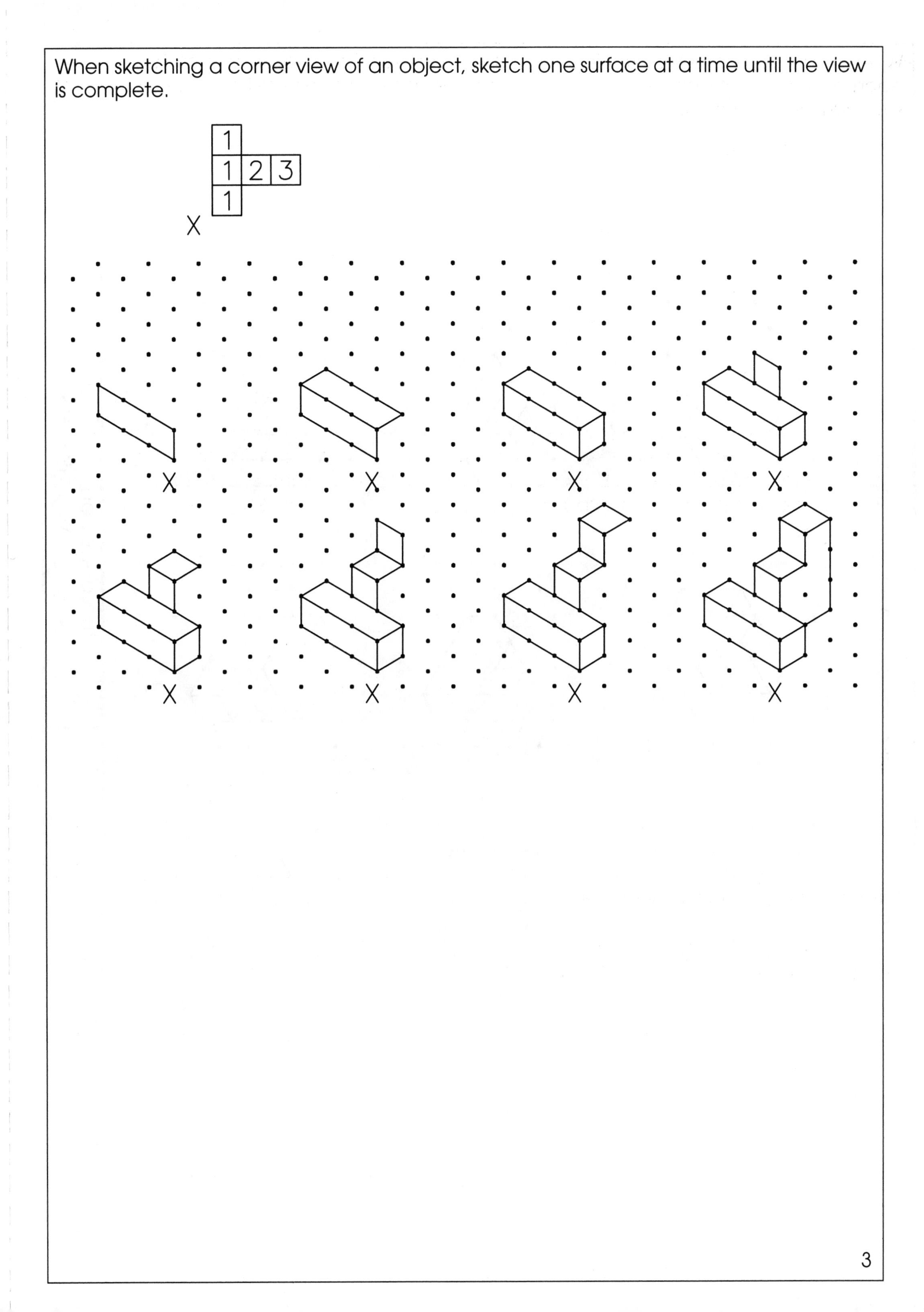

Circle the letter beneath the isometric sketch of the object that corresponds to the coded plan shown on the left.

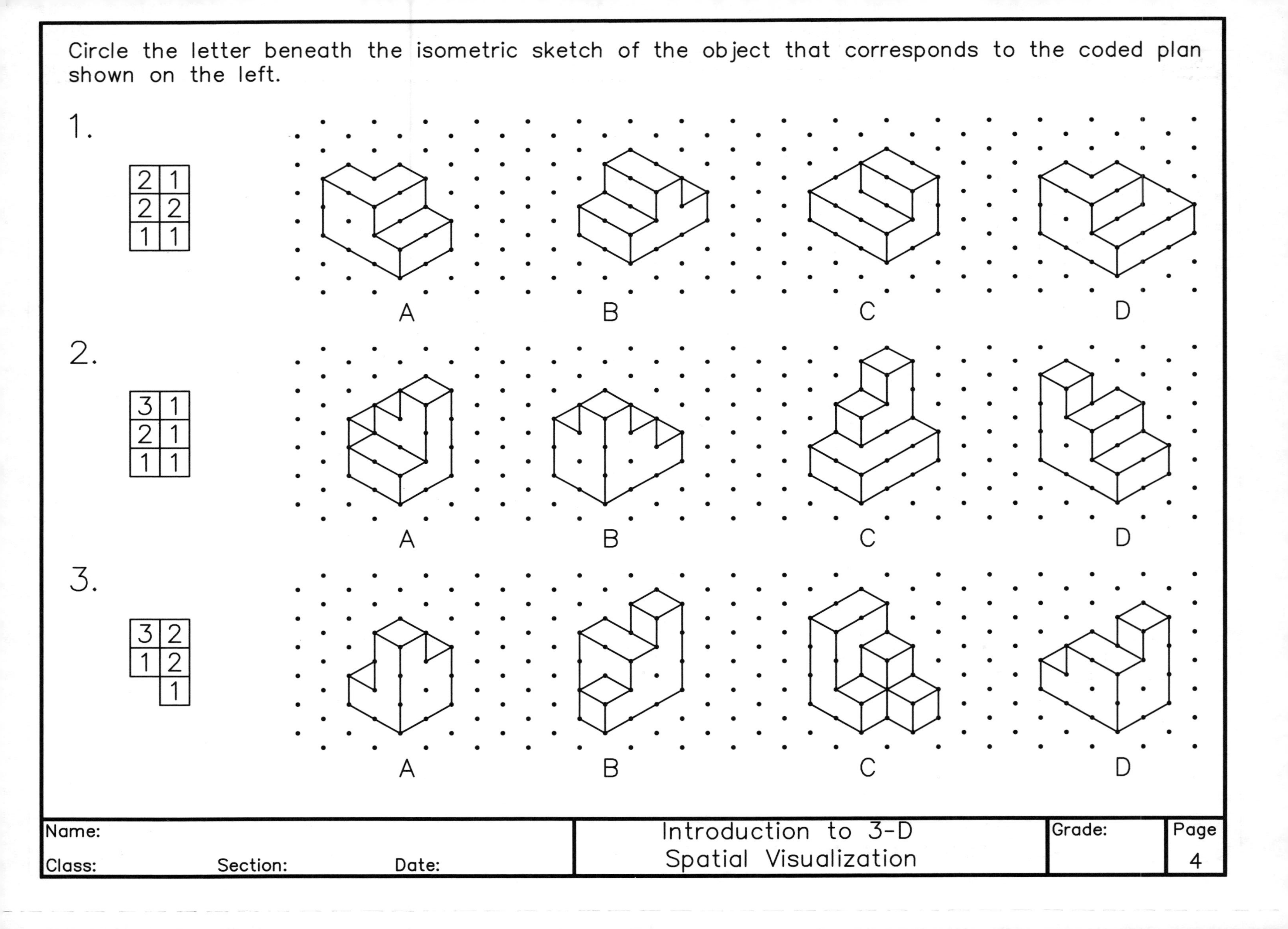

Circle the letter beneath the isometric sketch of the object that corresponds to the coded plan shown on the left.

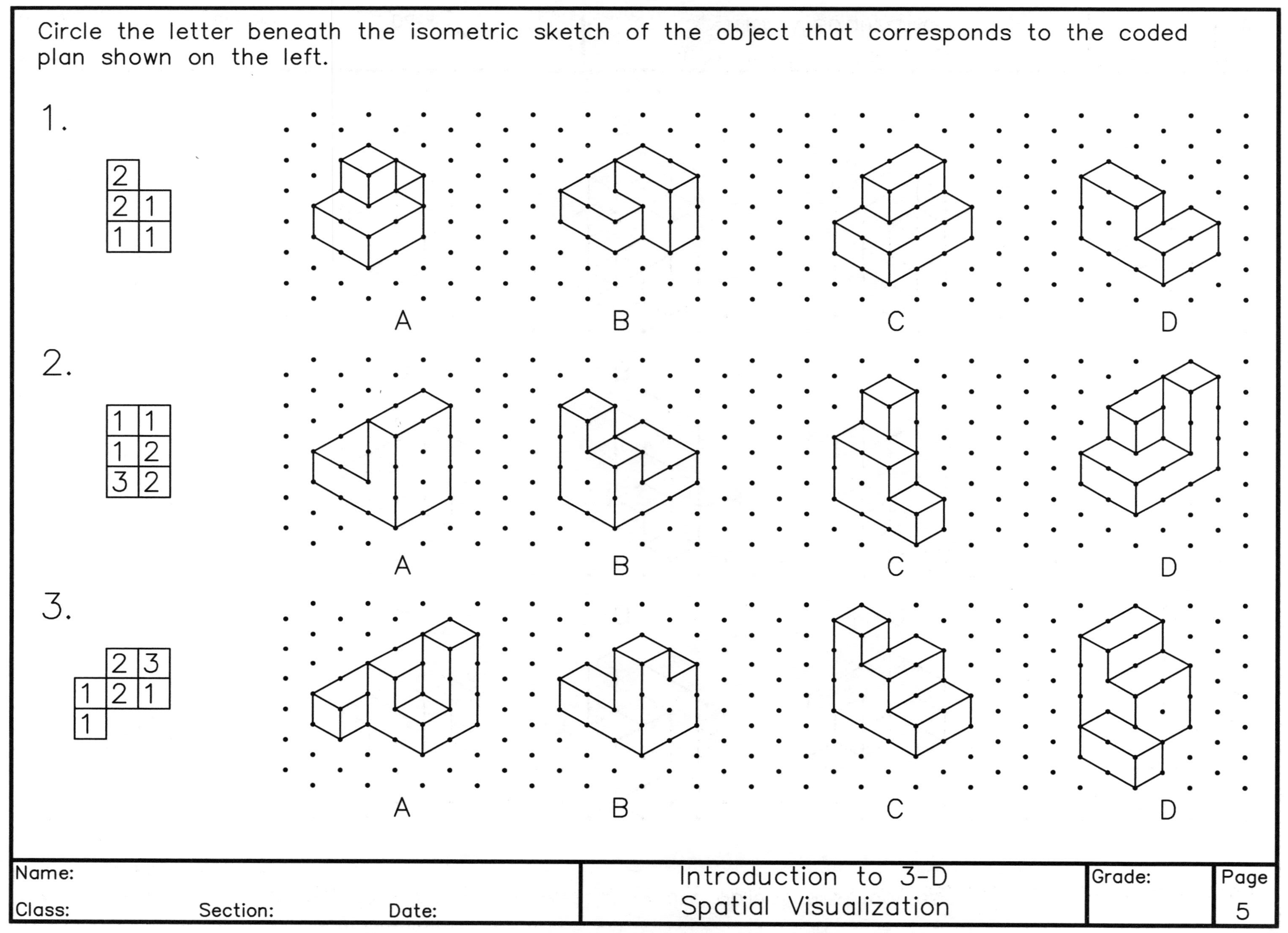

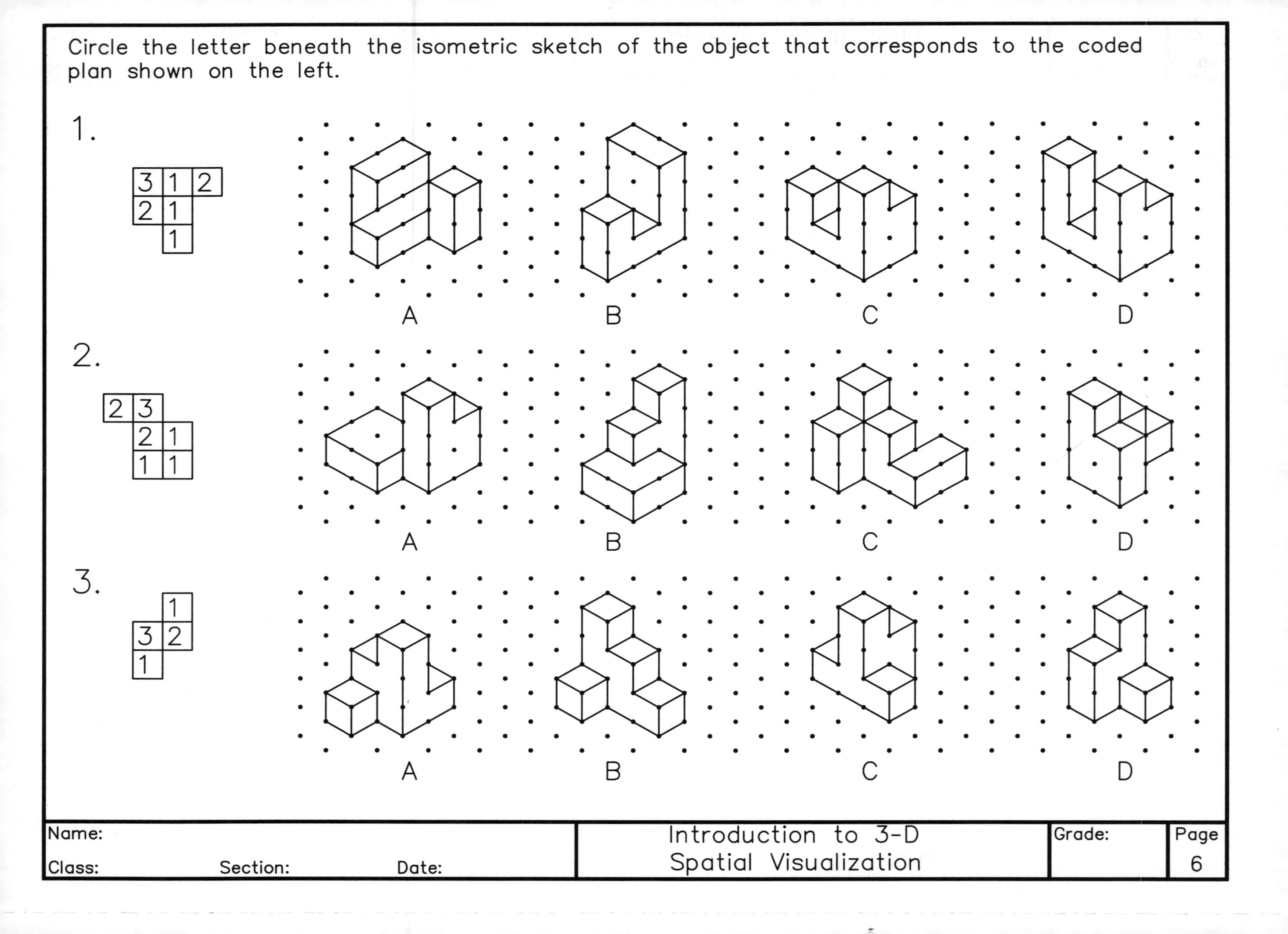
Circle the letter beneath the isometric sketch of the object that corresponds to the coded plan shown on the left.
1.
3 1 2
2 1
1
A
B
C
D
2.
2 3
2 1
1 1
A
B
C
D
3.
1
3 2
1
A
B
C
D
Name:
Class:
Section:
Date:
Introduction to 3-D Spatial Visualization
Grade:
Page
6

Complete the coded plan for the object shown in an isometric sketch on the right.

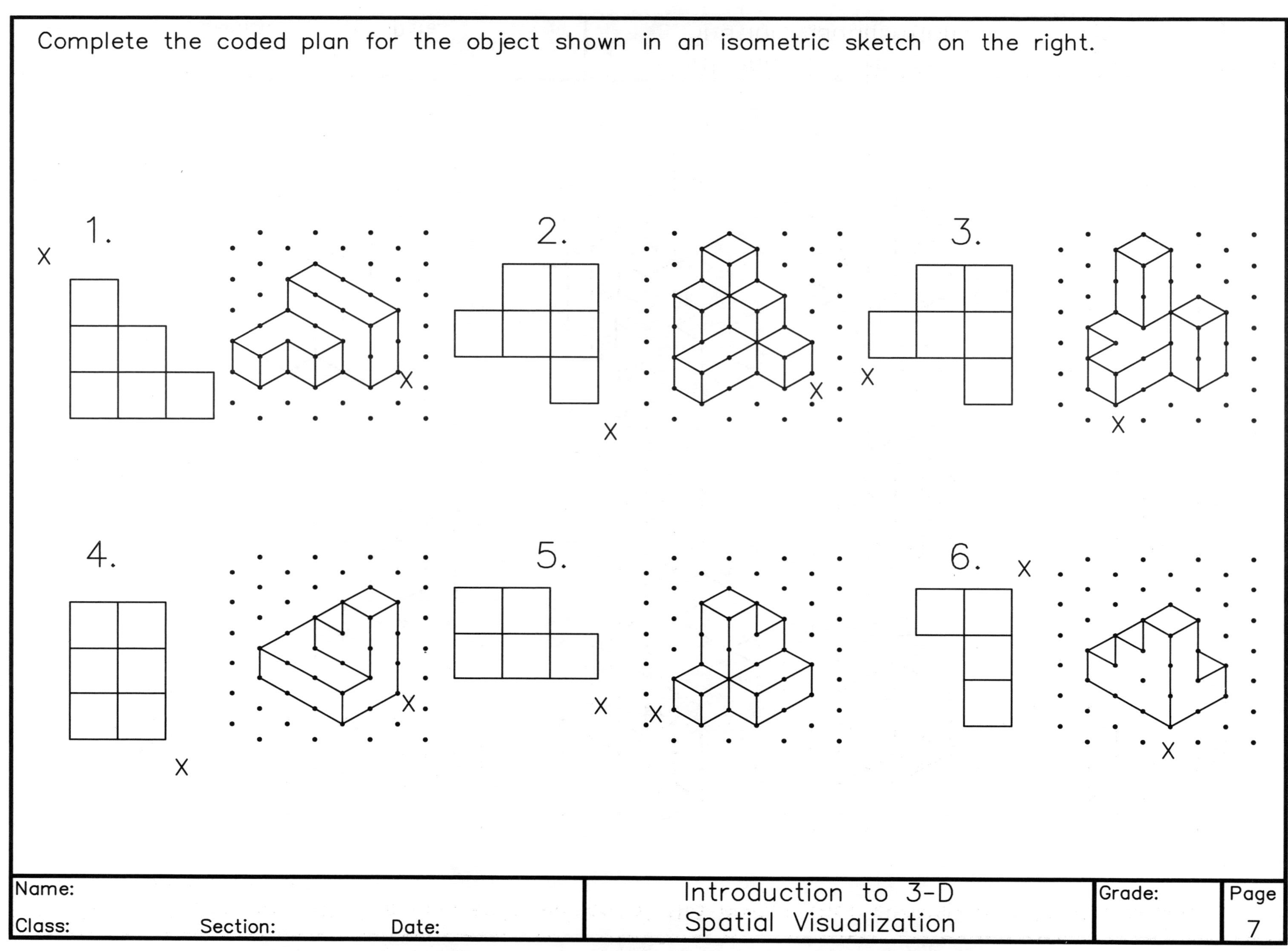

Complete the coded plan for the object shown in an isometric sketch on the right.

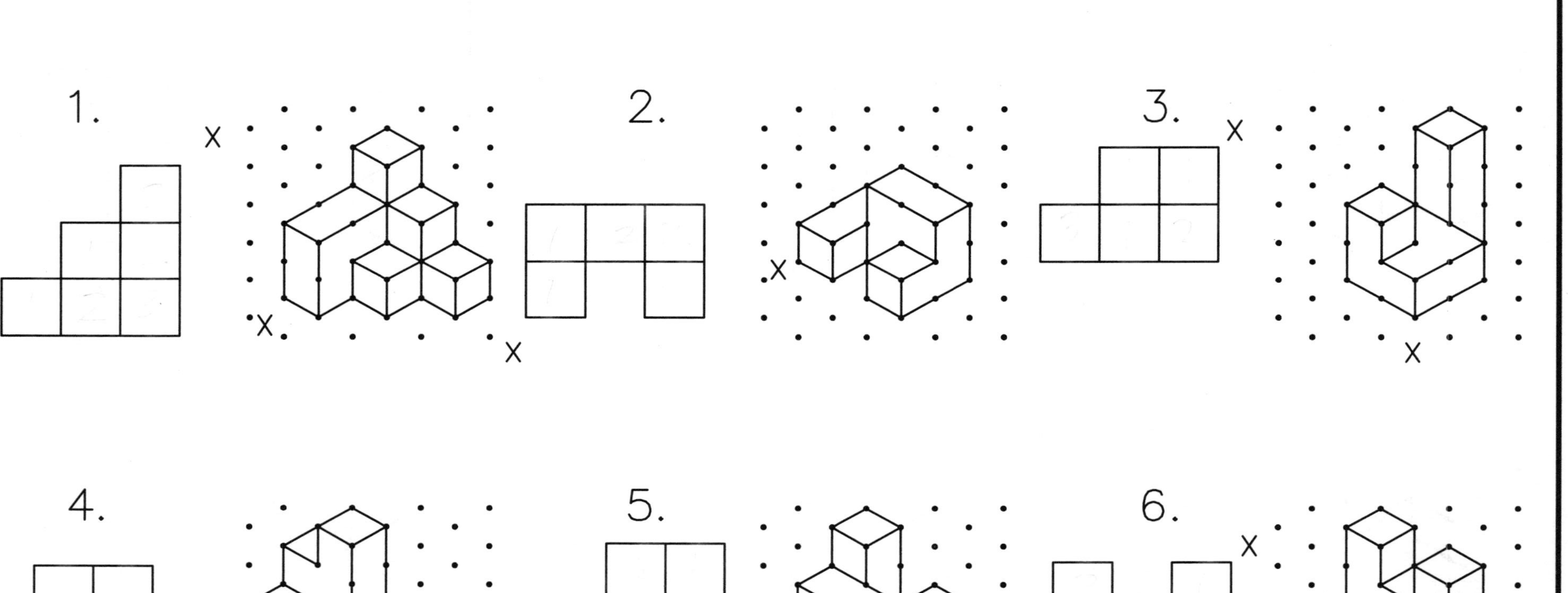

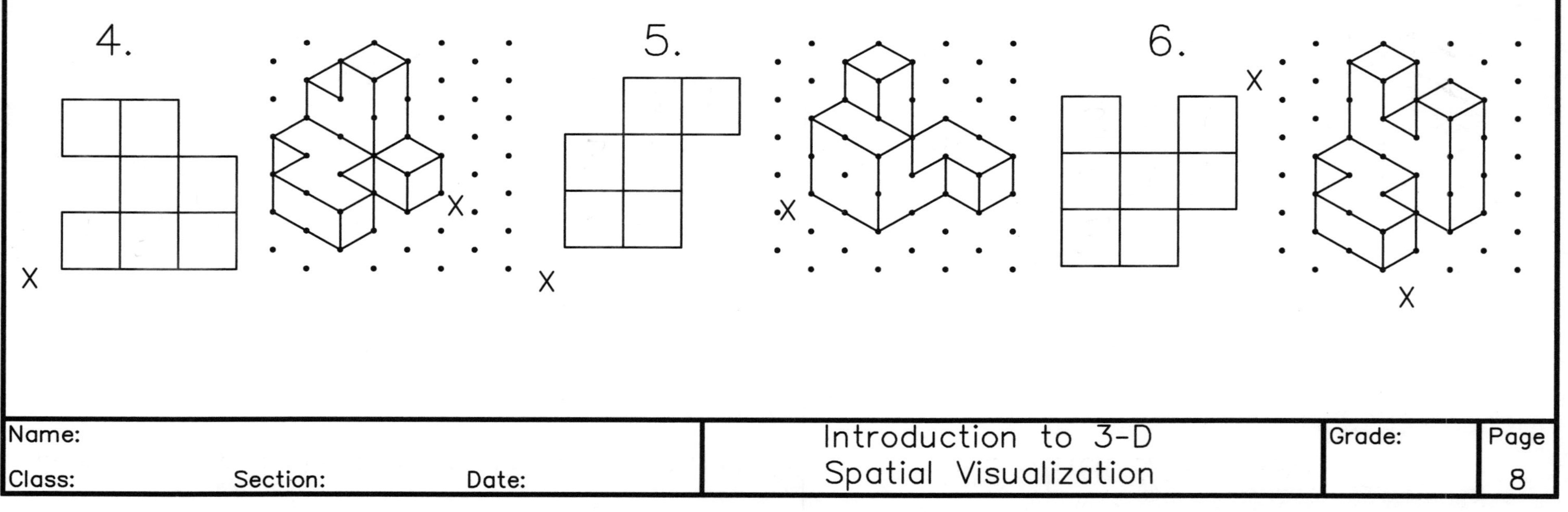

1.

2.

3.

4.

5.

6.

Name:

Class: Section: Date:

Introduction to 3-D
Spatial Visualization

Grade:

Complete the coded plan for the object shown in an isometric sketch on the right.

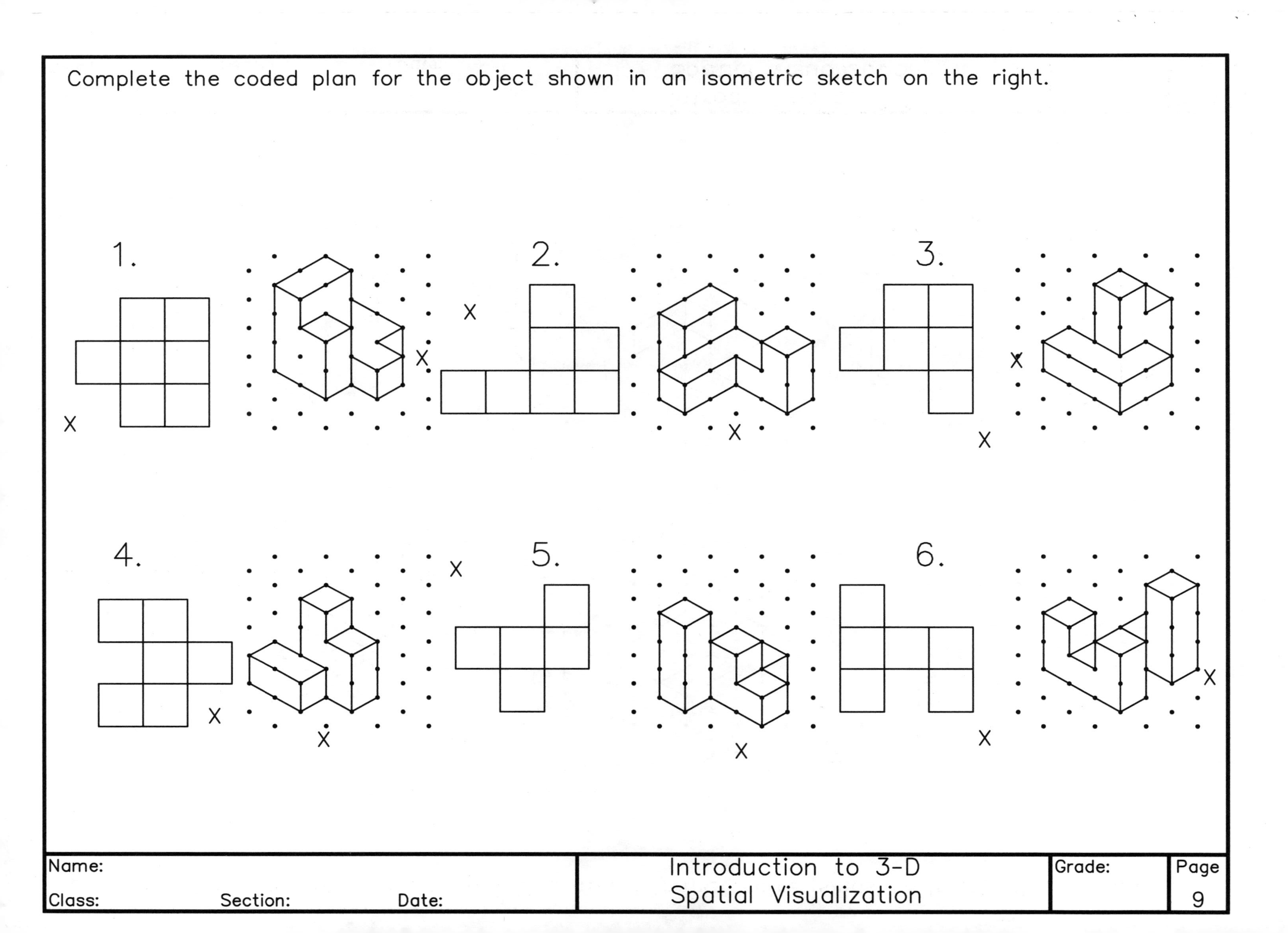

Circle the letter on the coded plan (W, X, Y, or Z) that corresponds to the given isometric sketch.

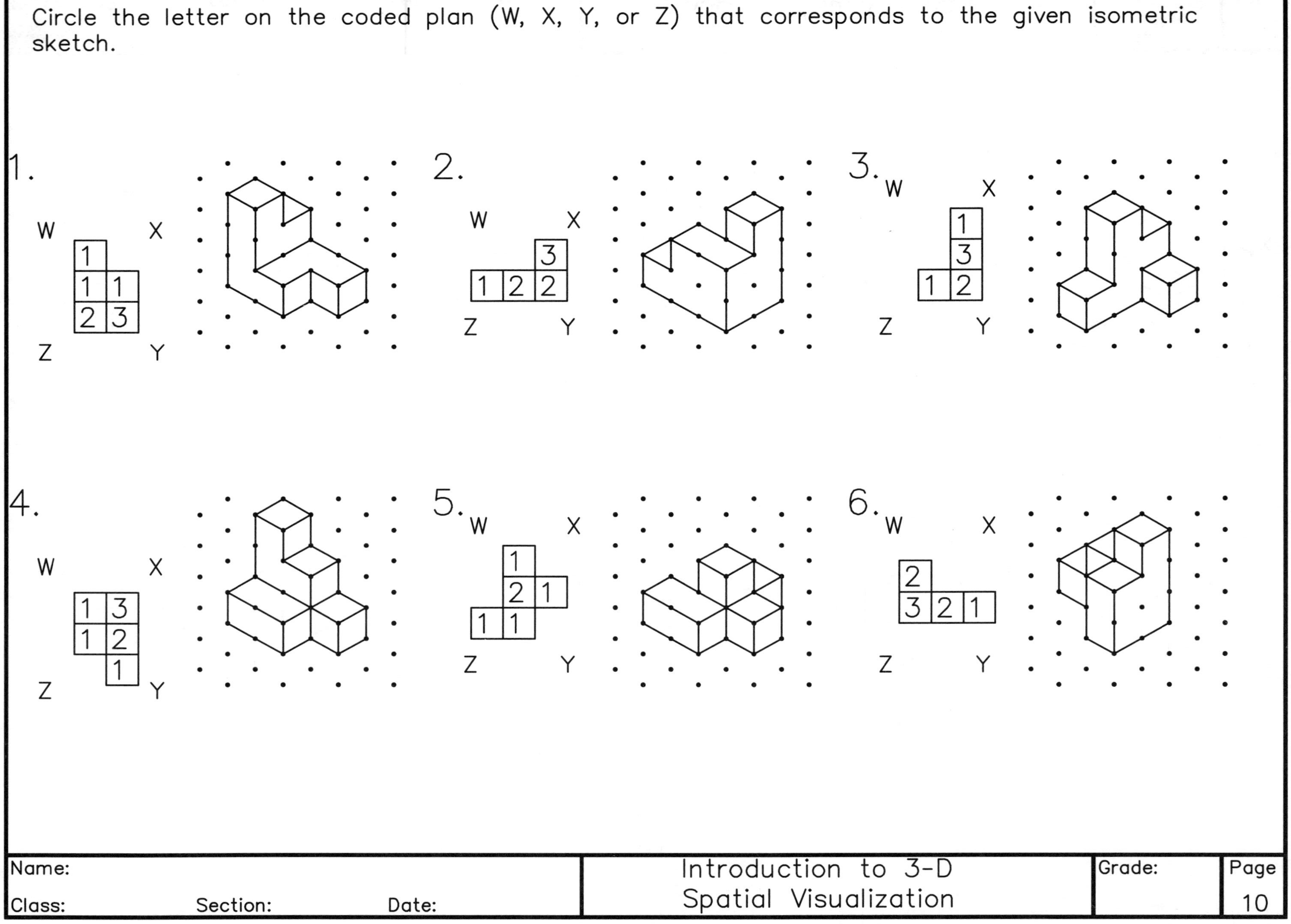


| Name: | Introduction to 3-D Spatial Visualization | Grade: | Page 10 |
|---|---|---|---|
| Class: Section: Date: | | | |

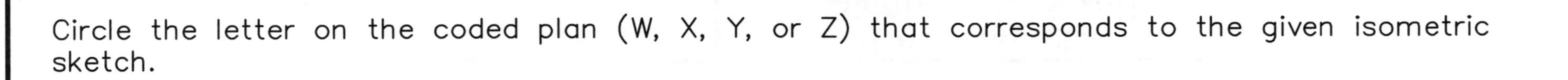

Circle the letter on the coded plan (W, X, Y, or Z) that corresponds to the given isometric sketch.

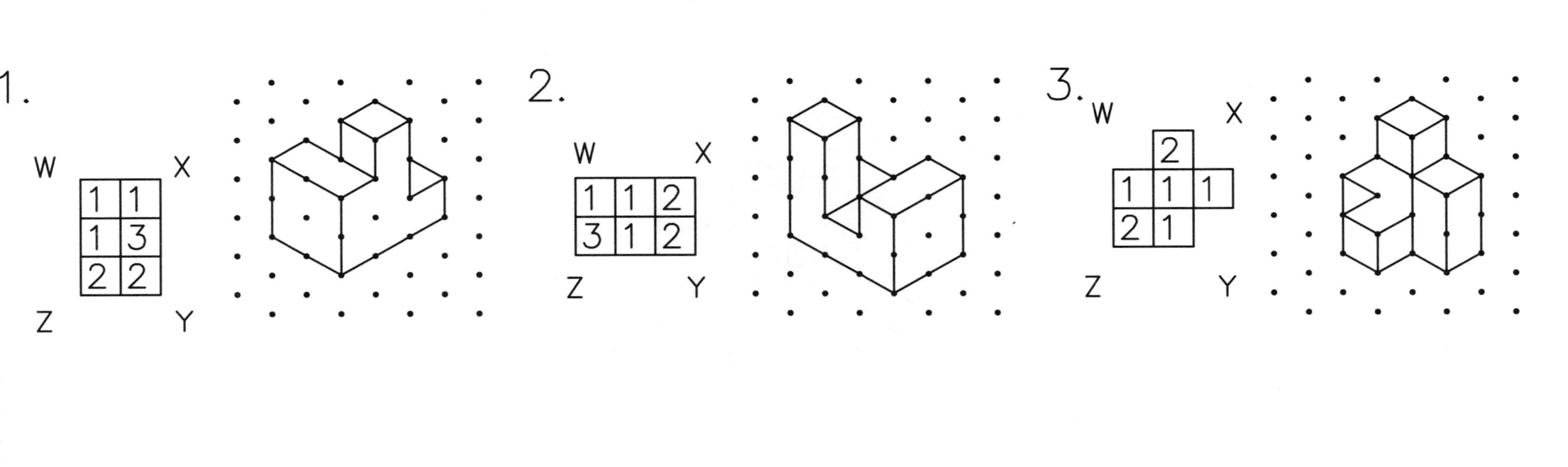


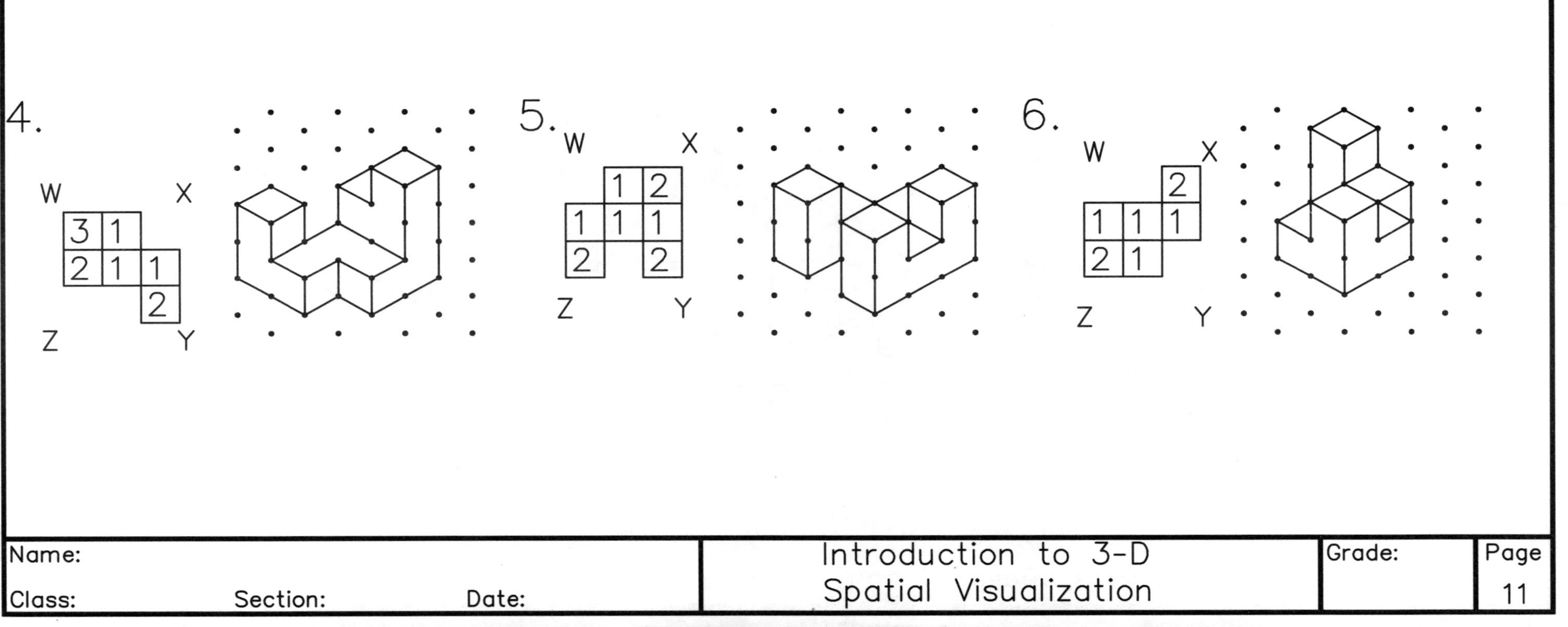


| Name: | Introduction to 3-D Spatial Visualization | Grade: | Page 11 |
|---|---|---|---|
| Class: Section: Date: | | | |

Circle the letter on the coded plan (W, X, Y, or Z) that corresponds to the given isometric sketch.

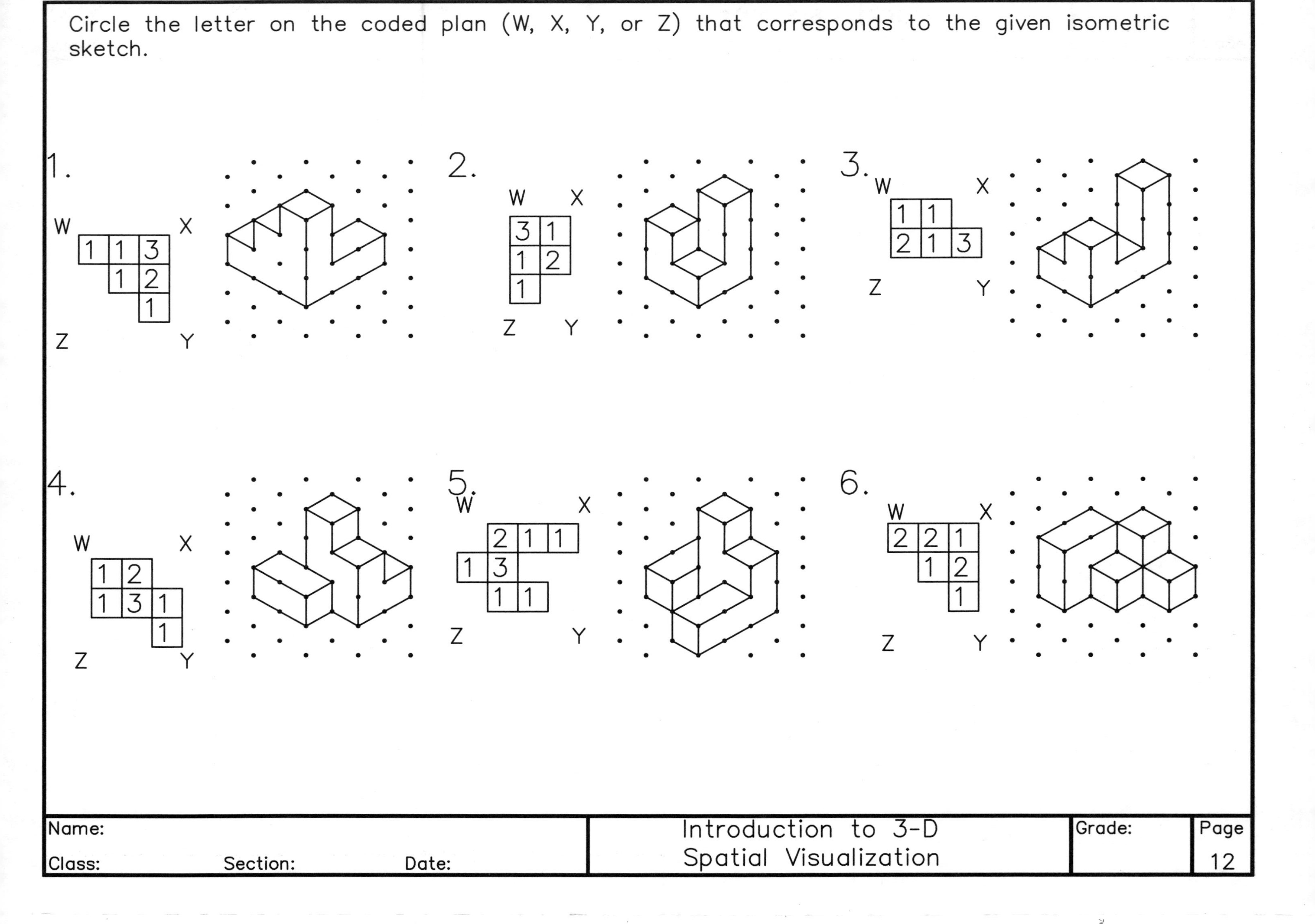

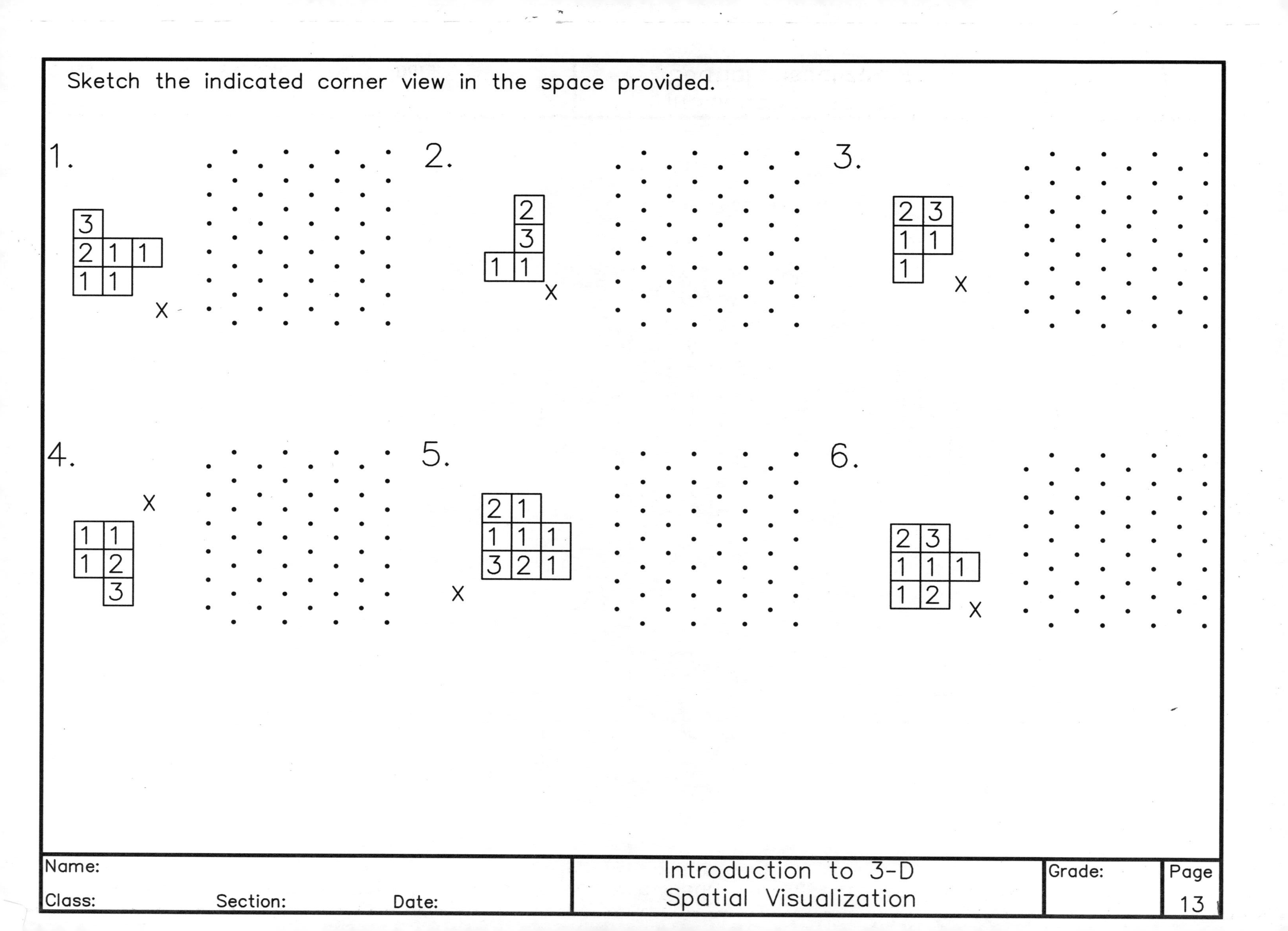

Sketch the indicated corner view in the space provided.
1.
3
2 1 1
1 1
X
2.
2
3
1 1
X
3.
2 3
1 1
1
X
4.
X
1 1
1 2
3
5.
2 1
1 1 1
3 2 1
X
6.
2 3
1 1 1
1 2
X
Name:
Class:
Section:
Date:
Introduction to 3-D
Spatial Visualization
Grade:
Page
13

Sketch the indicated corner view in the space provided.

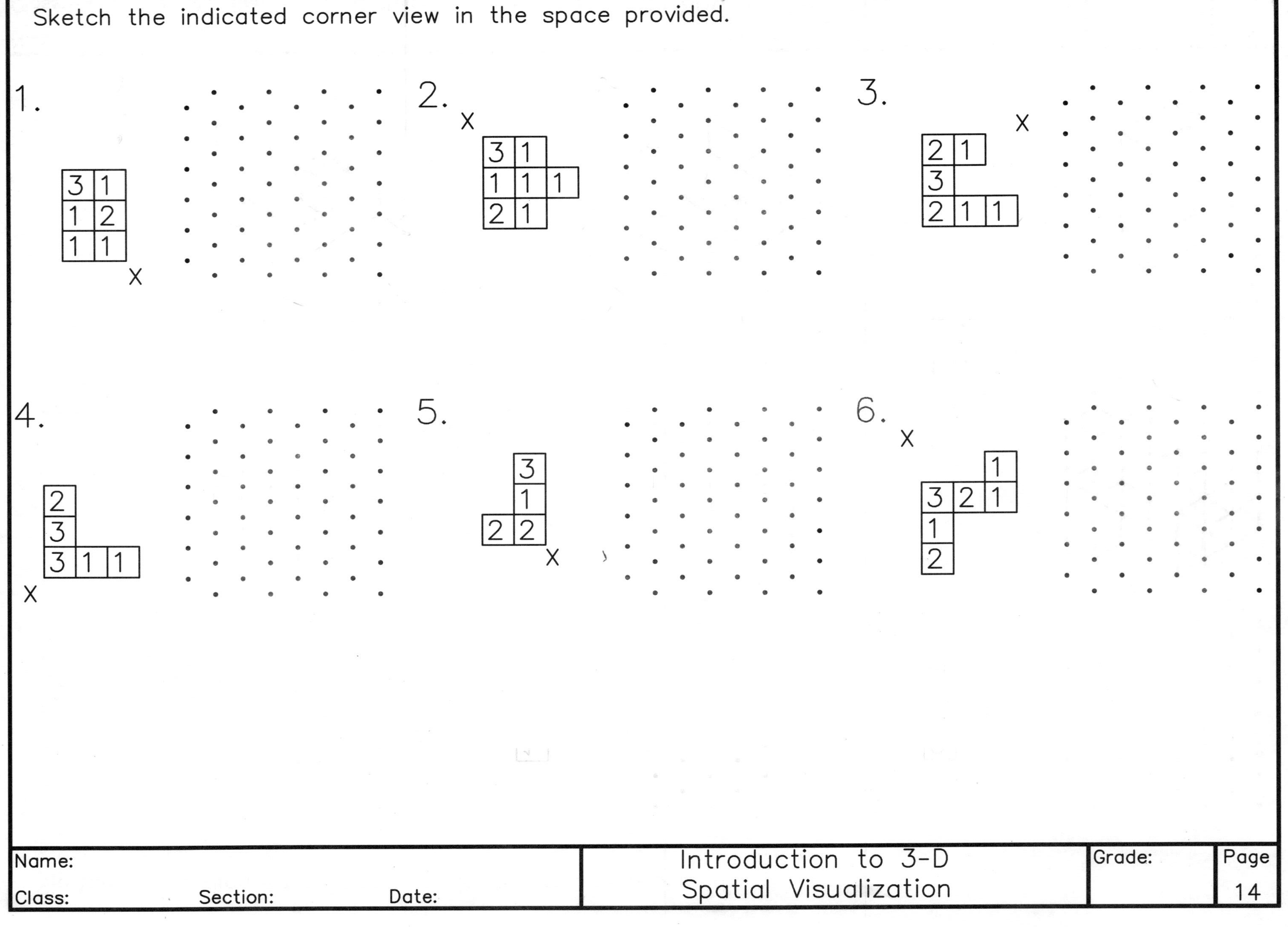

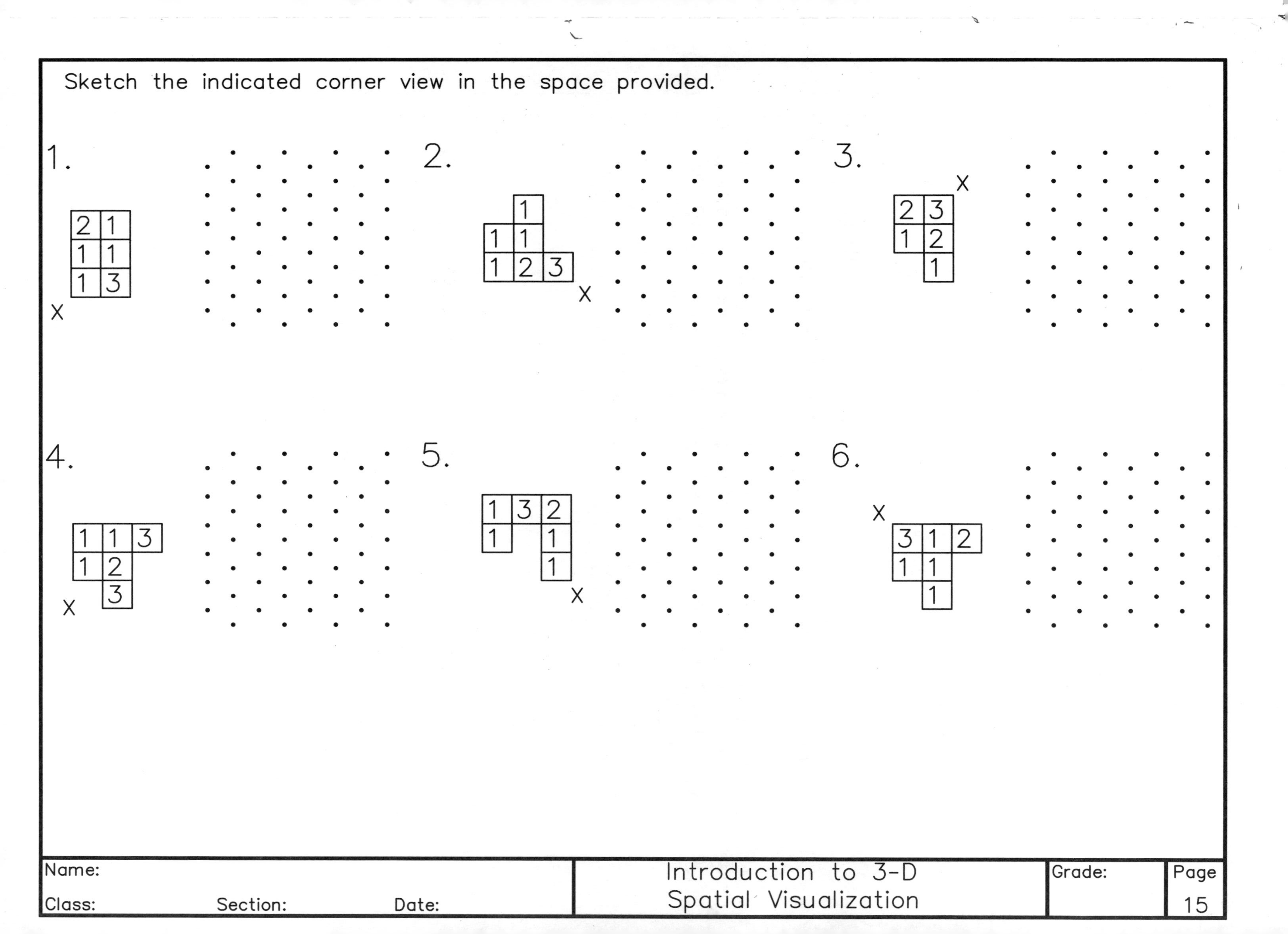

Sketch the indicated corner view in the space provided.
1.
2 1
1 1
1 3
X
2.
1
1 1
1 2 3
X
3.
X
2 3
1 2
1
4.
1 1 3
1 2
3
X
5.
1 3 2
1 1
1
X
6.
X
3 1 2
1 1
1
Name:
Class:
Section:
Date:
Introduction to 3-D
Spatial Visualization
Grade:
Page
15

## Orthographic Drawings

Isometric views show objects from their corners. Orthographic views show the faces of the object straight on or parallel to the viewing plane:

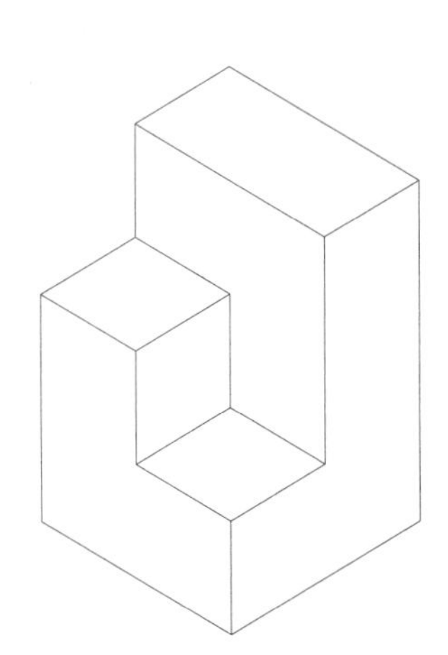

Isometric Drawing

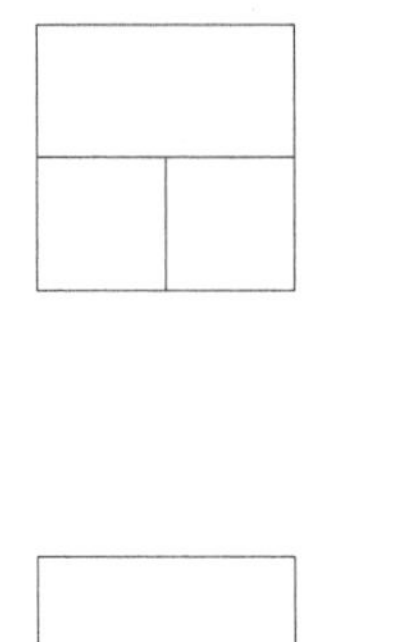

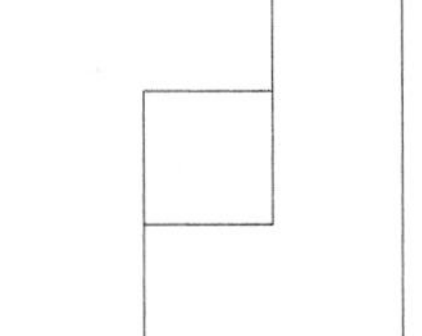

Orthographic Drawing

In creating orthographic views, you imagine that the object is surrounded by a transparent glass cube. The edges and surfaces of the object are projected onto the panes of the glass cube and the cube is unfolded so the panes of glass lie in one plane.

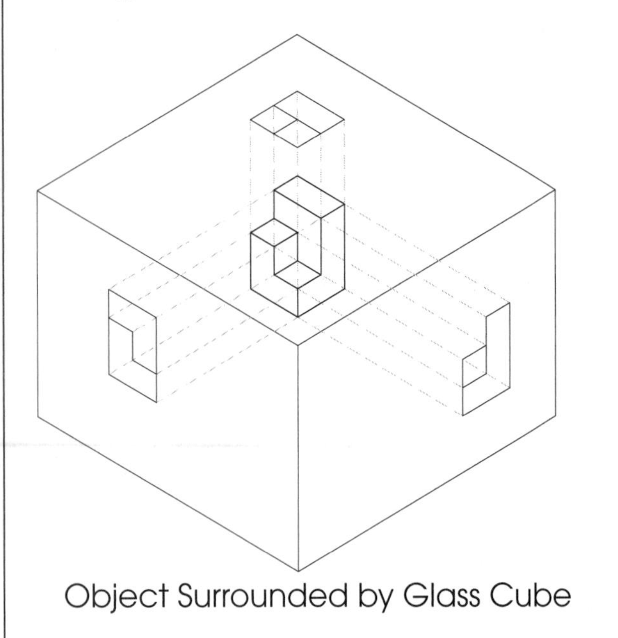

Object Surrounded by Glass Cube

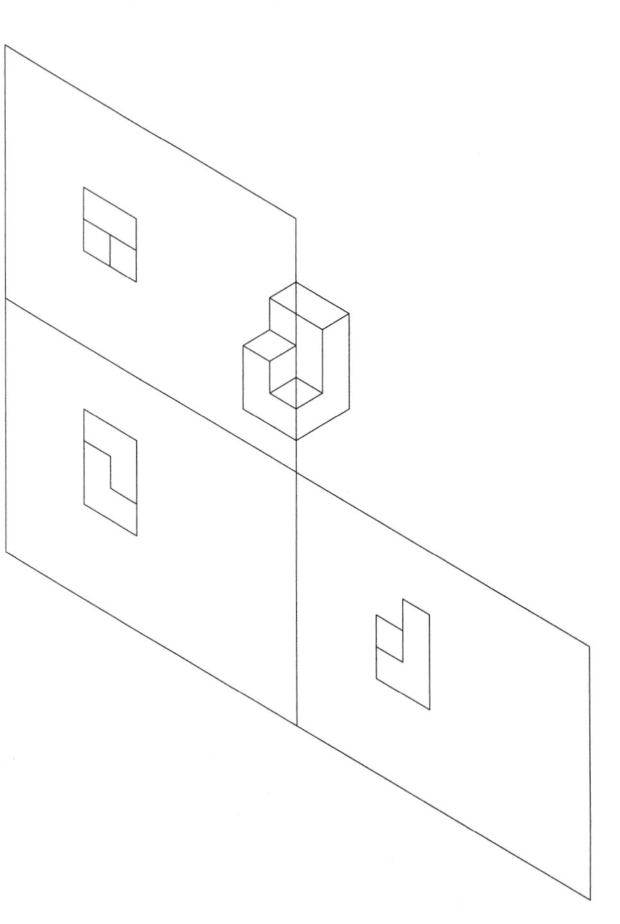

Glass Cube Unfolded

Normal surfaces are defined as being parallel to either the top, front, or side views. A normal surface is seen as a surface in the view to which it is parallel and is seen as an edge in the other views.

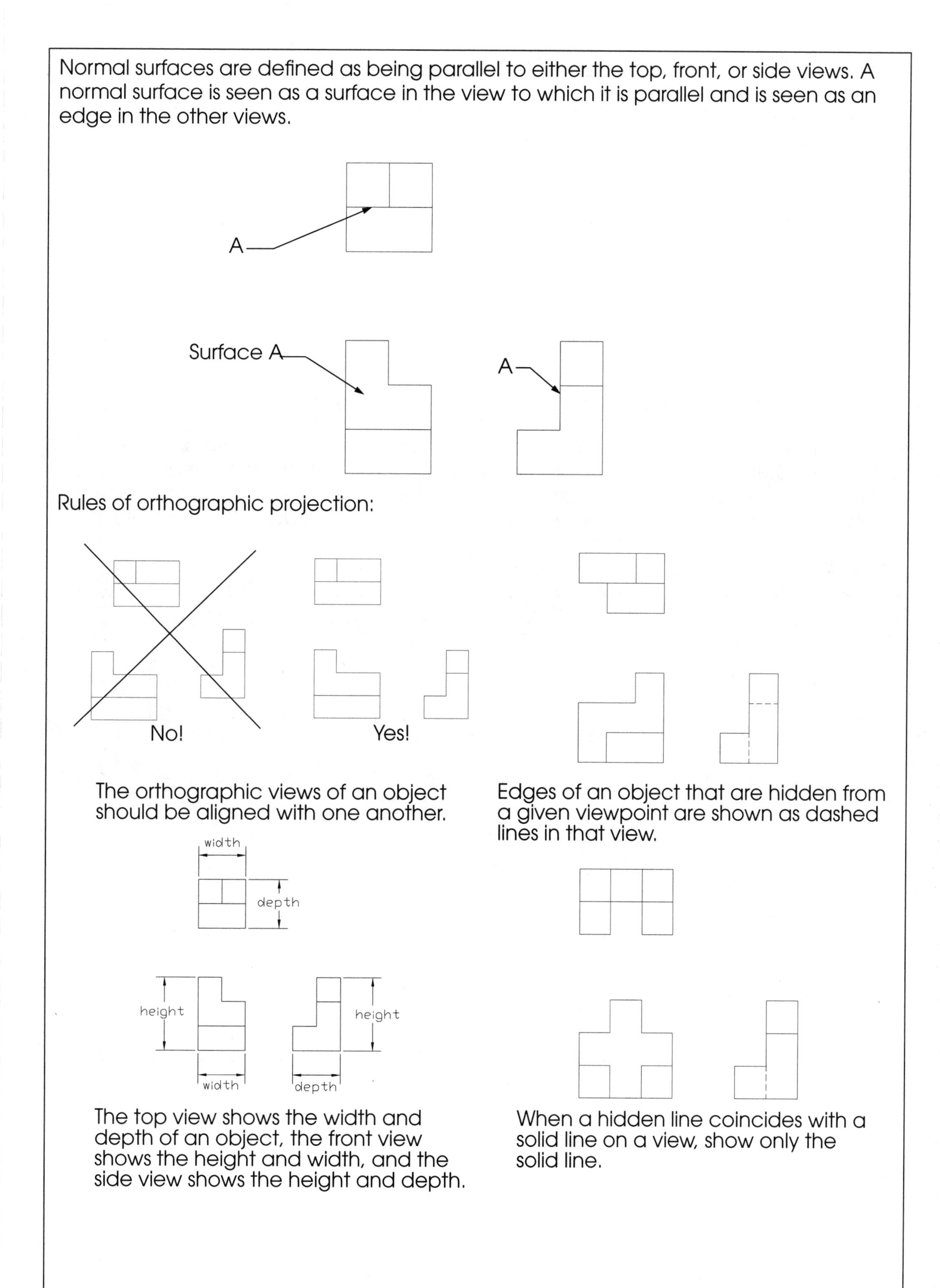


Rules of orthographic projection:

The orthographic views of an object should be aligned with one another.

Edges of an object that are hidden from a given viewpoint are shown as dashed lines in that view.

The top view shows the width and depth of an object, the front view shows the height and width, and the side view shows the height and depth.

When a hidden line coincides with a solid line on a view, show only the solid line.

Sometimes only two views are required to completely describe an object. However, most times, three views are shown (top, front, and right side).

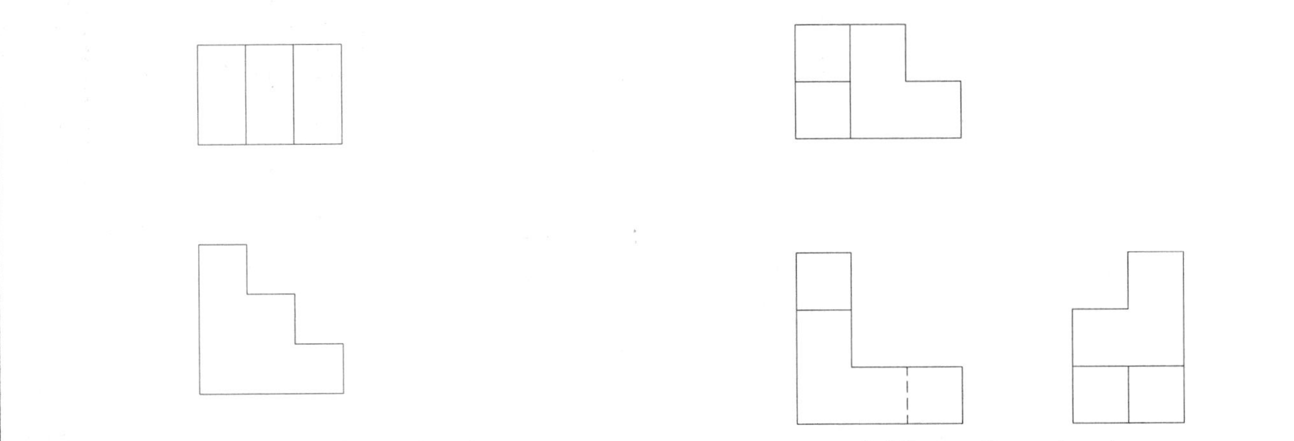

Only 2 Views Required

3 Views Required

The box method is sometimes used to create an isometric drawing from three orthographic views. Step 1: Construct an isometric drawing of a box that is the same overall size as the object as seen in the orthographic views. Step 2: Draw the top, front and side views on this box. Step 3: Add and erase lines from the isometric drawing to complete it.

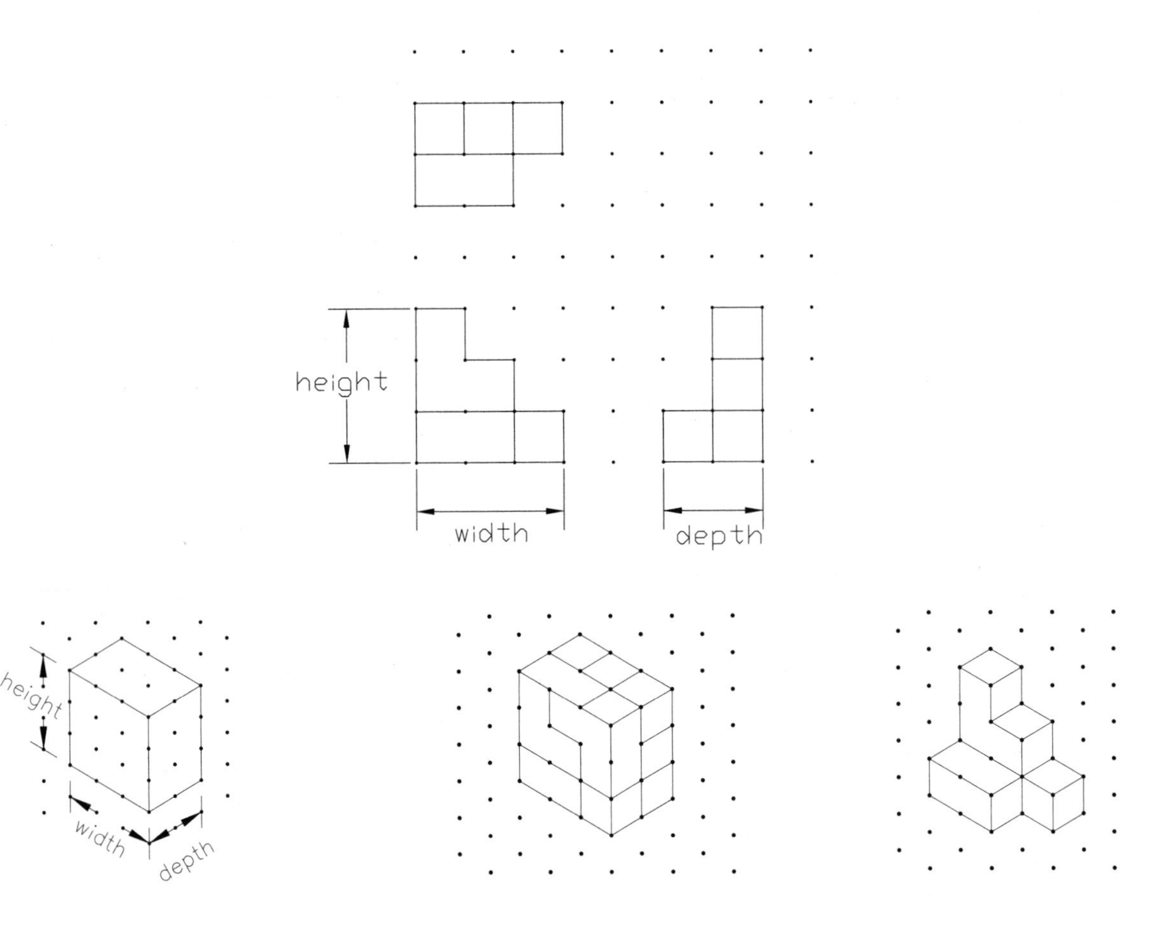

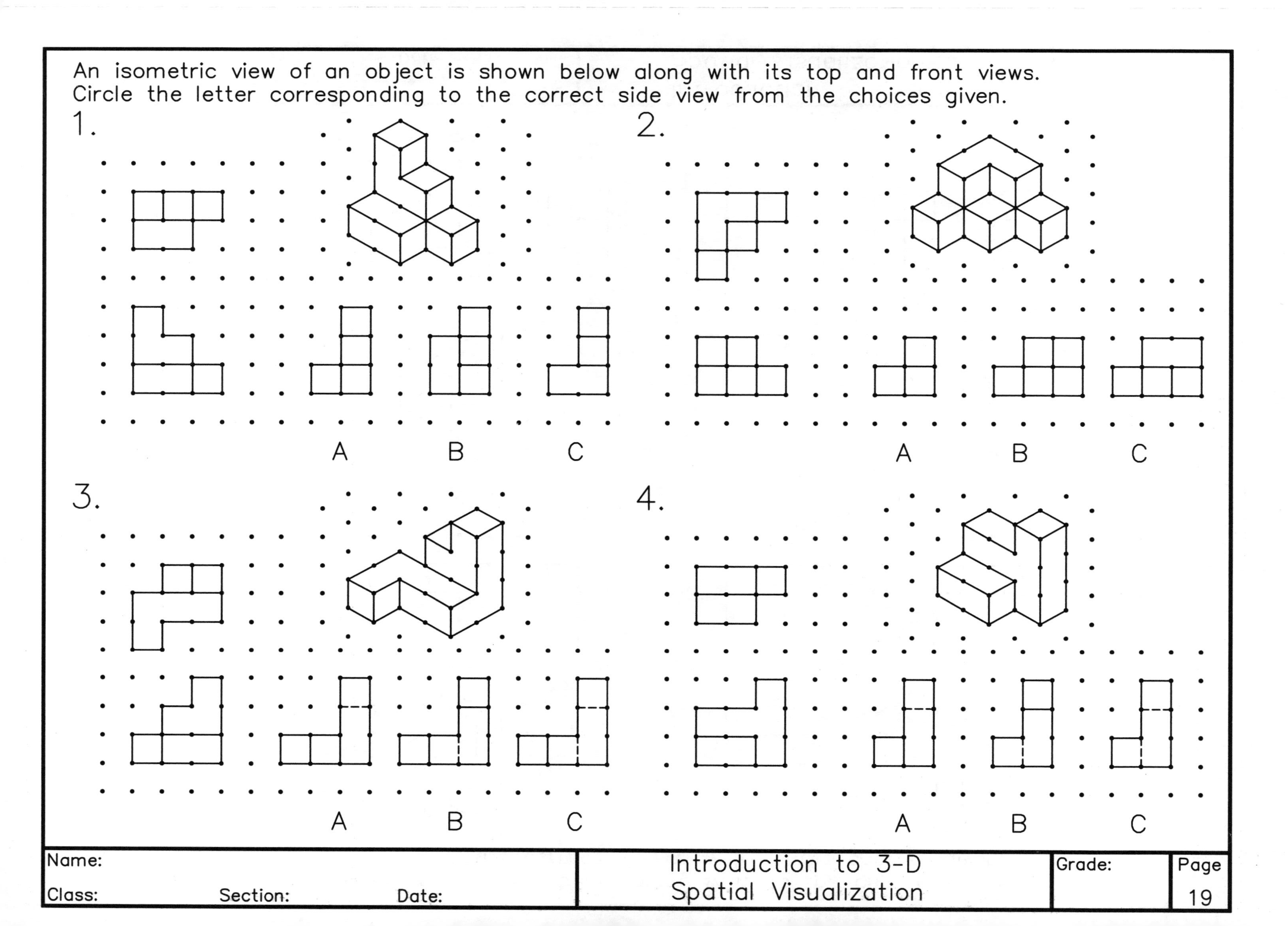

An isometric view of an object is shown below along with its top and front views.
Circle the letter corresponding to the correct side view from the choices given.

An isometric view of an object is shown below along with its top and front views. Circle the letter corresponding to the correct side view from the choices given.

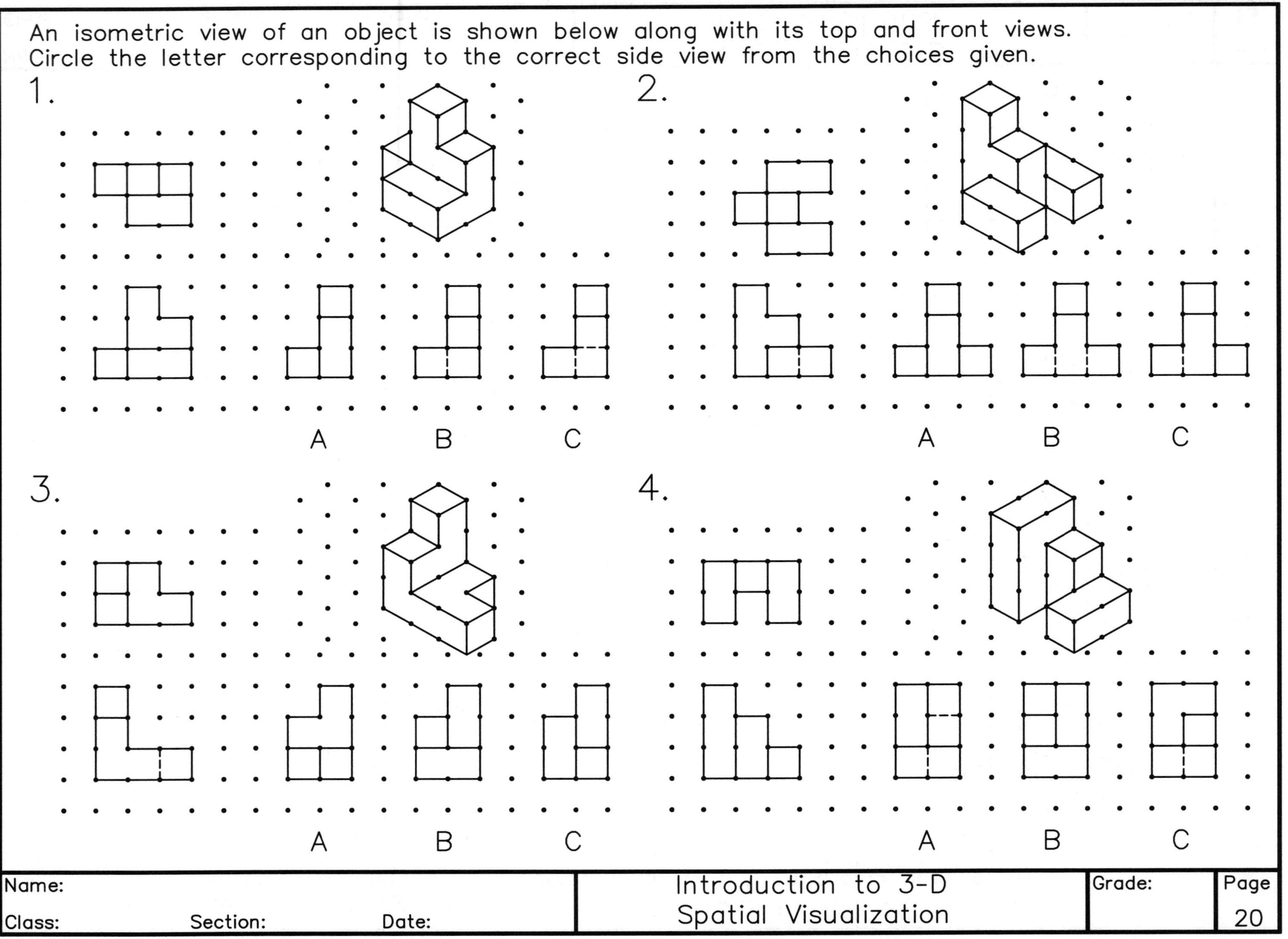

An isometric view of an object is shown below along with its top and front views. Circle the letter corresponding to the correct side view from the choices given.

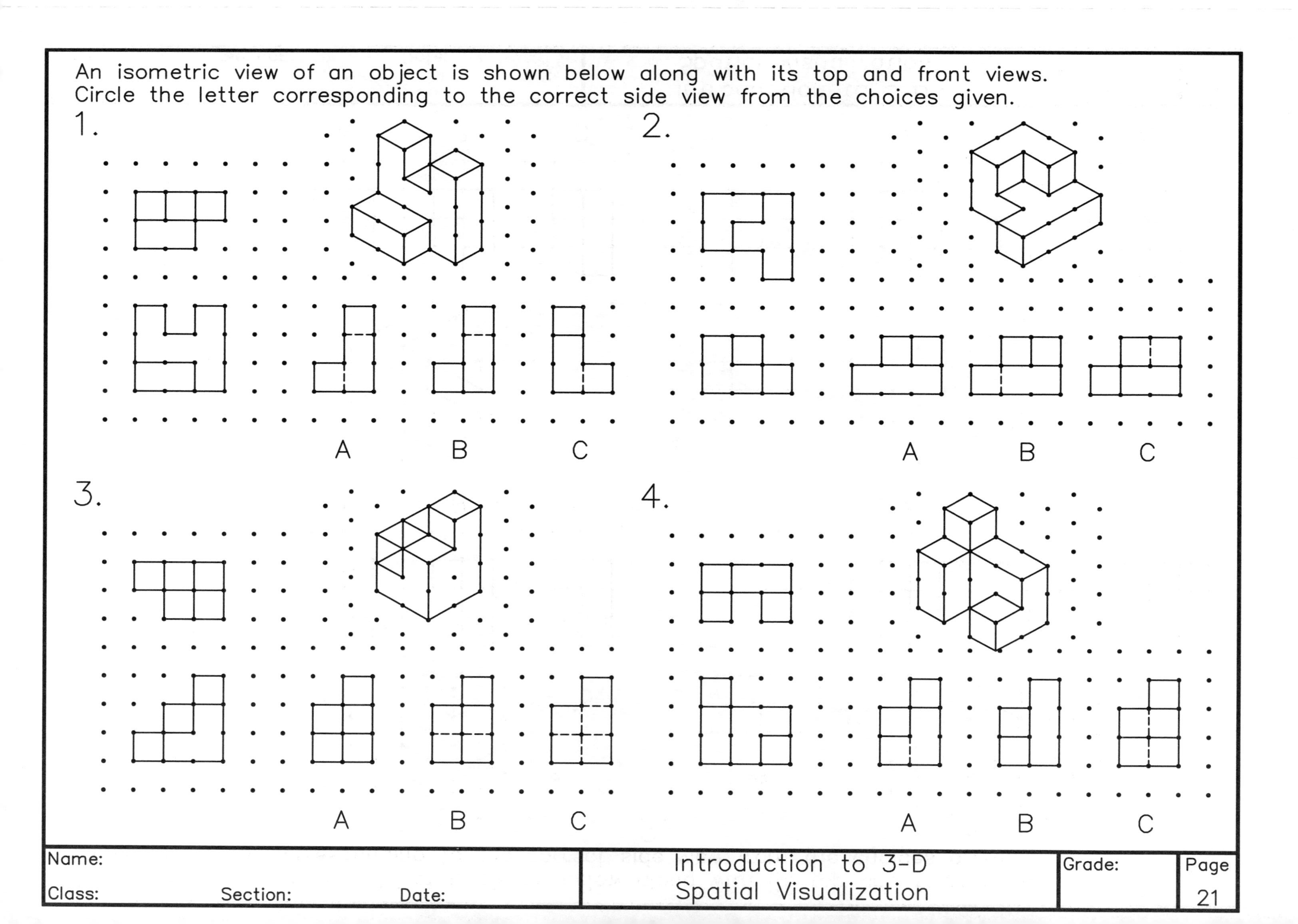

An isometric view of an object is shown below along with its top and front views. Circle the letter corresponding to the correct side view from the choices given.

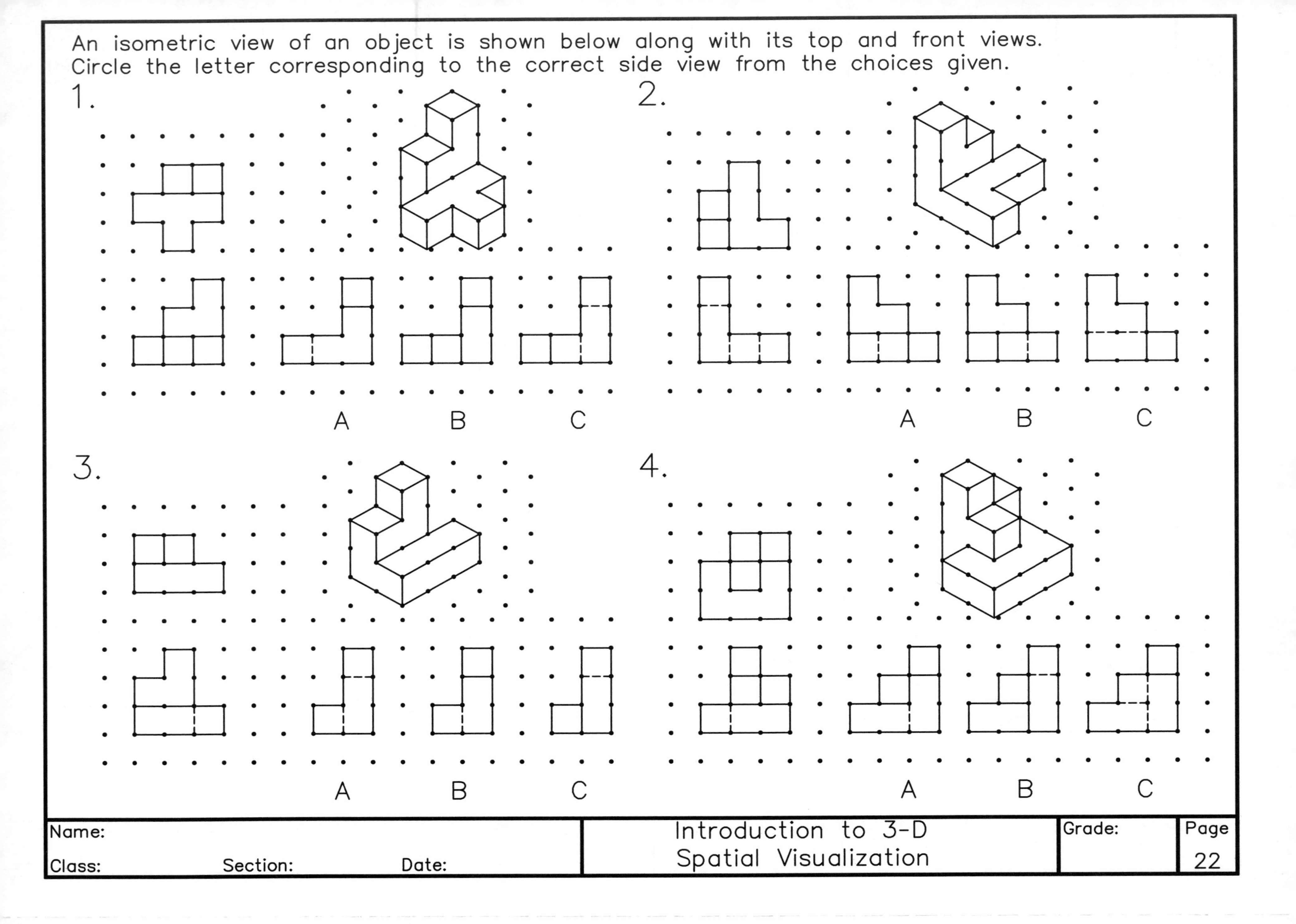

For the object shown in orthographic projection on the left, circle the letter of the correct corresponding isometric view.

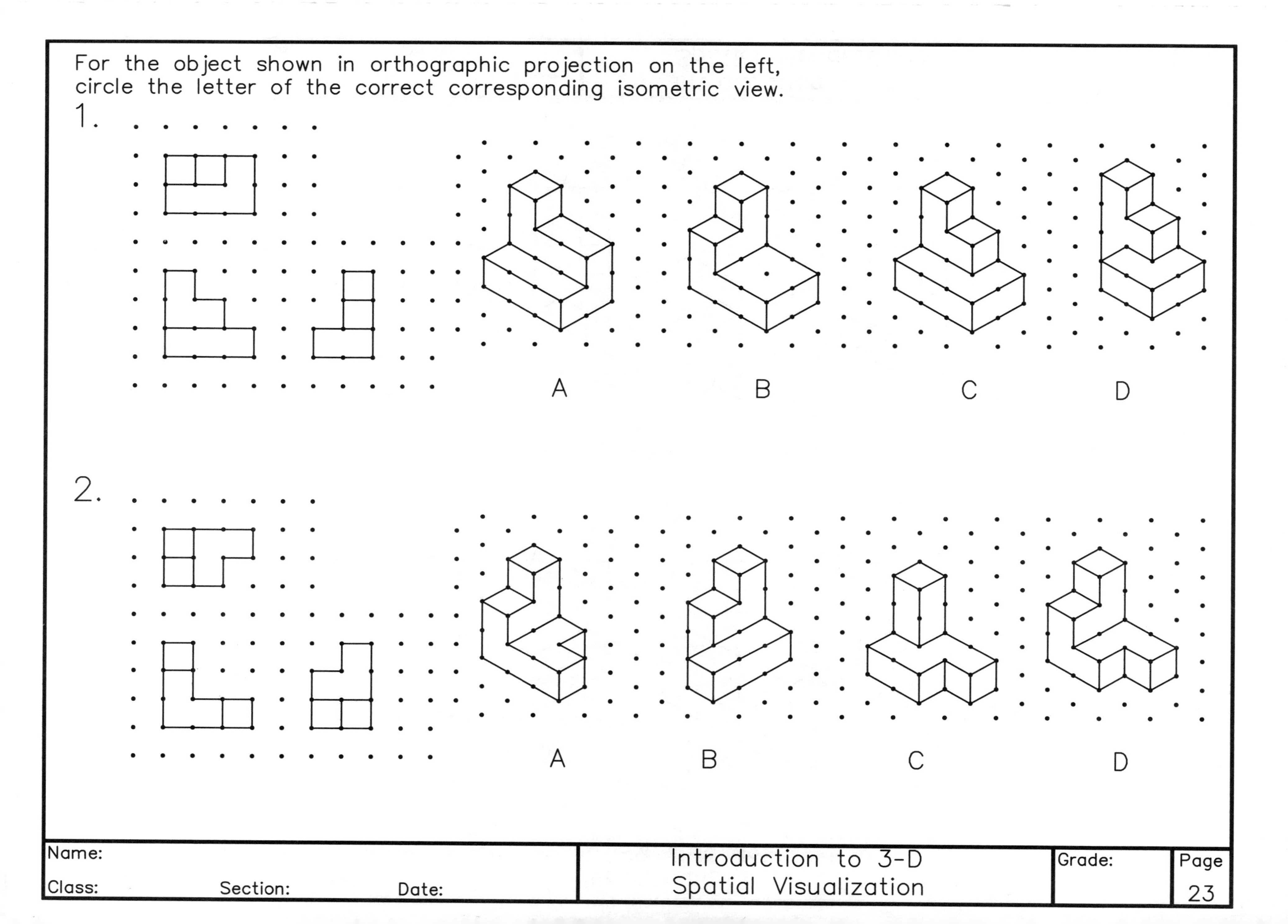

For the object shown in orthographic projection on the left, circle the letter of the correct corresponding isometric view.

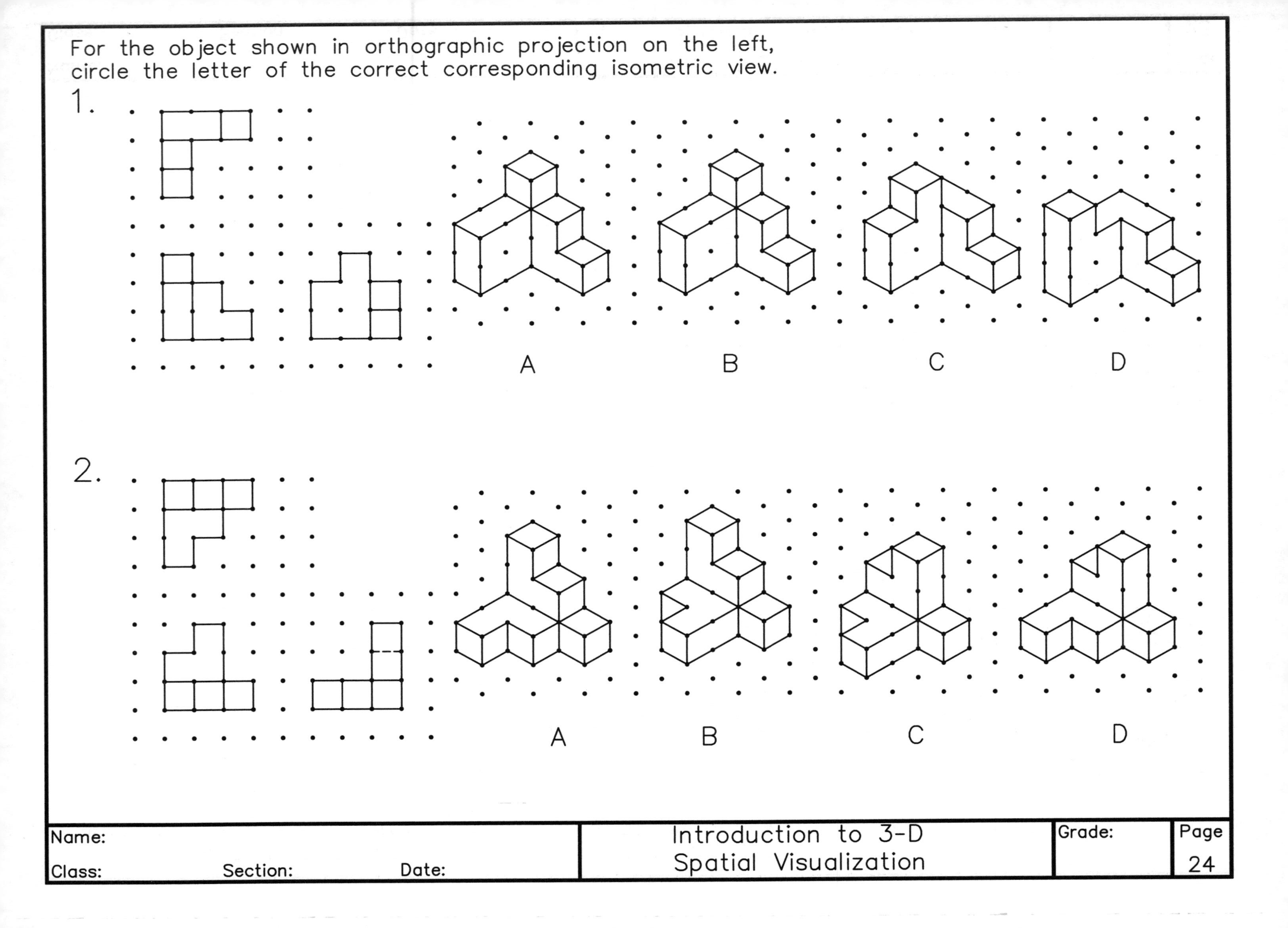


| Name: | Introduction to 3-D Spatial Visualization | Grade: | Page 24 |
|---|---|---|---|
| Class: Section: Date: | | | |

For the object shown in orthographic projection on the left, circle the letter of the correct corresponding isometric view.

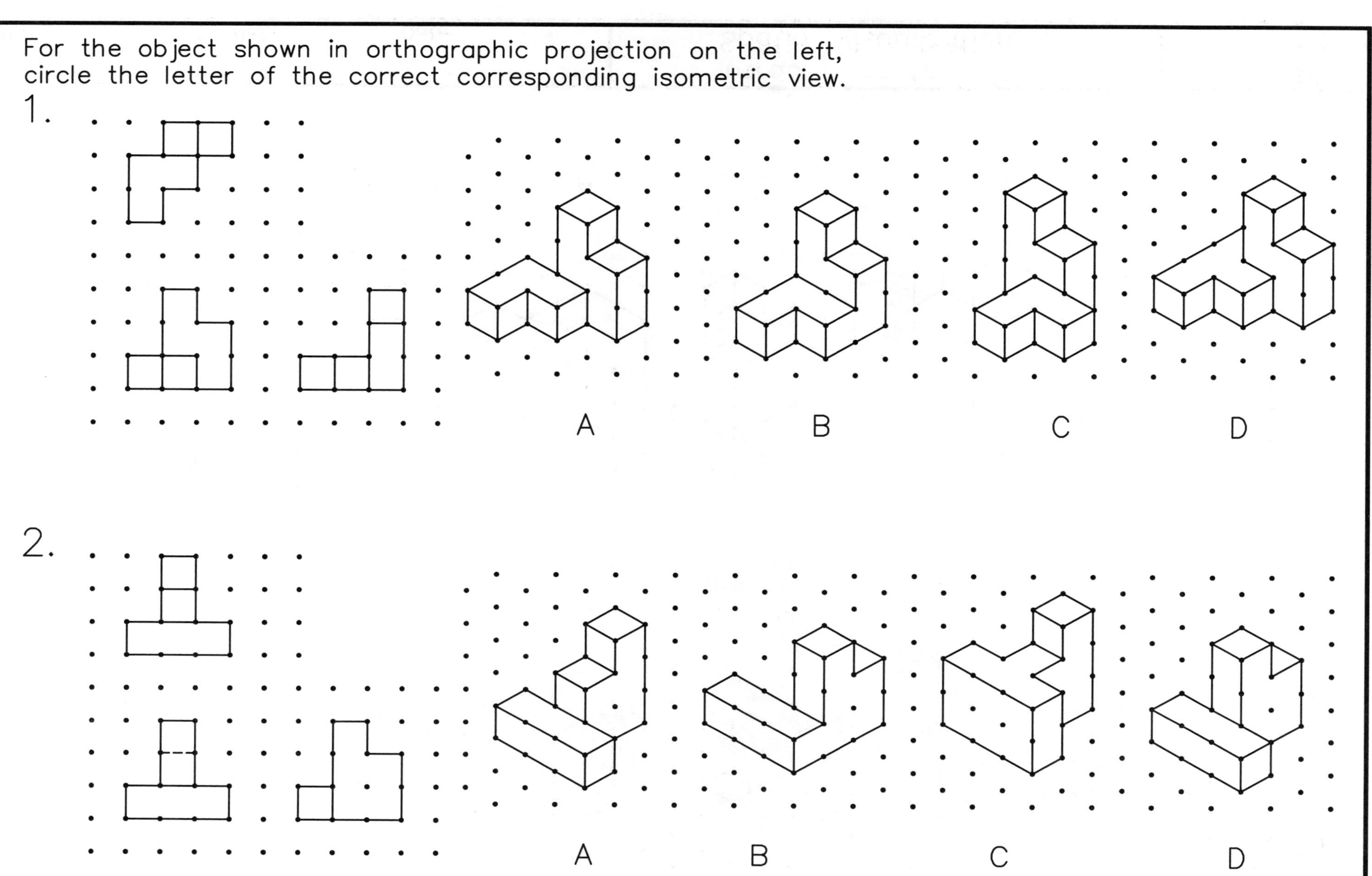


| Name: | Introduction to 3-D Spatial Visualization | Grade: | Page 25 |
|---|---|---|---|
| Class: Section: Date: | | | |

For the object shown in orthographic projection on the left, circle the letter of the correct corresponding isometric view.

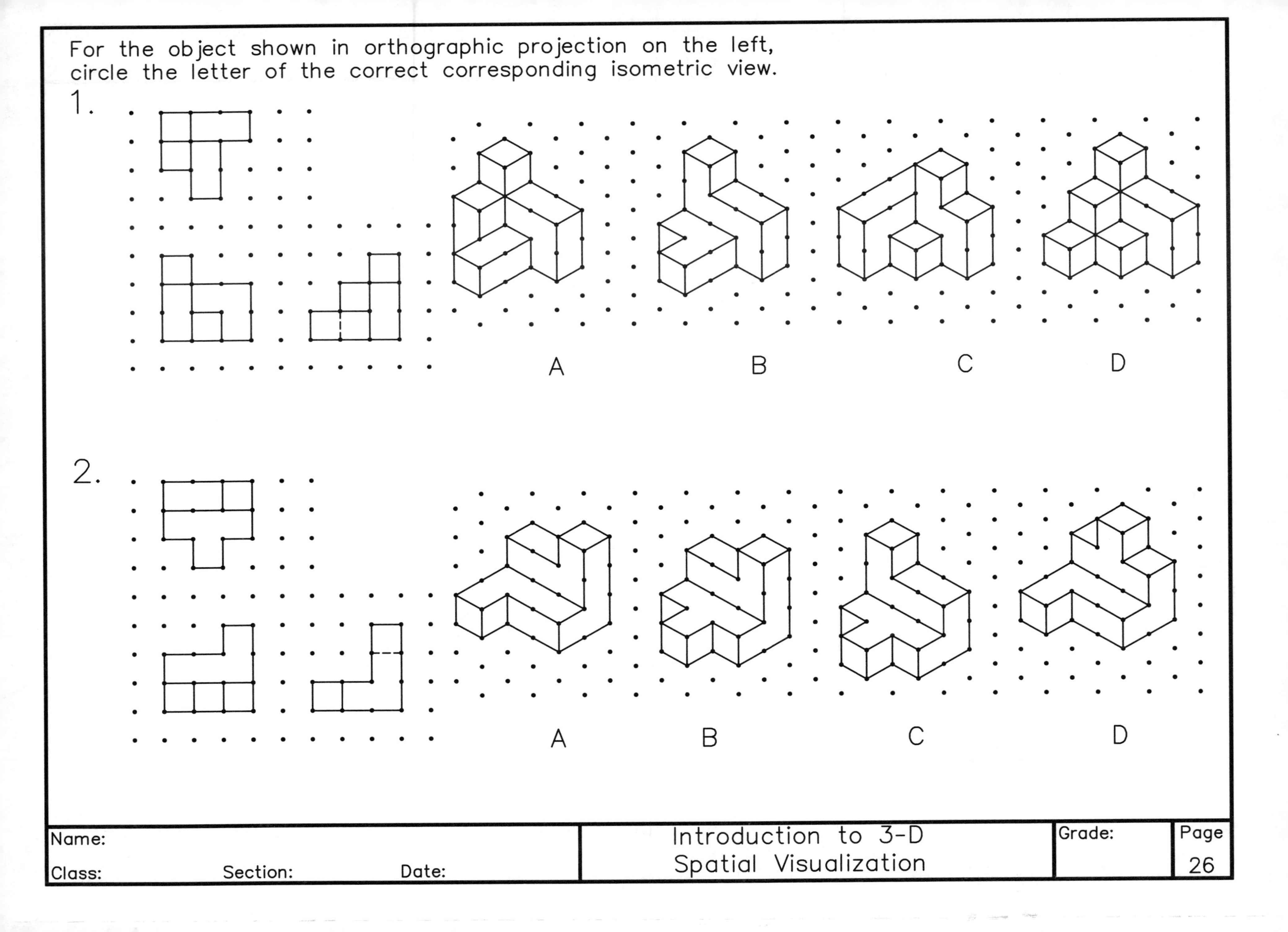


| Name: | Introduction to 3-D Spatial Visualization | Grade: | Page |
|---|---|---|---|
| Class: Section: Date: | | | 26 |

For the object shown in orthographic projection on the left, circle the letter of the correct corresponding isometric view.

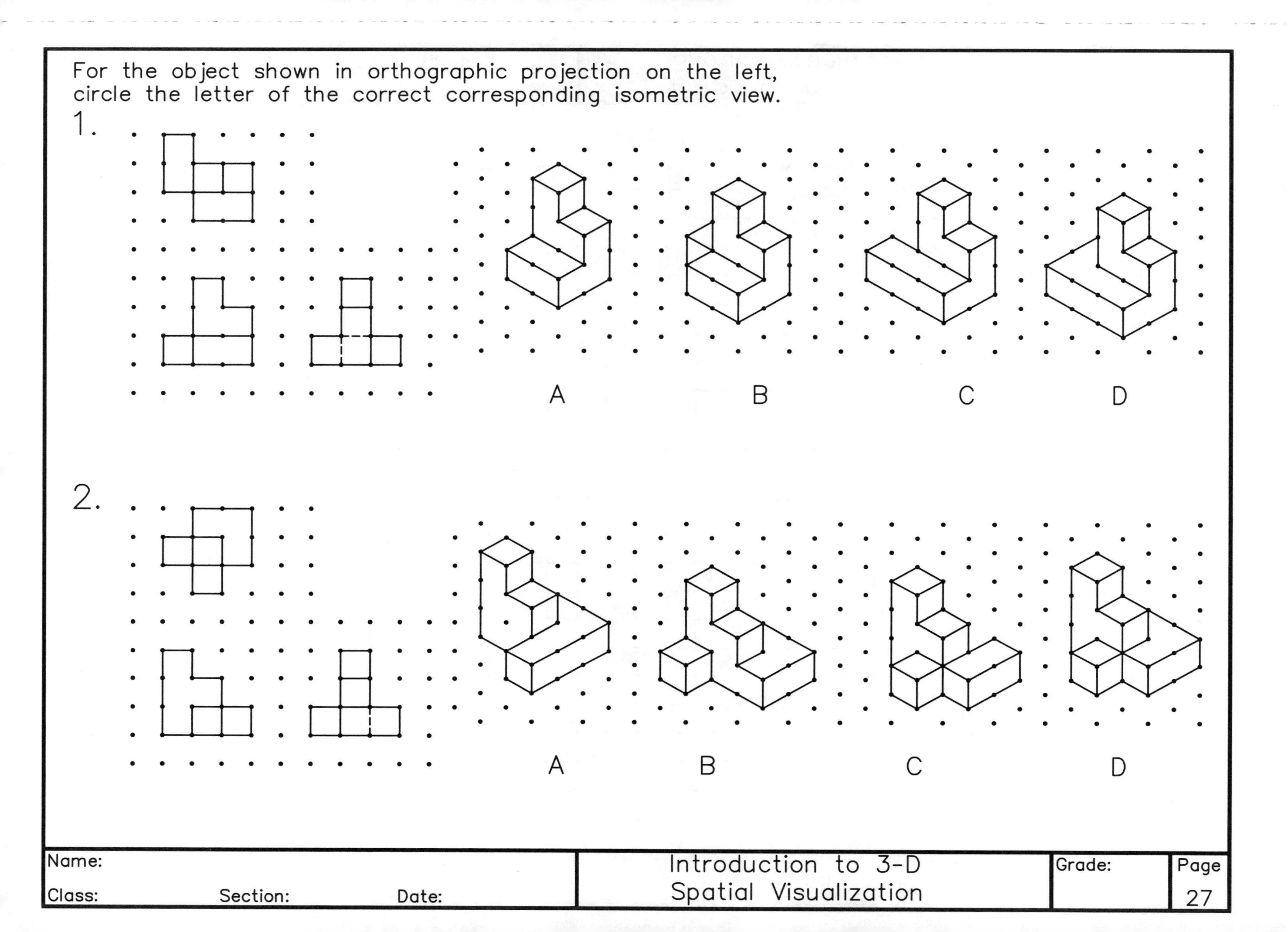


| Name: | Introduction to 3-D Spatial Visualization | Grade: | Page 27 |
|---|---|---|---|
| Class: Section: Date: | | | |

Circle the letter corresponding to the correctly aligned set of orthographic views.

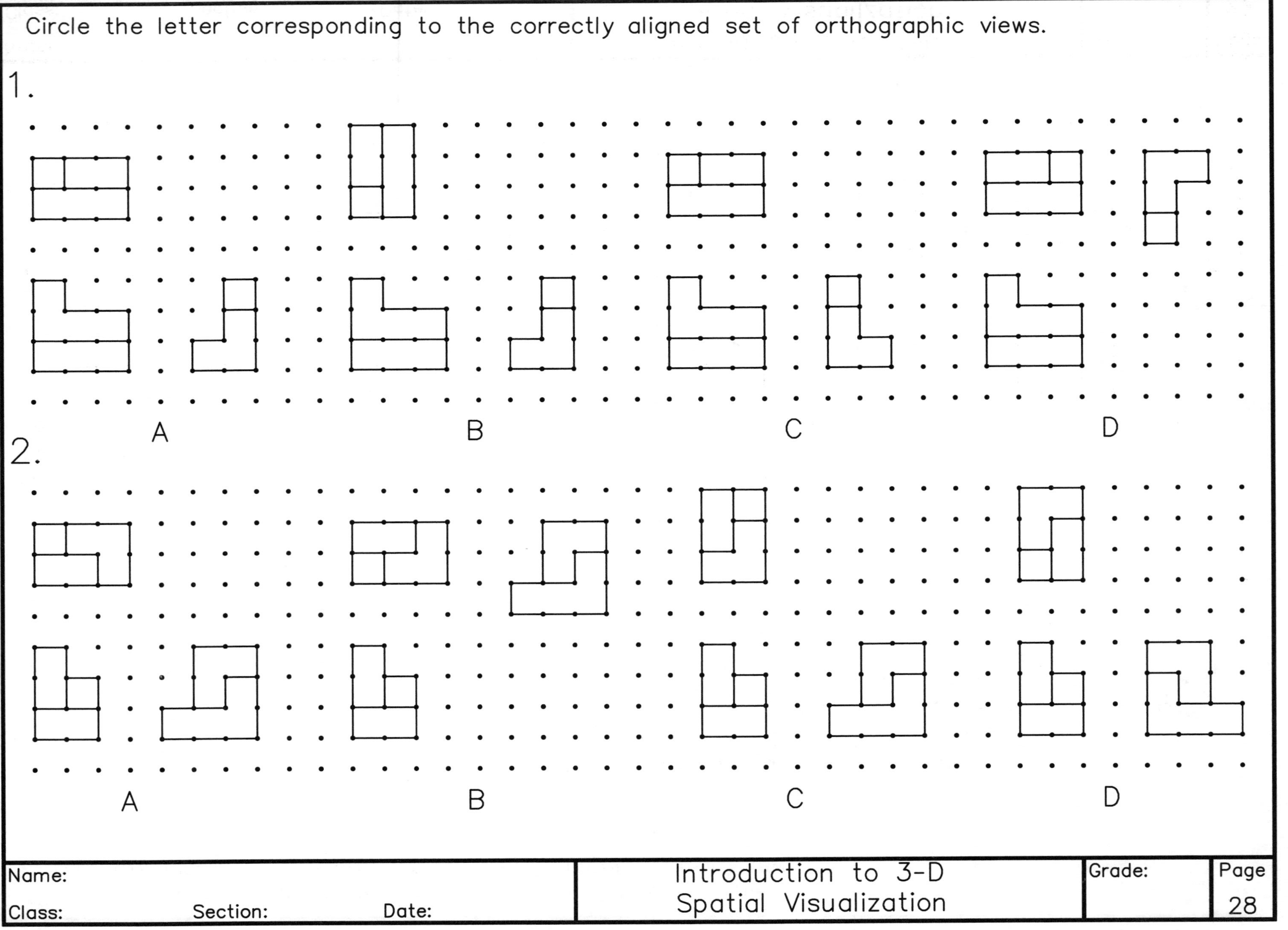

Circle the letter corresponding to the correctly aligned set of orthographic views.

1.

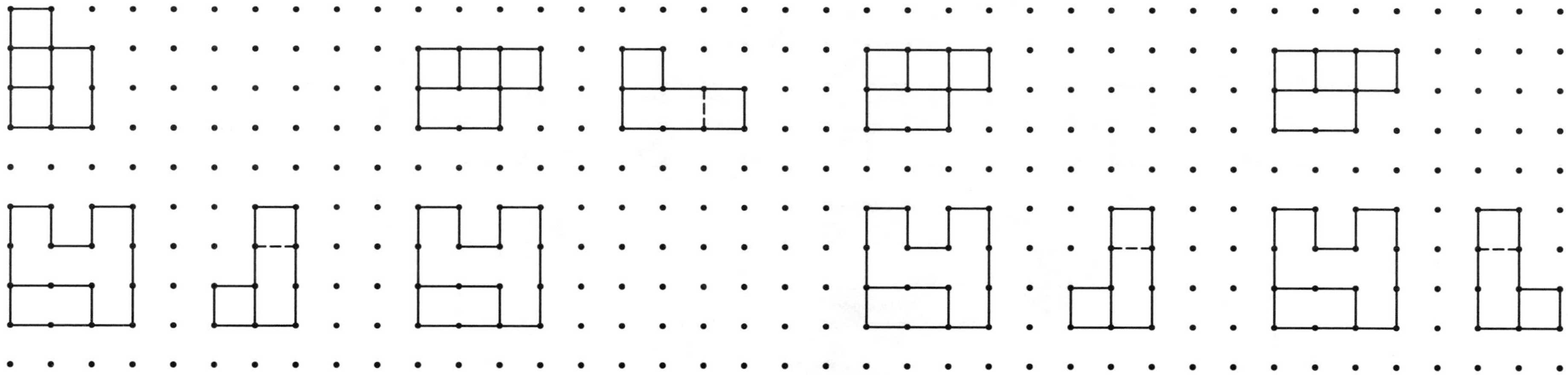

A B C D

2.

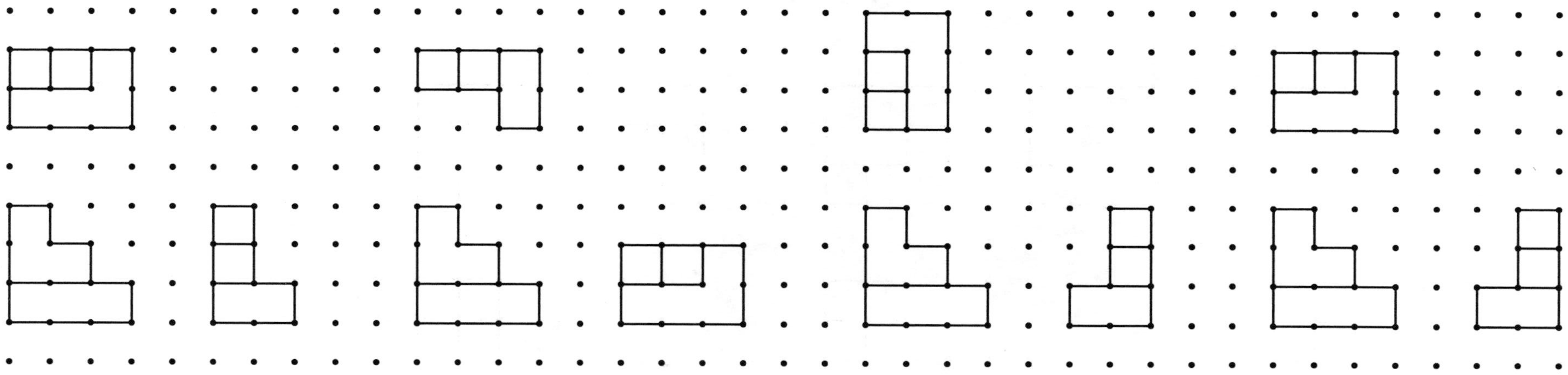

A B C D

Circle the letter corresponding to the correctly aligned set of orthographic views.

1.

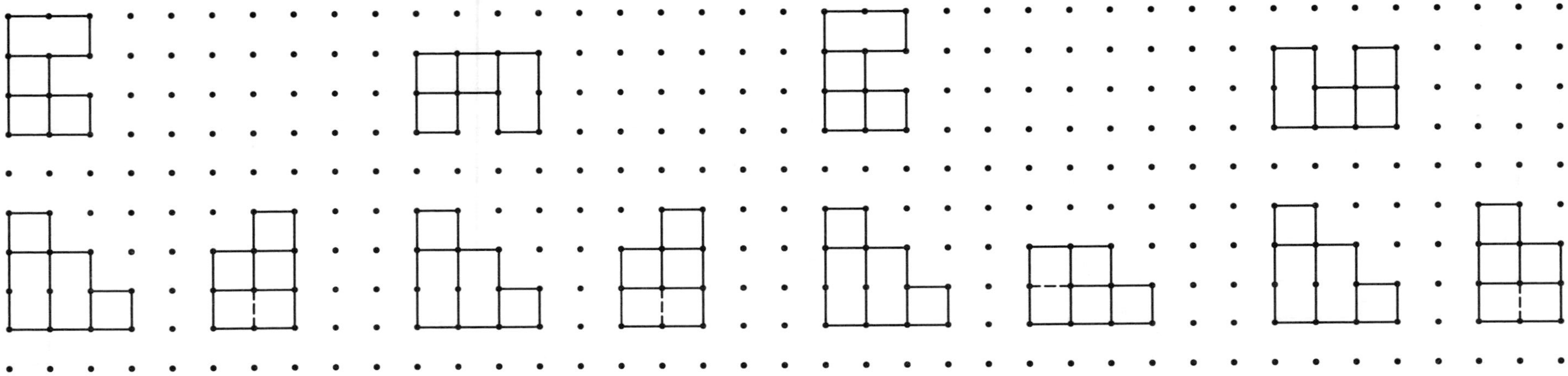

A B C D

2.

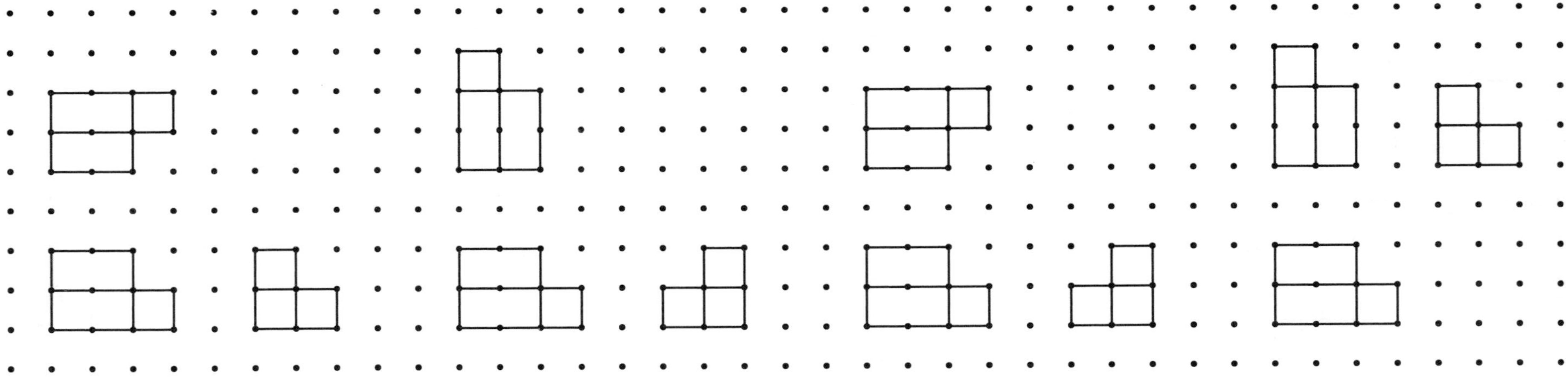

A B C D

| Name: | Introduction to 3-D Spatial Visualization | Grade: | Page |
|---|---|---|---|
| Class: Section: Date: | | | 30 |

Circle the letter corresponding to the correctly aligned set of orthographic views.

1.

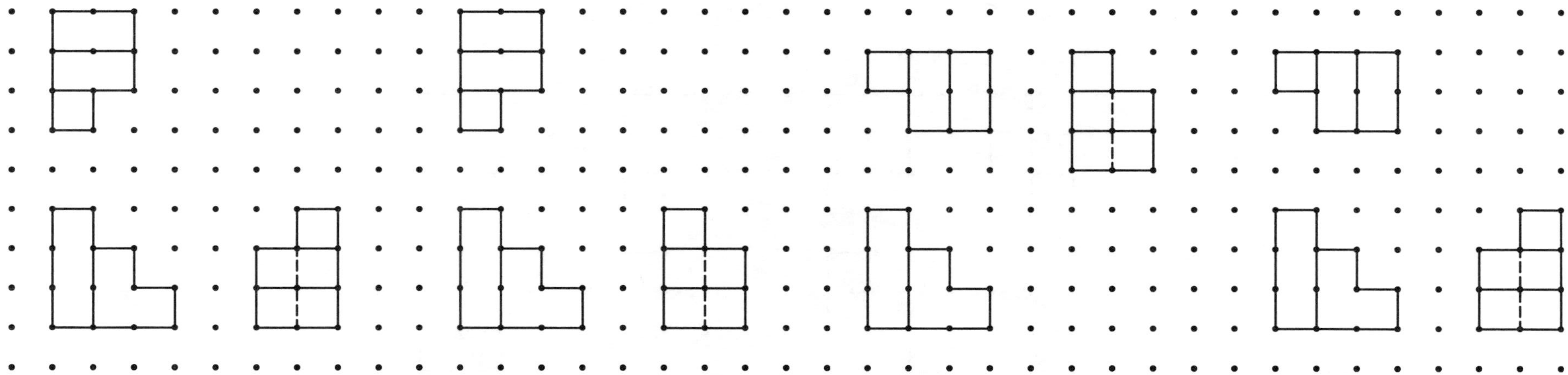

A B C D

2.

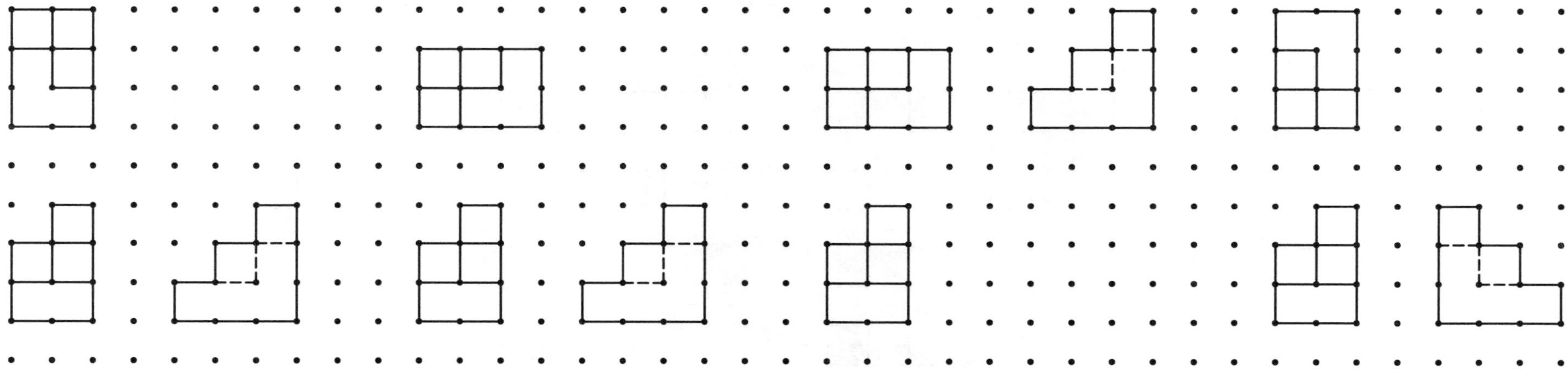

A B C D

| Name: | Introduction to 3-D Spatial Visualization | Grade: | Page |
|---|---|---|---|
| Class: Section: Date: | | | 31 |

Circle the letter corresponding to the correctly aligned set of orthographic views.

1.

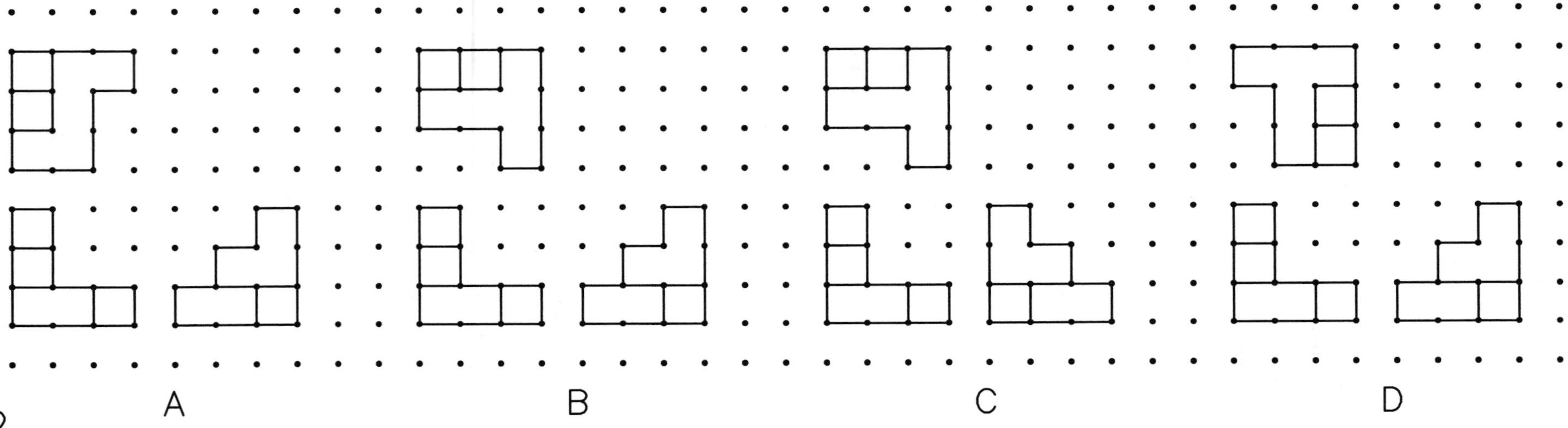

A B C D

2.

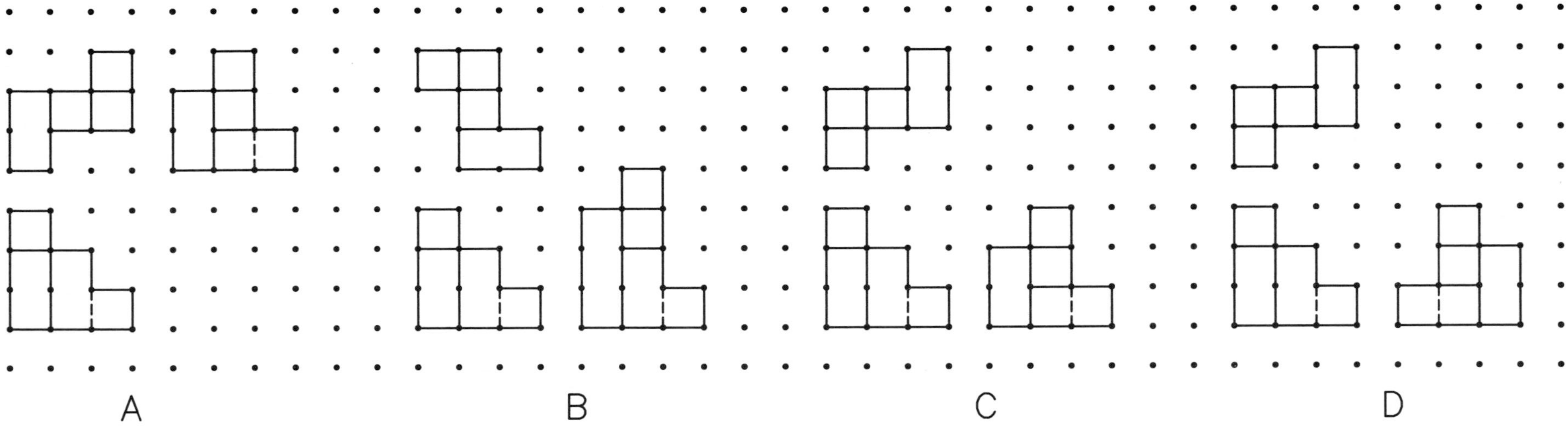

A B C D

For the objects shown in isometric below, sketch the top, front, and right side views in the space provided. Make sure that your views are properly aligned.

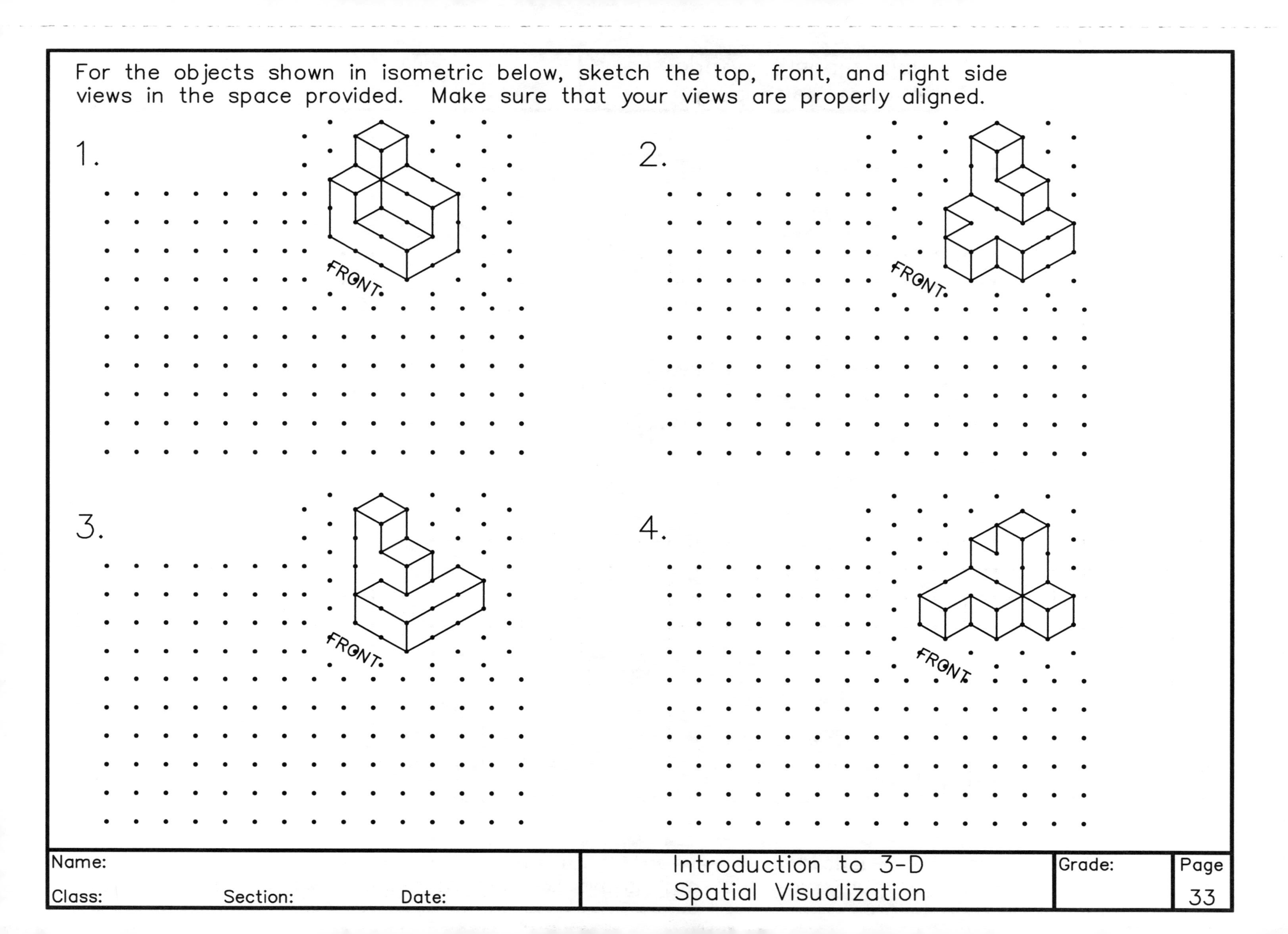


| Name: | Introduction to 3-D Spatial Visualization | Grade: | Page |
|---|---|---|---|
| Class: Section: Date: | | | 33 |

For the objects shown in isometric below, sketch the top, front, and right side views in the space provided. Make sure that your views are properly aligned.

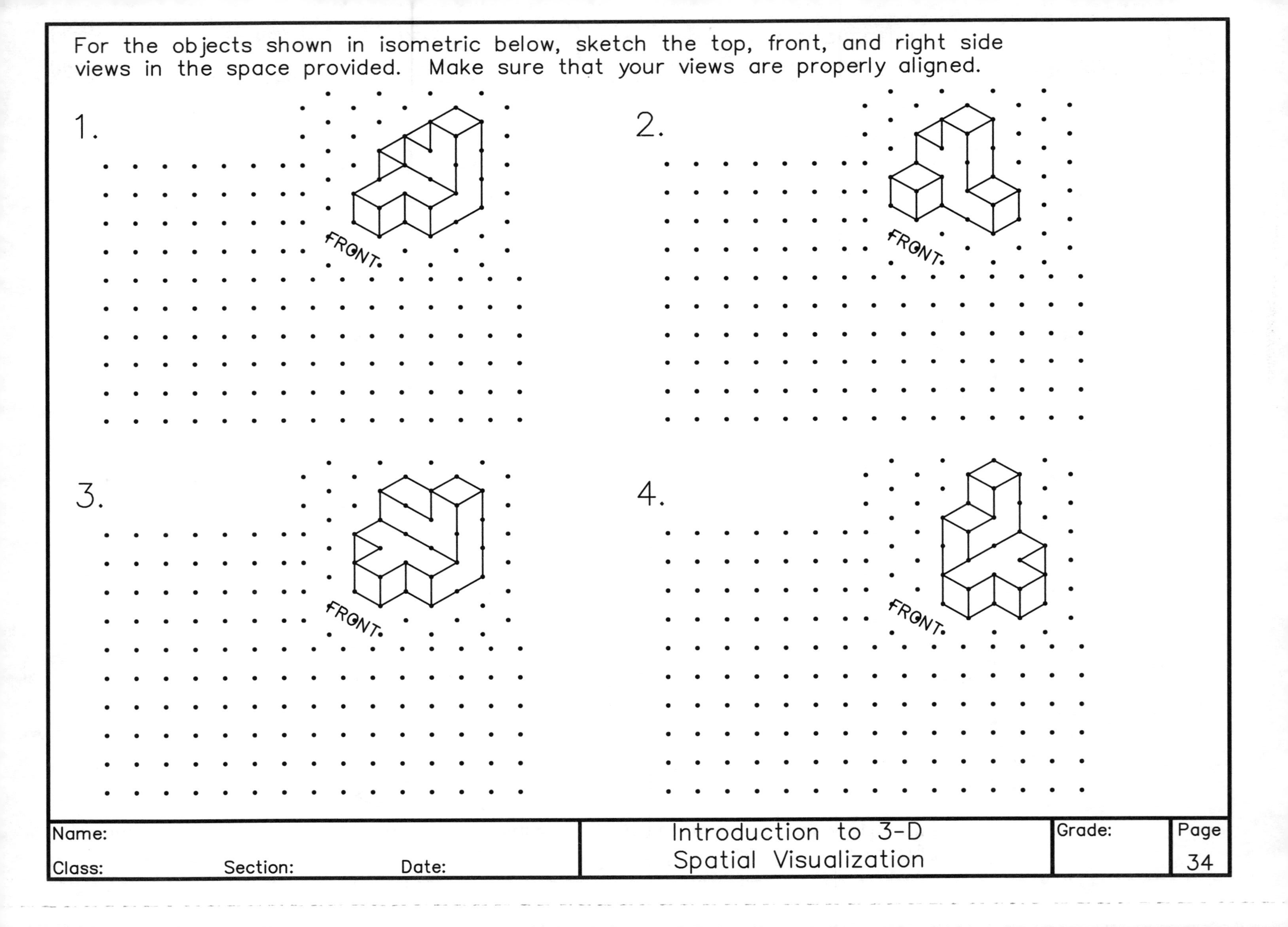

For the objects shown in isometric below, sketch the top, front, and right side views in the space provided. Make sure that your views are properly aligned.

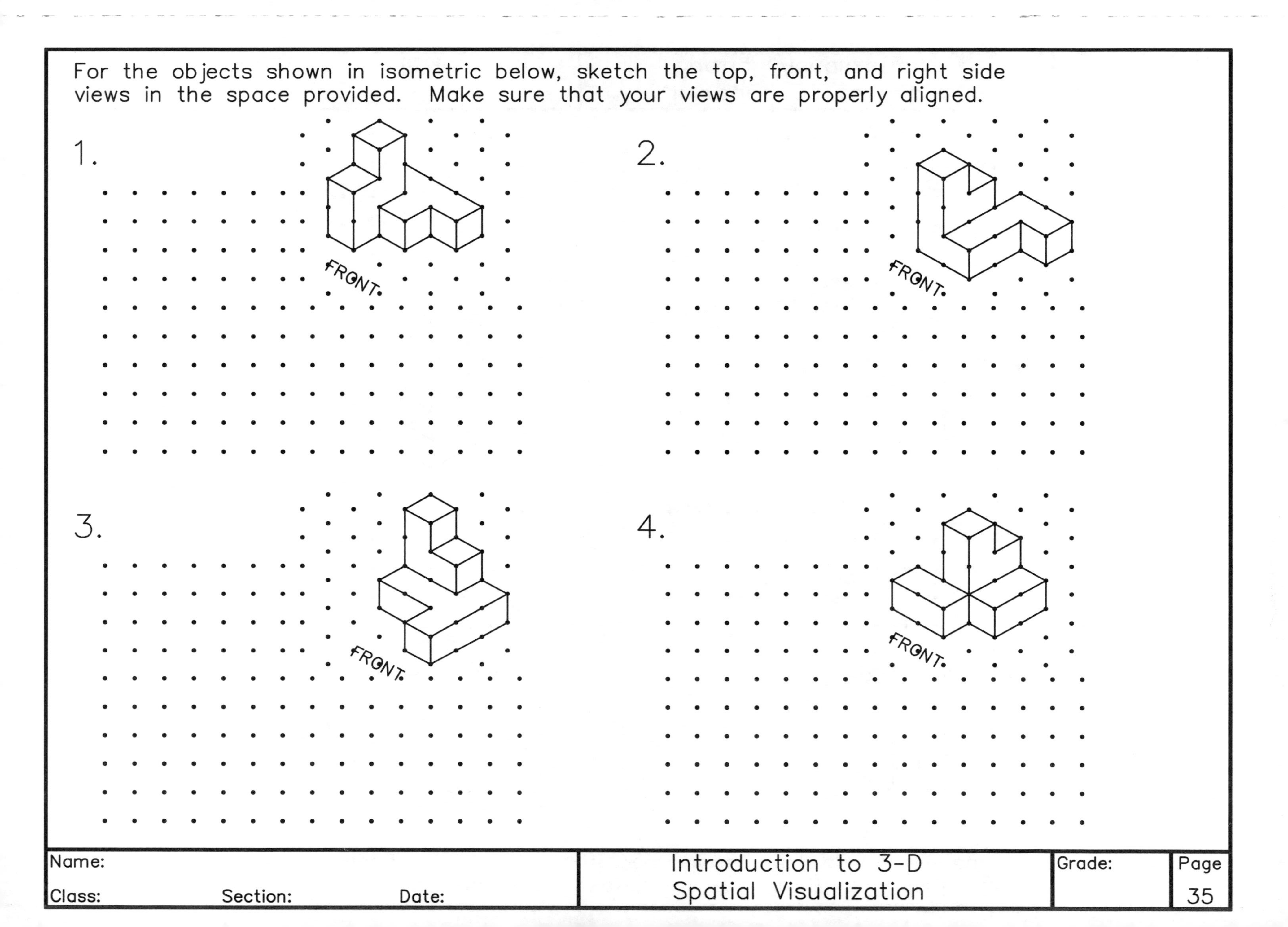

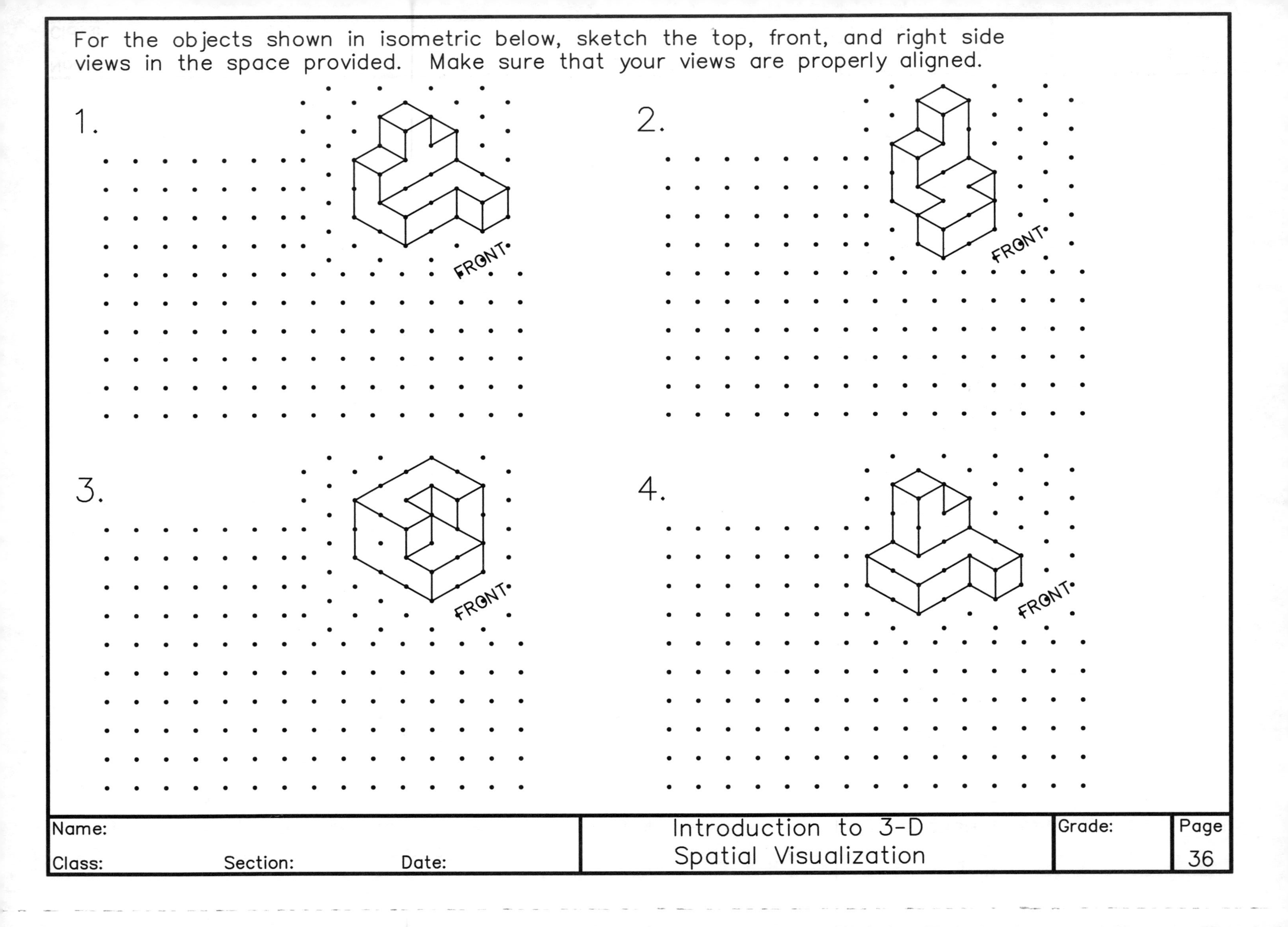
For the objects shown in isometric below, sketch the top, front, and right side views in the space provided. Make sure that your views are properly aligned.

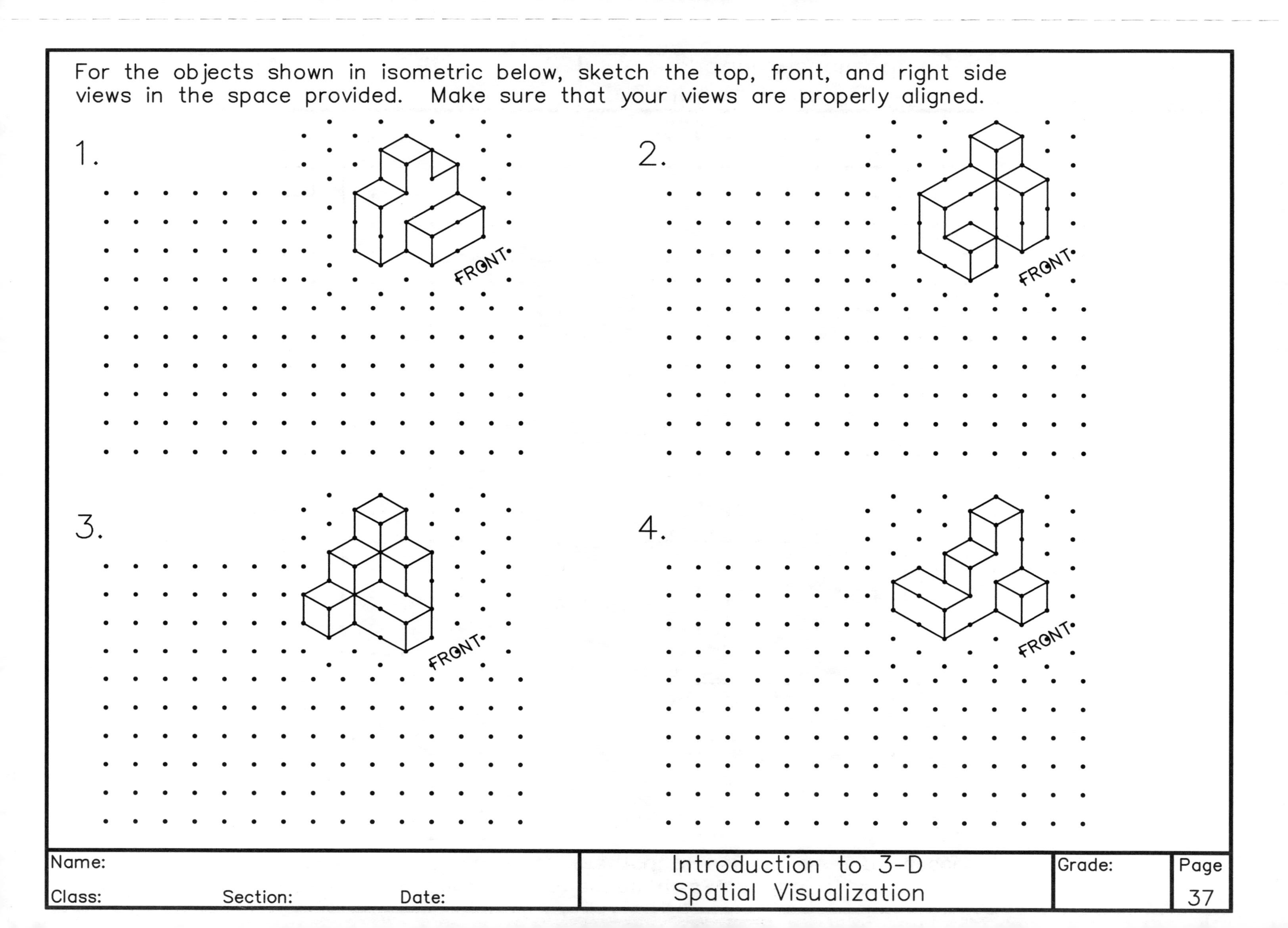

For the objects shown in isometric below, sketch the top, front, and right side views in the space provided. Make sure that your views are properly aligned.

1.



2.



3.



4.

For the objects shown in orthographic projection below, construct an isometric view in the space provided. Use the box method to assist you if necessary.

1.

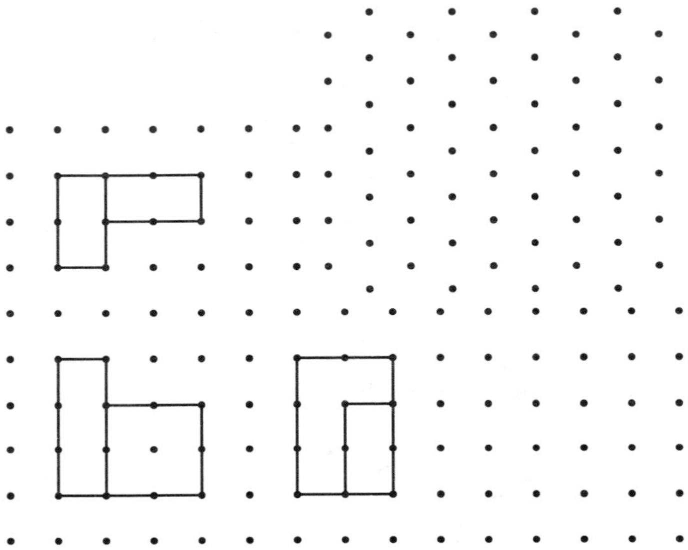

2.

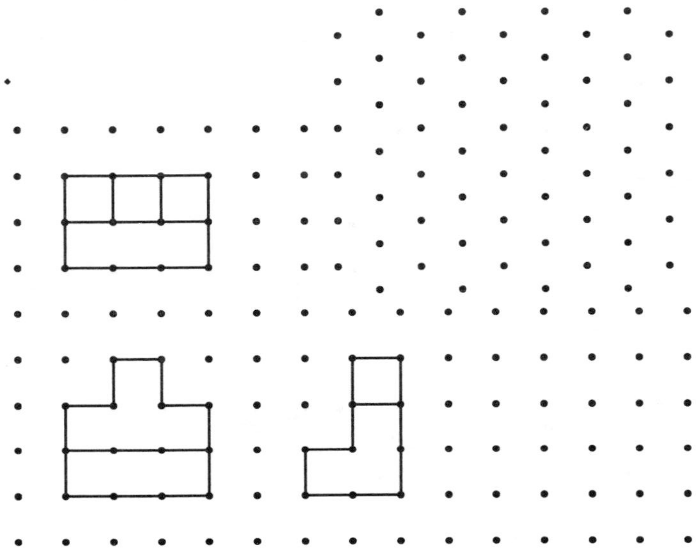

3.

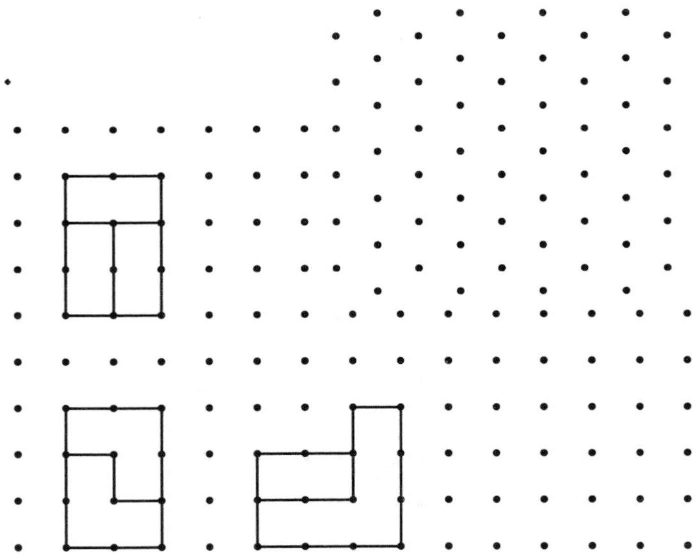

4.

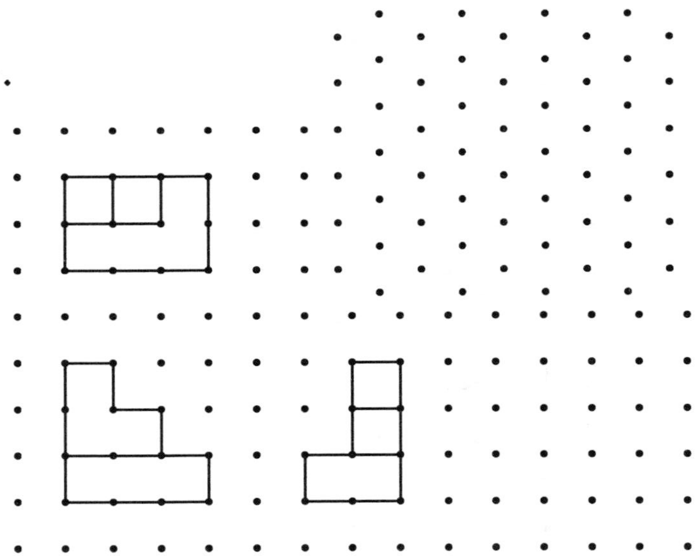

| | | | |
|---|---|---|---|
| Name:<br>Class: Section: Date: | Introduction to 3-D<br>Spatial Visualization | Grade: | Page<br>38 |

For the objects shown in orthographic projection below, construct an isometric view in the space provided. Use the box method to assist you if necessary.

1.

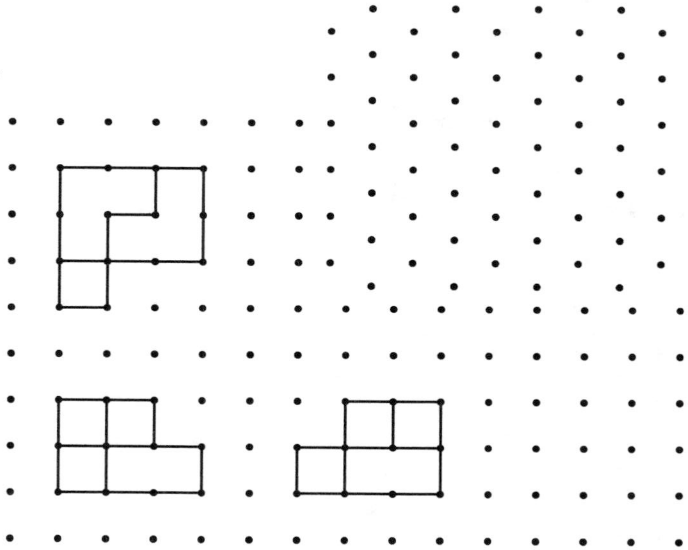

2.

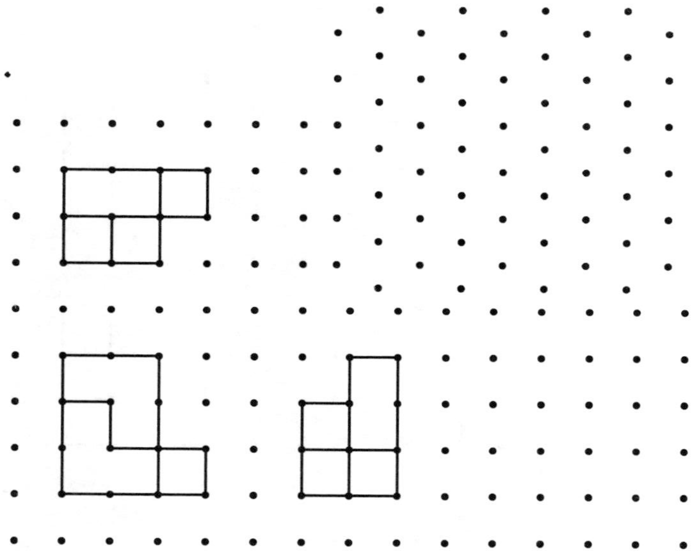

3.

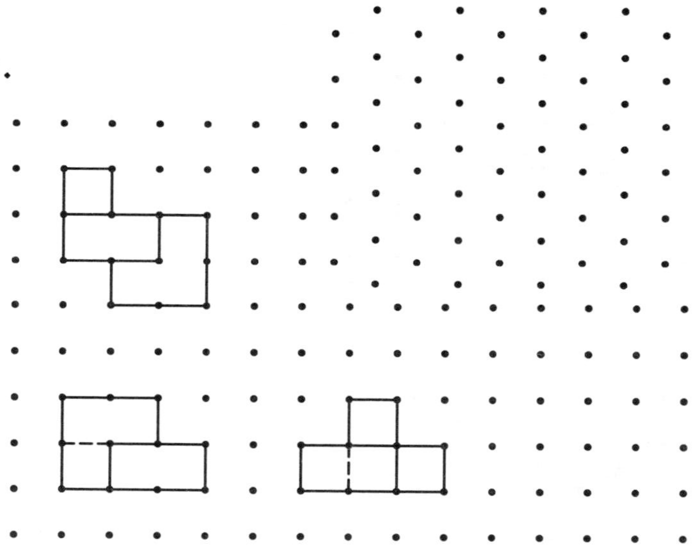

4.

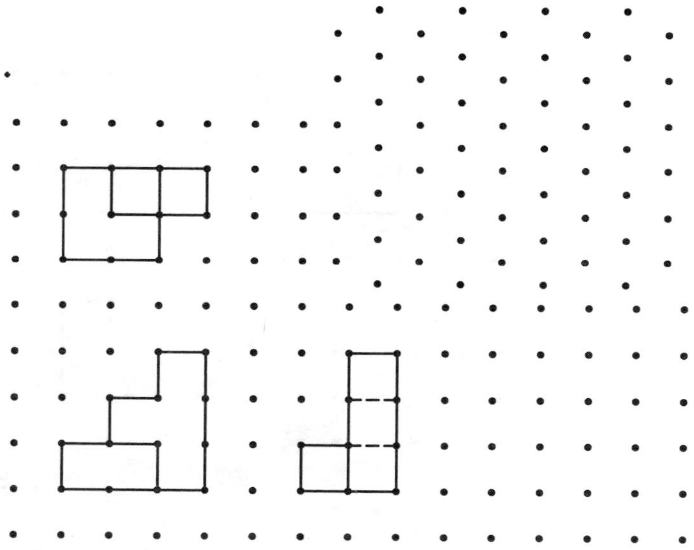

For the objects shown in orthographic projection below, construct an isometric view in the space provided. Use the box method to assist you if necessary.

1.

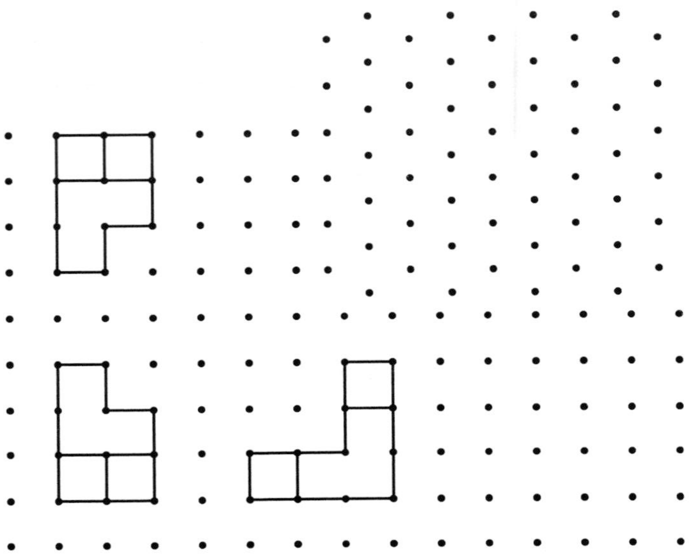

2.

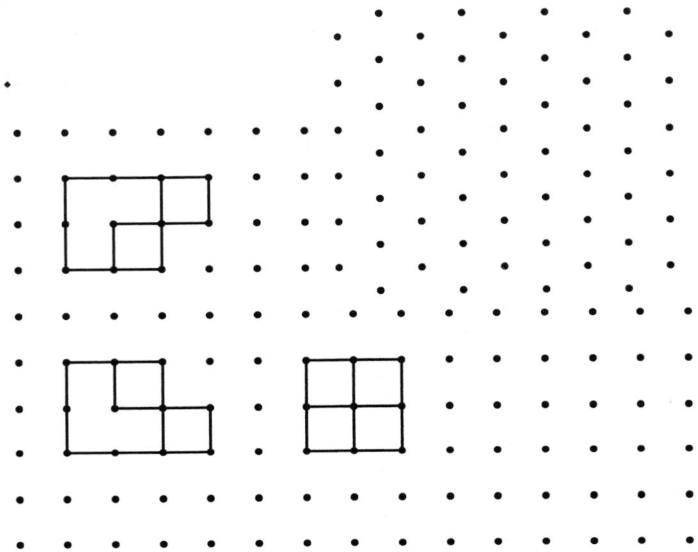

3.

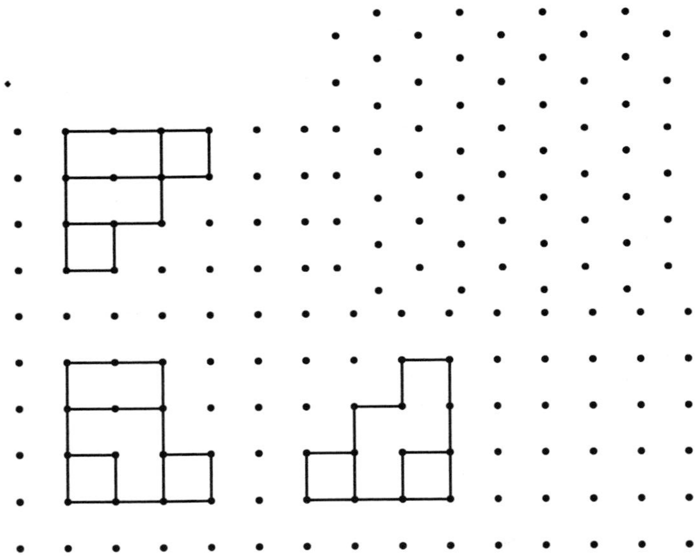

4.

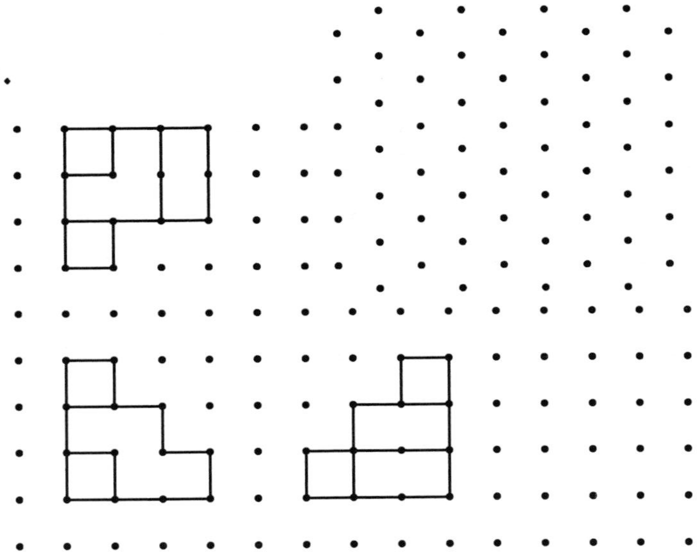

For the objects shown in orthographic projection below, construct an isometric view in the space provided. Use the box method to assist you if necessary.

1.

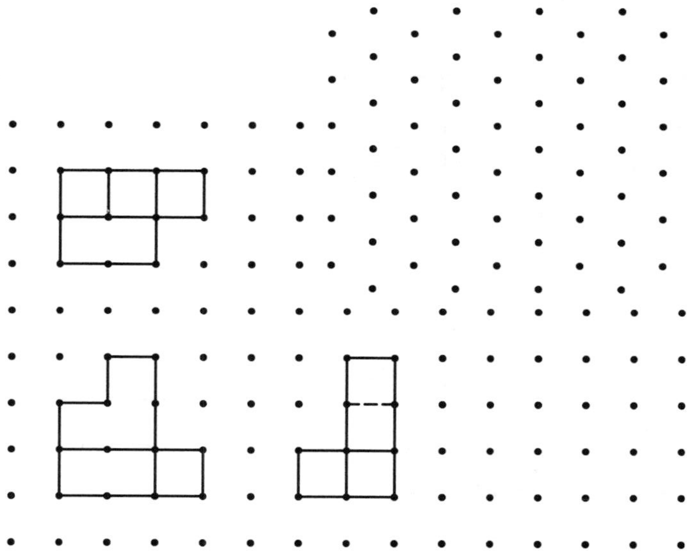

2.

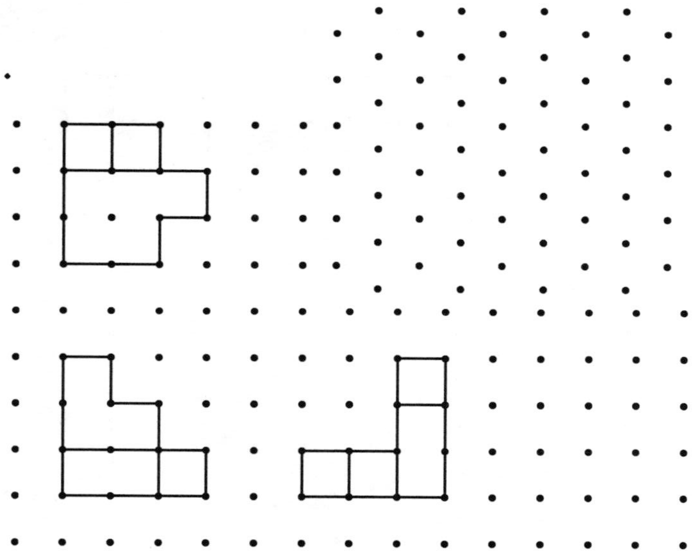

3.

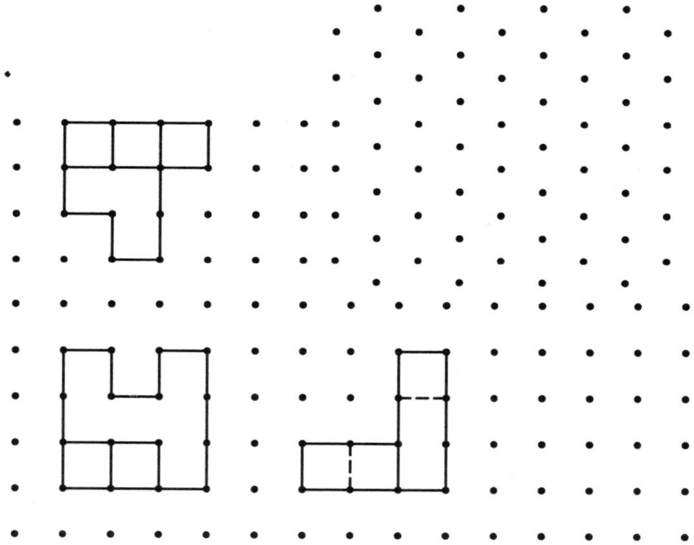

4.

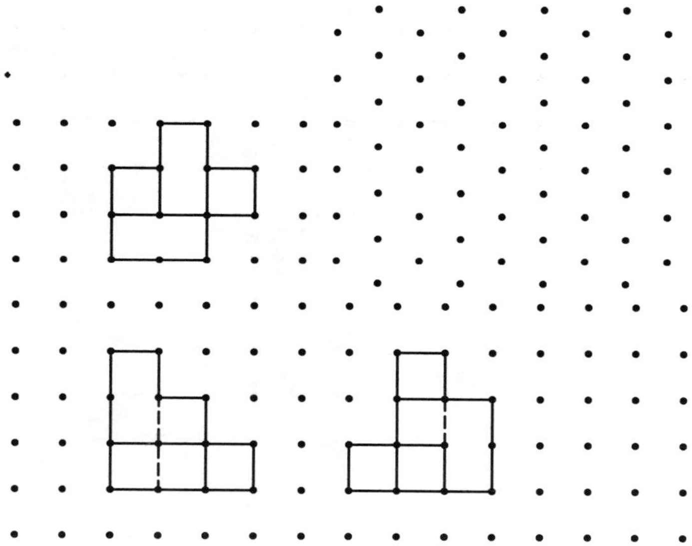

| Name: | Introduction to 3-D | Grade: | Page |
|---|---|---|---|
| Class: Section: Date: | Spatial Visualization | | 41 |

For the objects shown in orthographic projection below, construct an isometric view in the space provided. Use the box method to assist you if necessary.

1.

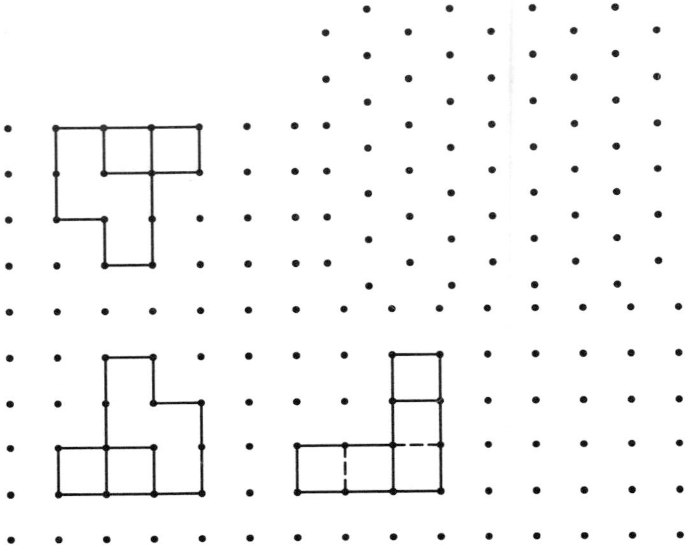

2.

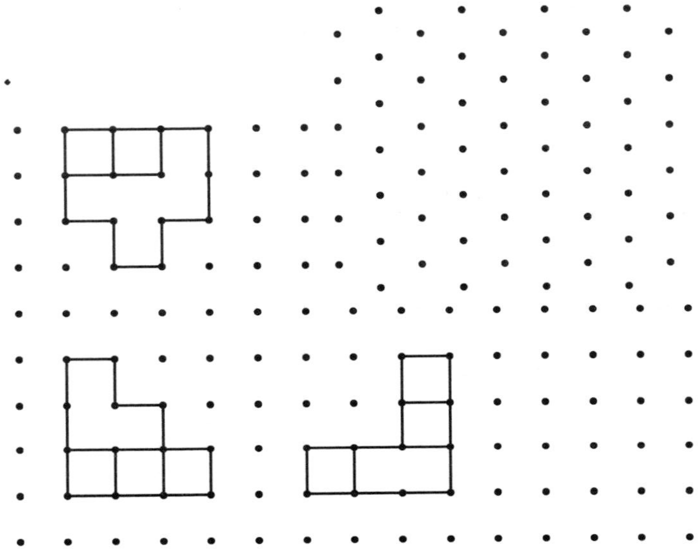

3.

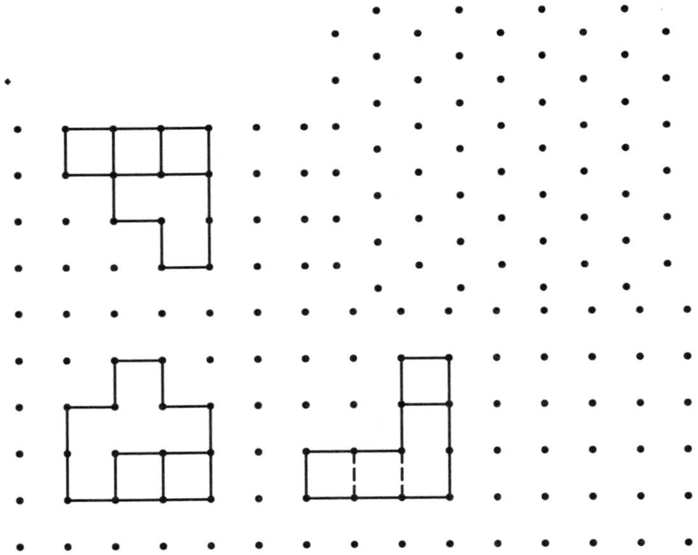

4.

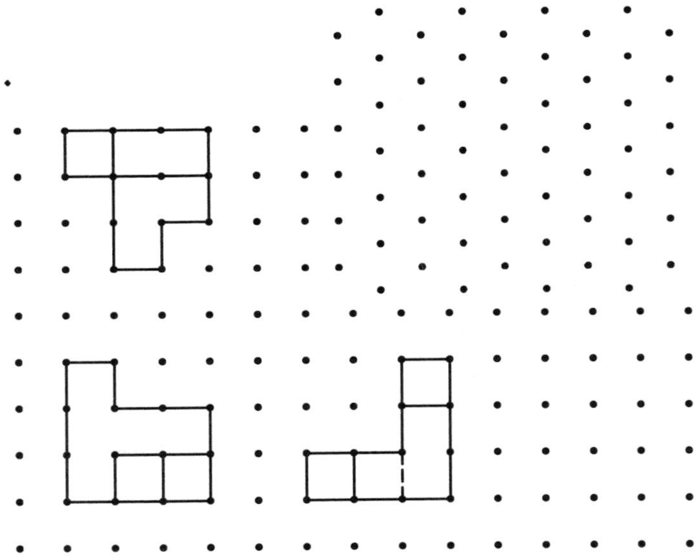

| | | | |
|---|---|---|---|
| Name:<br>Class: Section: Date: | Introduction to 3-D<br>Spatial Visualization | Grade: | Page<br>42 |

# Flat Patterns

Many 3-D objects can be formed by folding up a 2-D flat pattern:

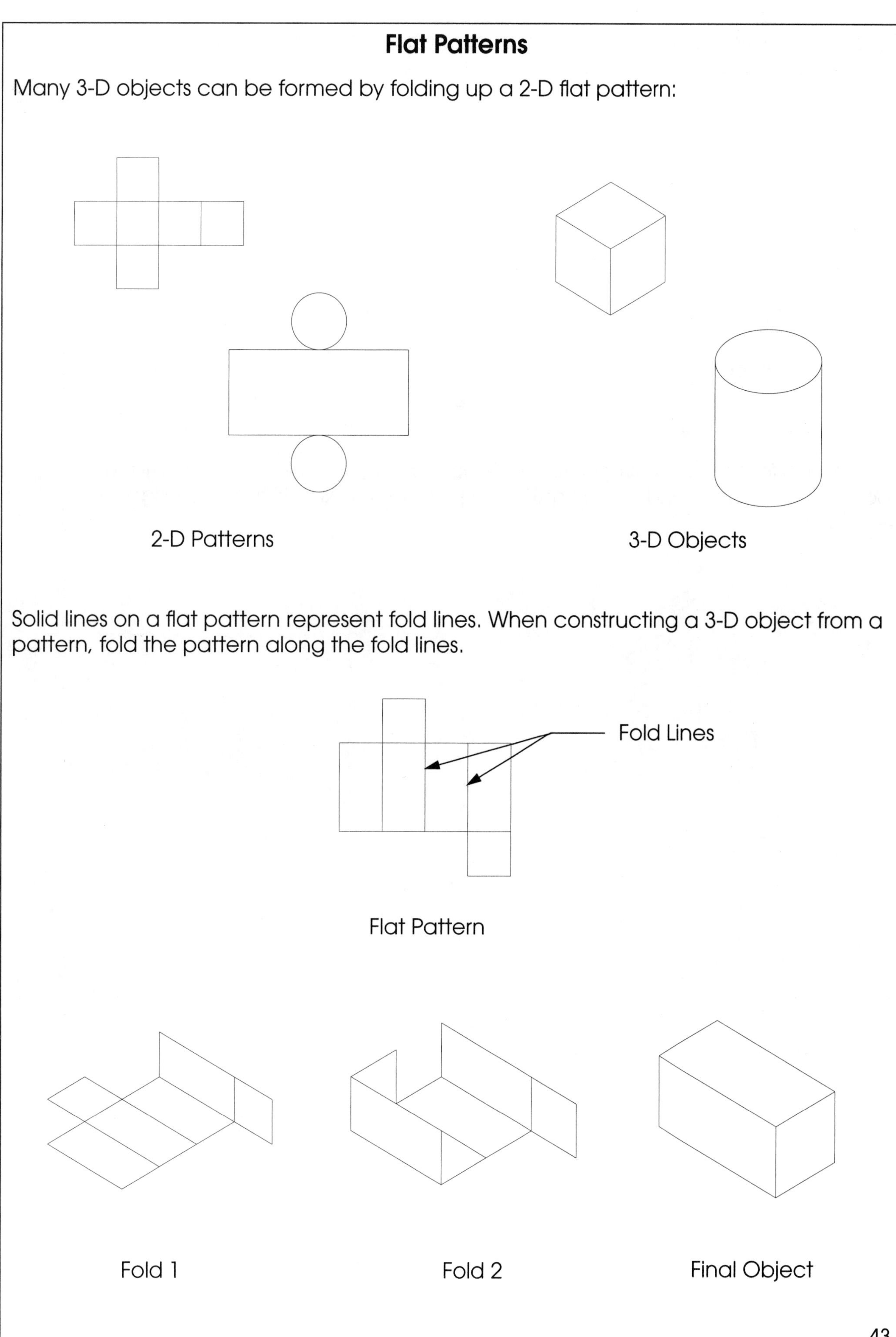


Solid lines on a flat pattern represent fold lines. When constructing a 3-D object from a pattern, fold the pattern along the fold lines.

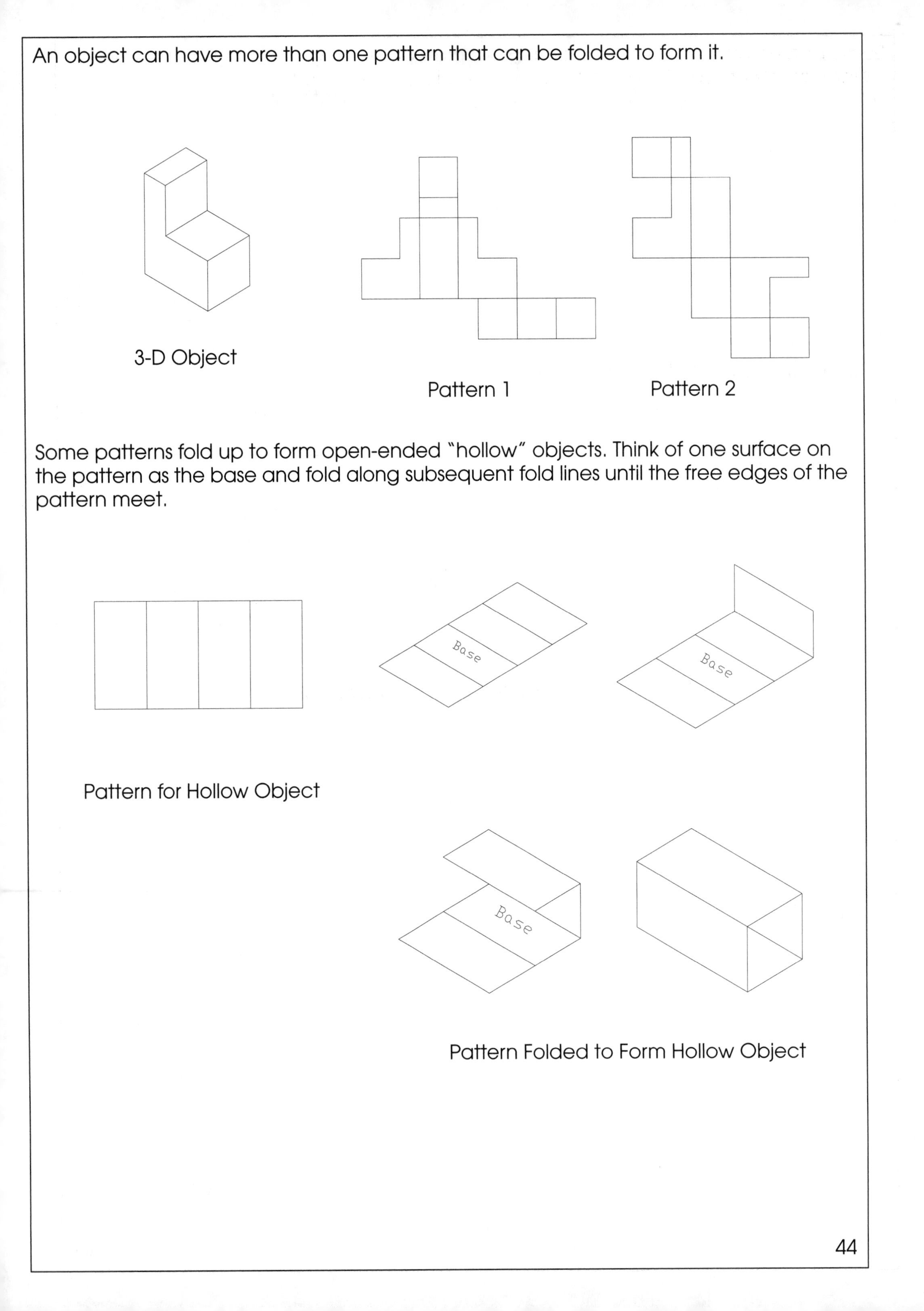

An object can have more than one pattern that can be folded to form it.

3-D Object

Pattern 1

Pattern 2

Some patterns fold up to form open-ended "hollow" objects. Think of one surface on the pattern as the base and fold along subsequent fold lines until the free edges of the pattern meet.



Pattern for Hollow Object

Pattern Folded to Form Hollow Object

Some patterns fold up to form objects with closed ends. Imagine folding up the sides as before, and then fold the ends into place to "close" the object.

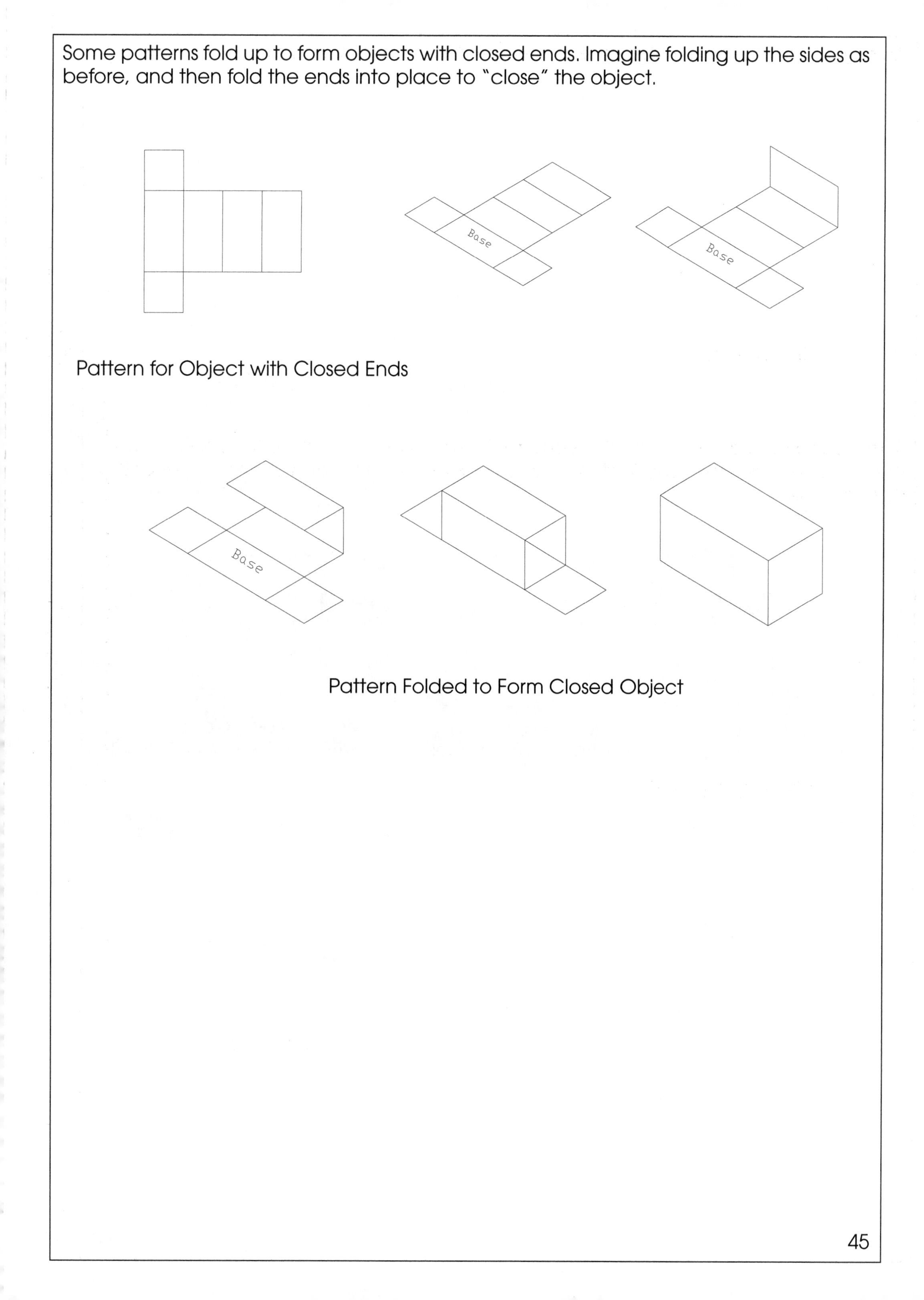


Pattern for Object with Closed Ends

Pattern Folded to Form Closed Object

Sometimes there are markings on a flat pattern. When visualizing folding up a pattern with markings on it, the markings must be oriented on the object the same way they are on the pattern. Also, surfaces that are next to each other on the pattern must be next to each other on the object.

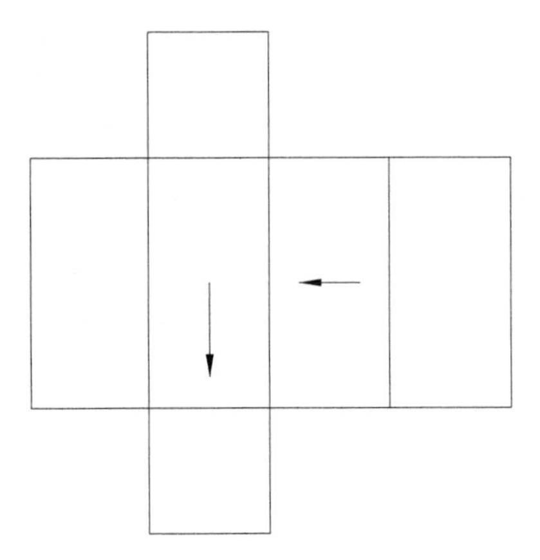

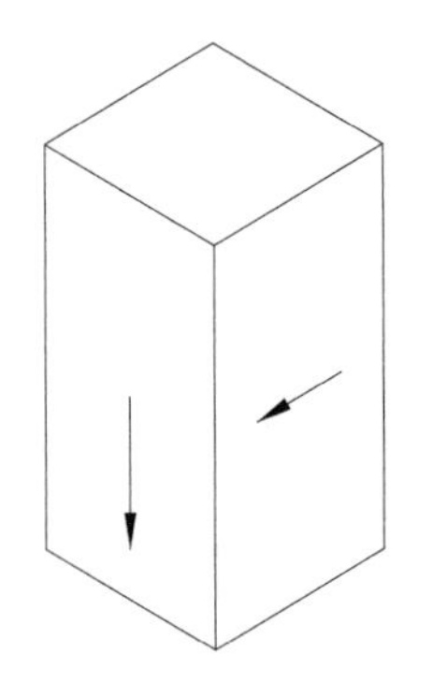

Correct 3-D Object

Markings not on Adjacent Surfaces

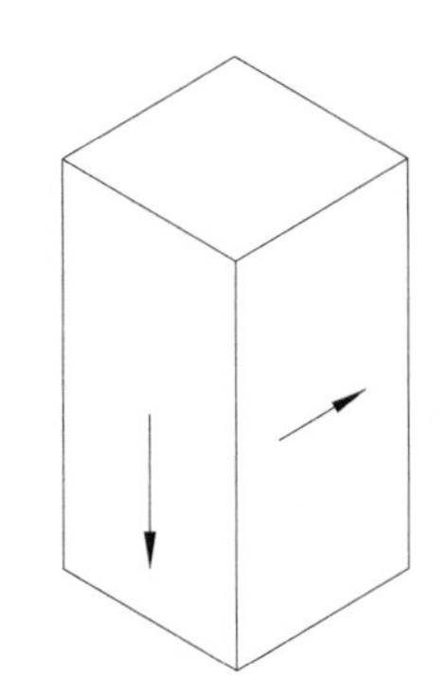

Markings on Adjacent Surfaces in Wrong Orientation

The patterns shown below fold up to form a cube with the word "CUBE" spelled around its four sides. Complete each pattern by placing the "B" on it in the correct orientation.

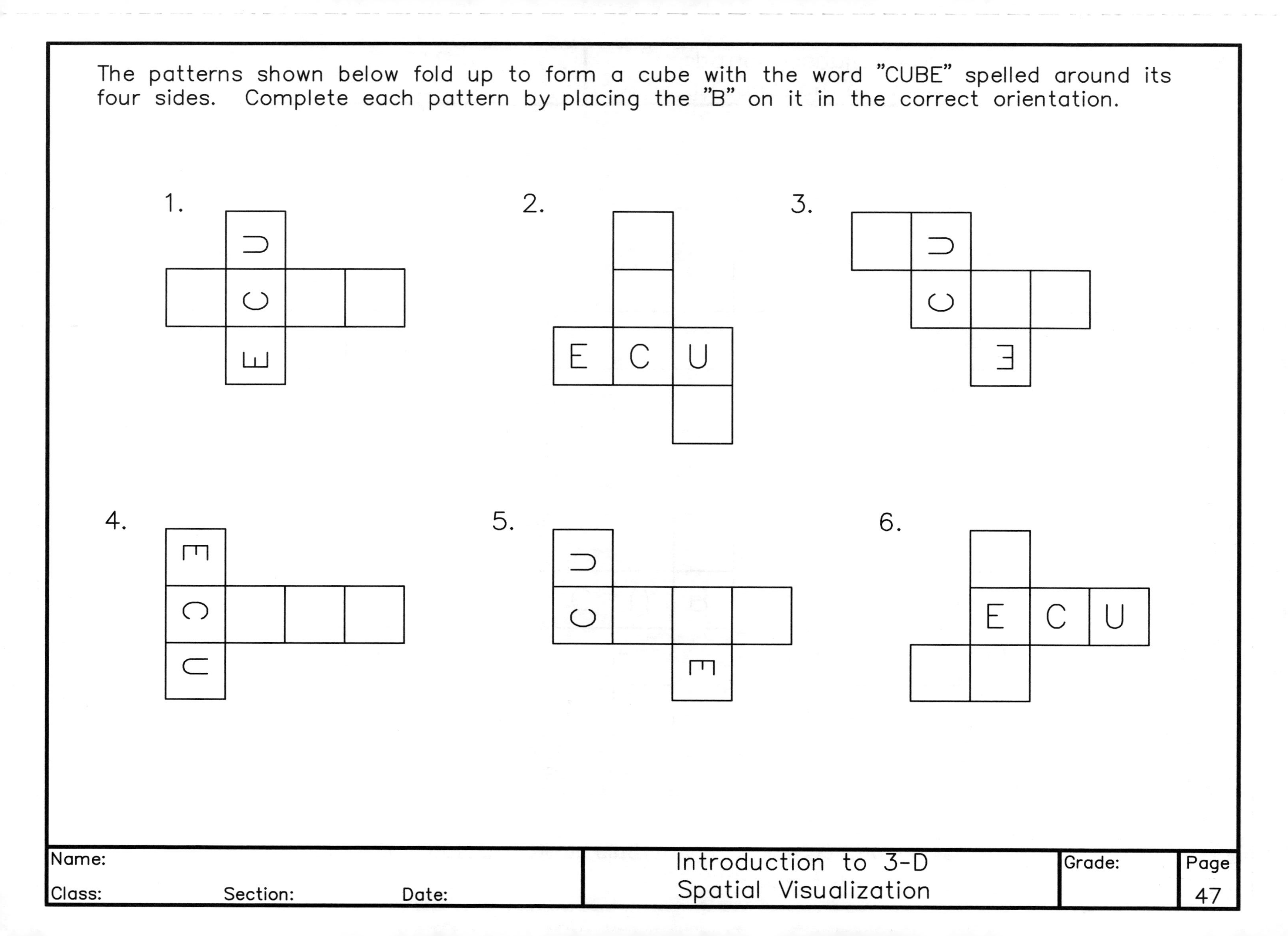

The patterns shown below fold up to form a cube with the word "CUBE" spelled around its four sides. Complete each pattern by placing the "E" on it in the correct orientation.

1. C U B

2. C U B

3. C U B

4. B U C

5. B C U

6. C U B

The patterns shown below fold up to form a cube with the word "CUBE" spelled around its four sides. Complete each pattern by placing the "E" on it in the correct orientation.

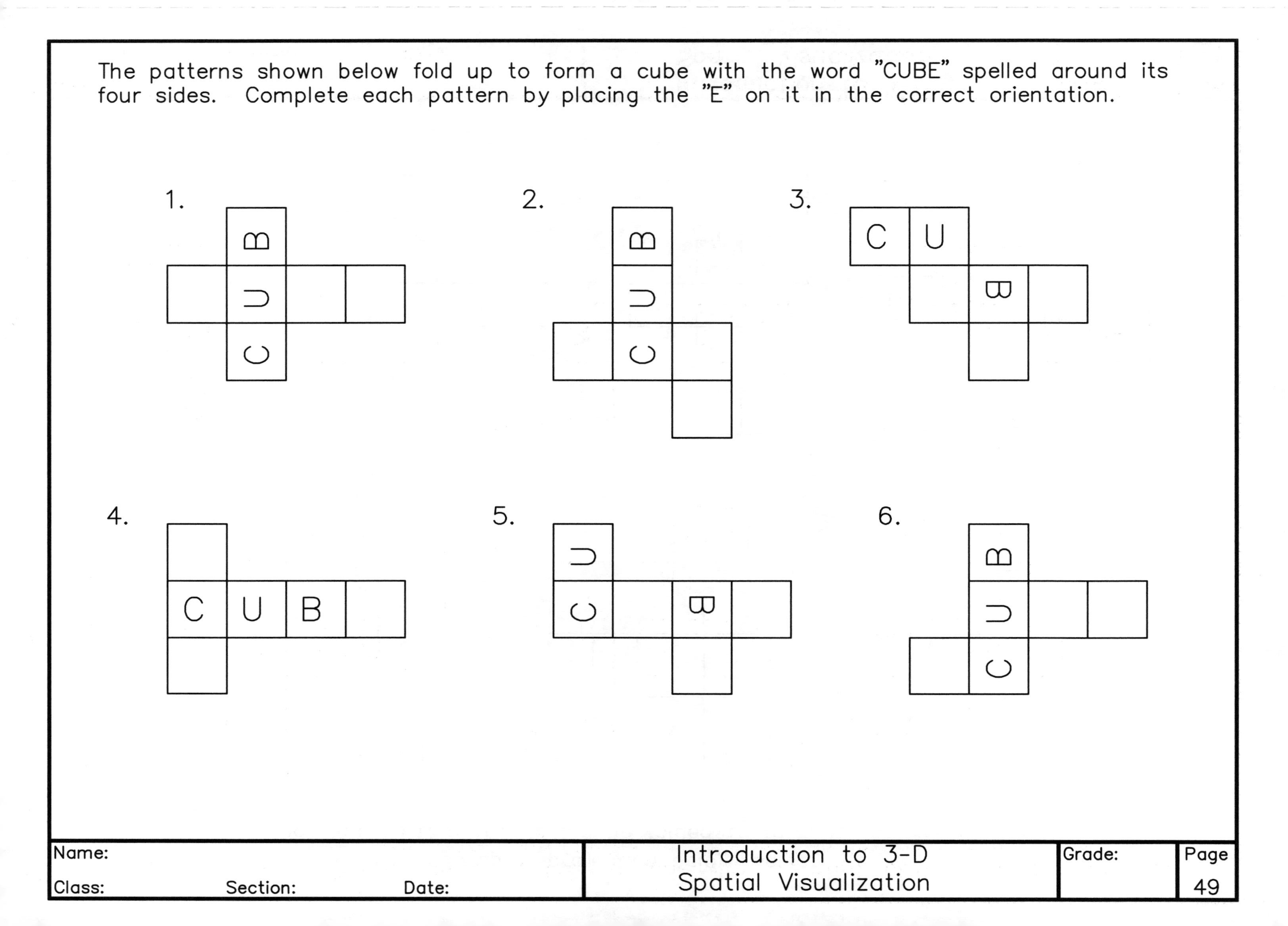


Name:

Class: Section: Date:

Given the object shown in orthographic projection below, select the letter of the correct flat pattern that could be folded to form it.

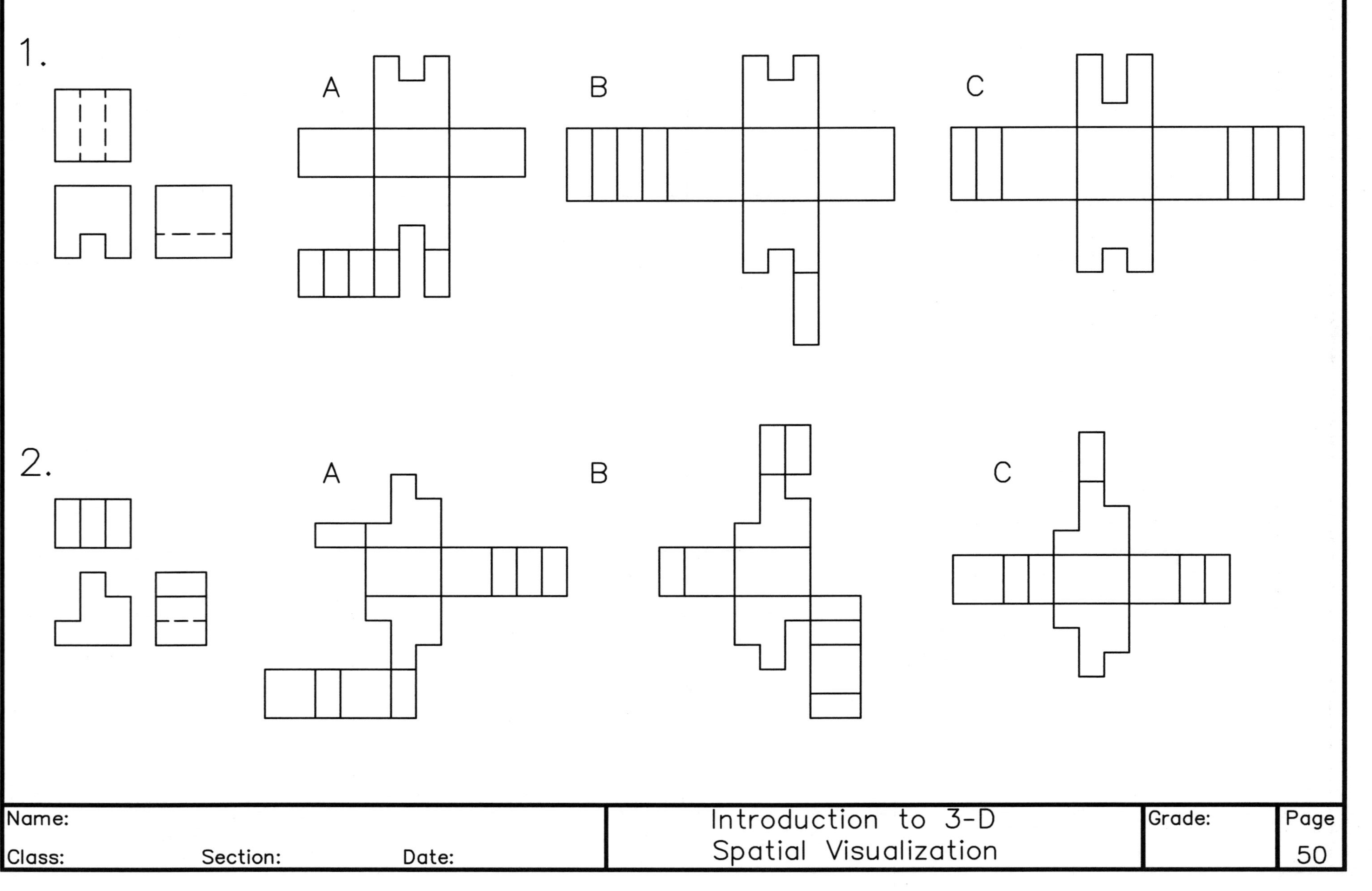

Given the object shown in orthographic projection below, select the letter of the correct flat pattern that could be folded to form it.

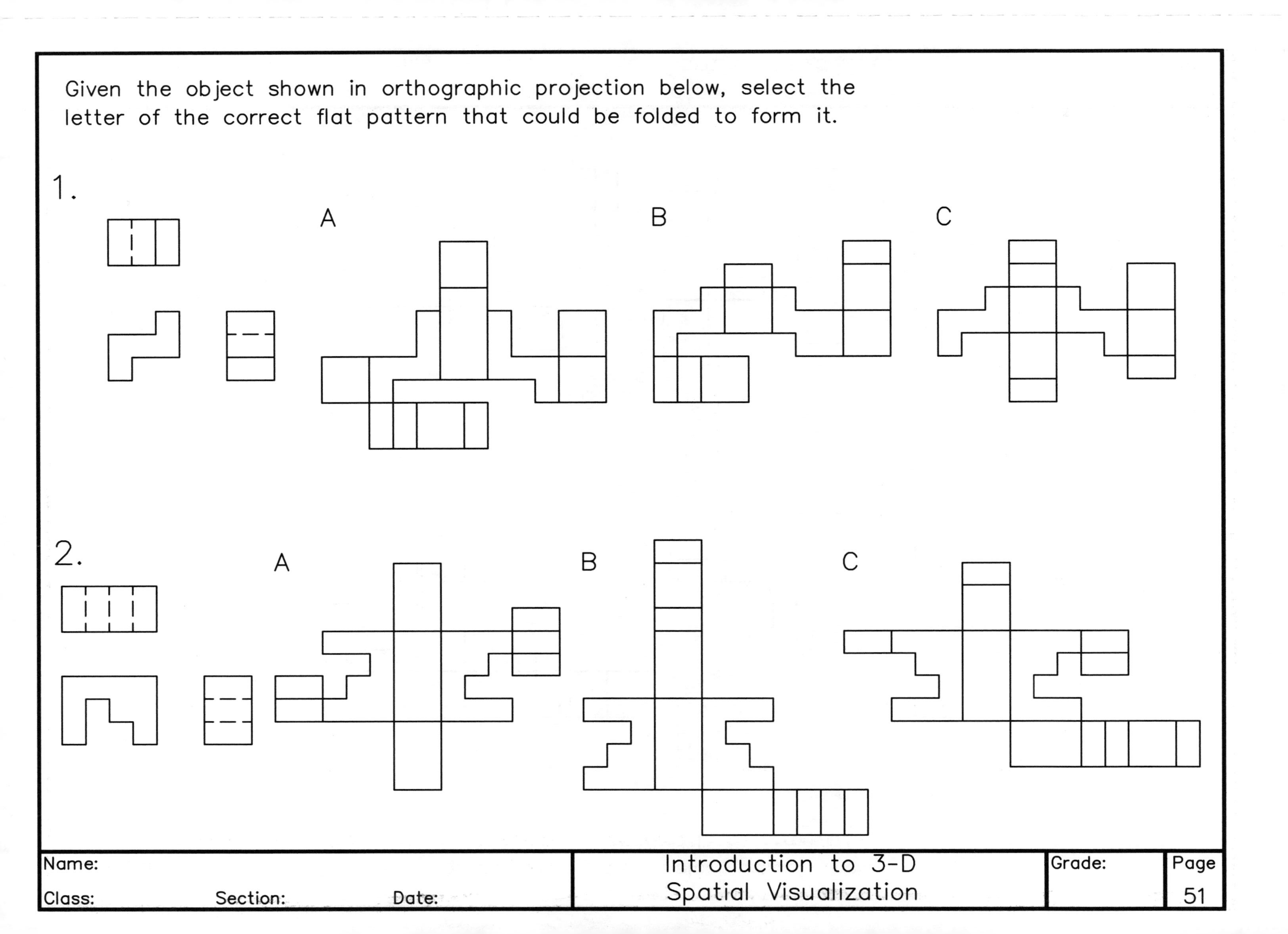


| Name: | Introduction to 3-D Spatial Visualization | Grade: | Page |
|---|---|---|---|
| Class: Section: Date: | | | 51 |

Given the object shown in orthographic projection below, select the letter of the correct flat pattern that could be folded to form it.

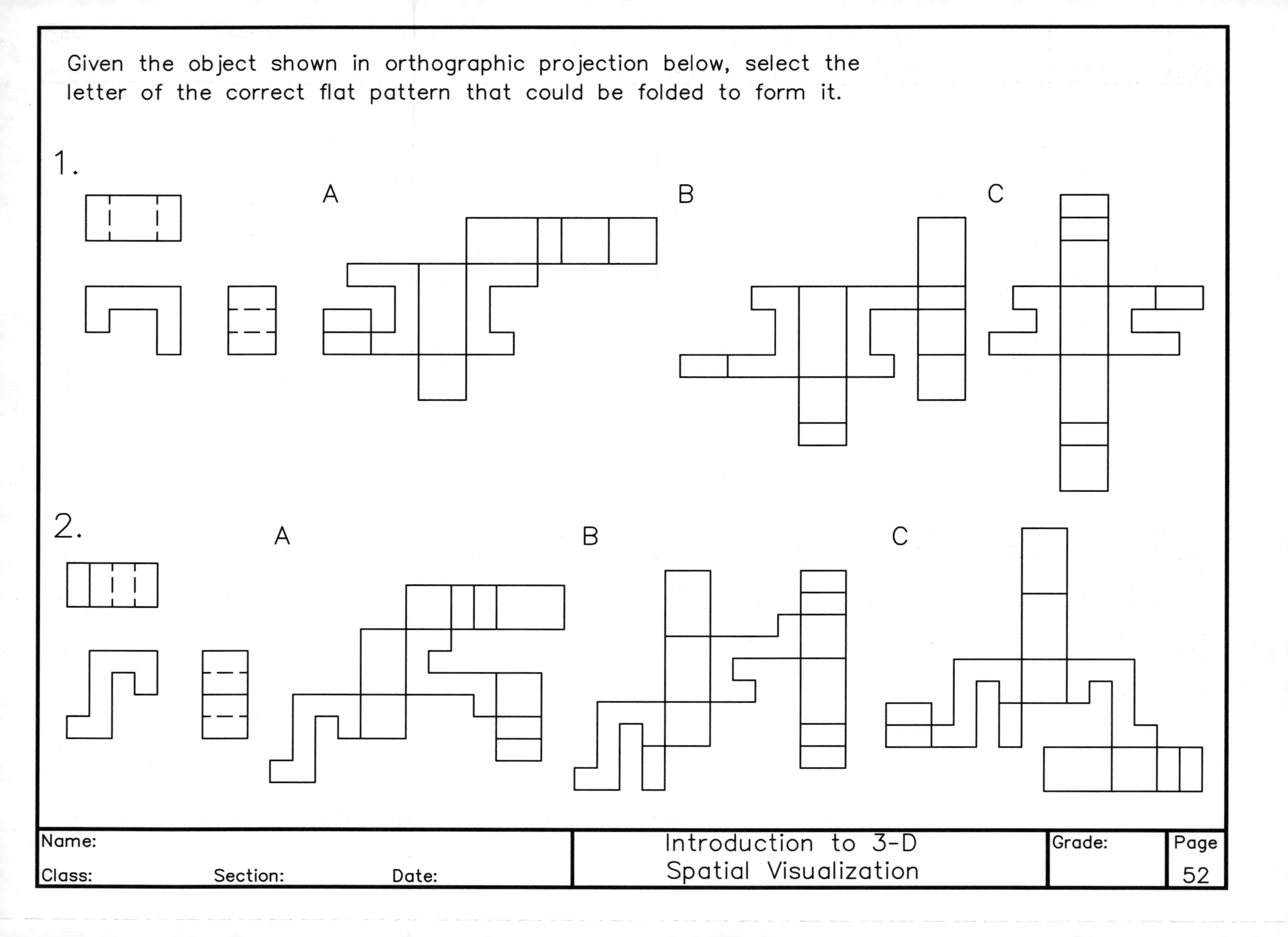

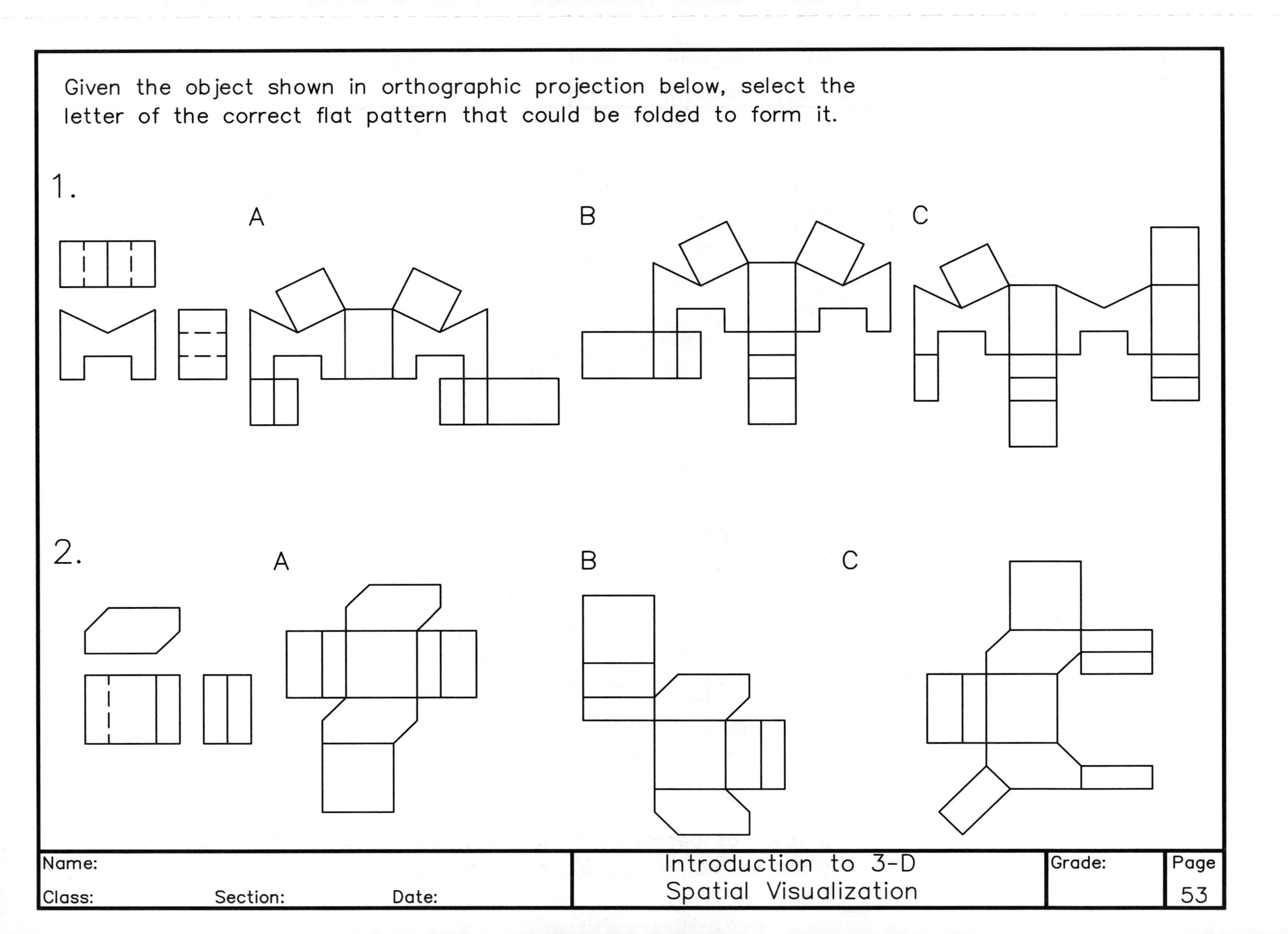
Given the object shown in orthographic projection below, select the letter of the correct flat pattern that could be folded to form it.
1.
A
B
C
2.
A
B
C
Name:
Class:
Section:
Date:
Introduction to 3-D Spatial Visualization
Grade:
Page
53

Given the object shown in orthographic projection below, select the letter of the correct flat pattern that could be folded to form it.

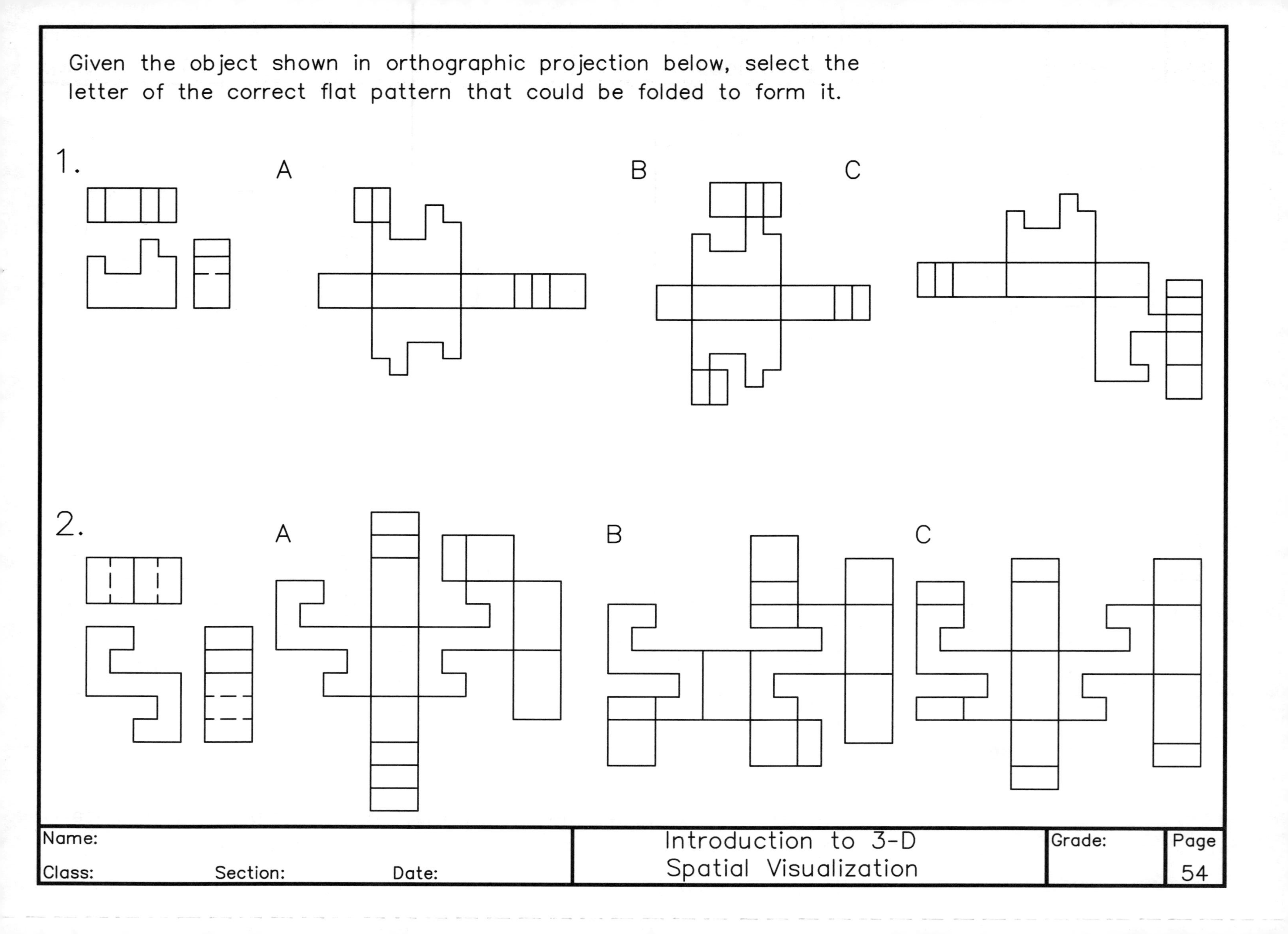

Select the object that results from folding up the pattern shown on the left below.

1.

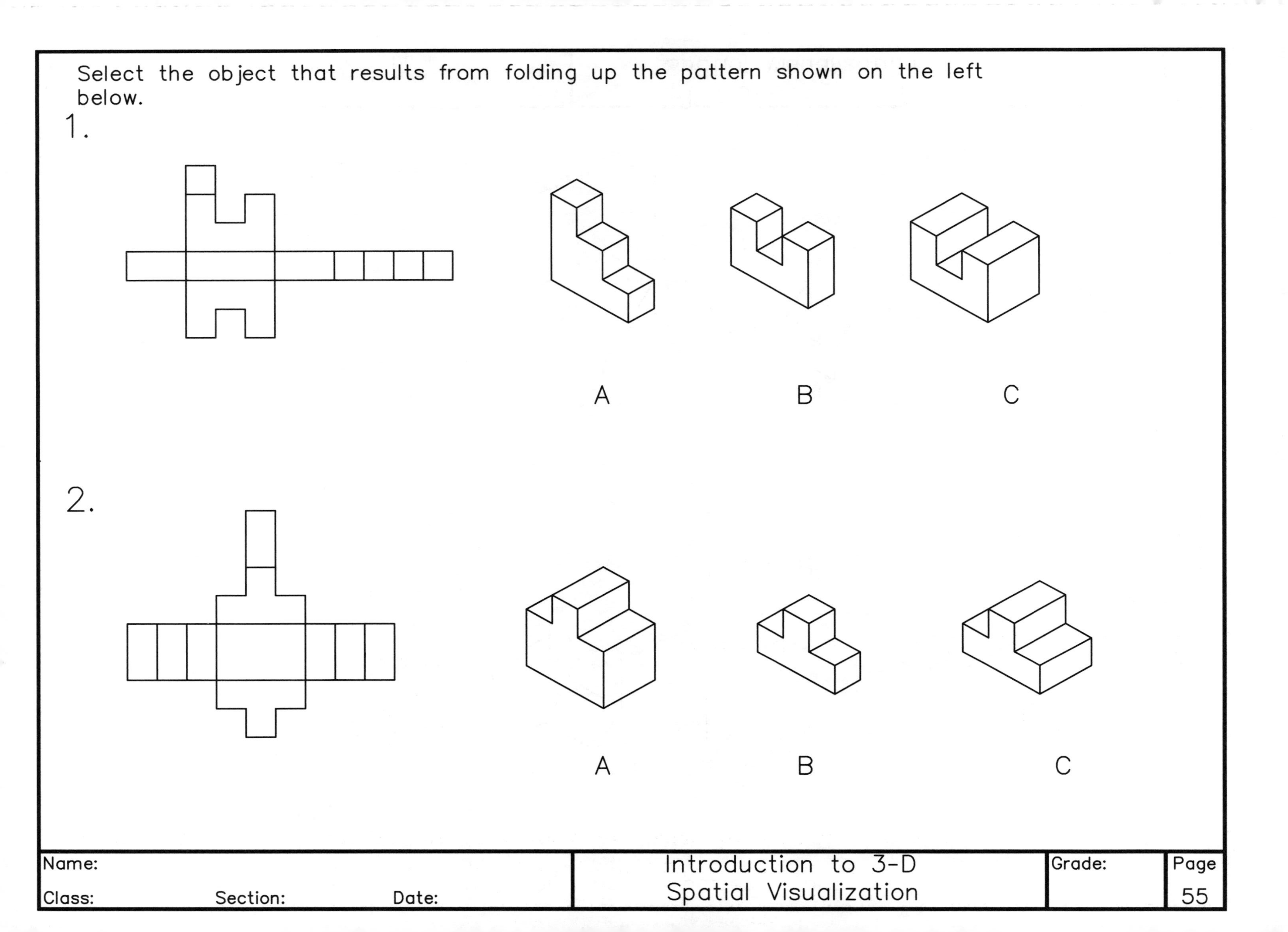

Select the object that results from folding up the pattern shown on the left below.

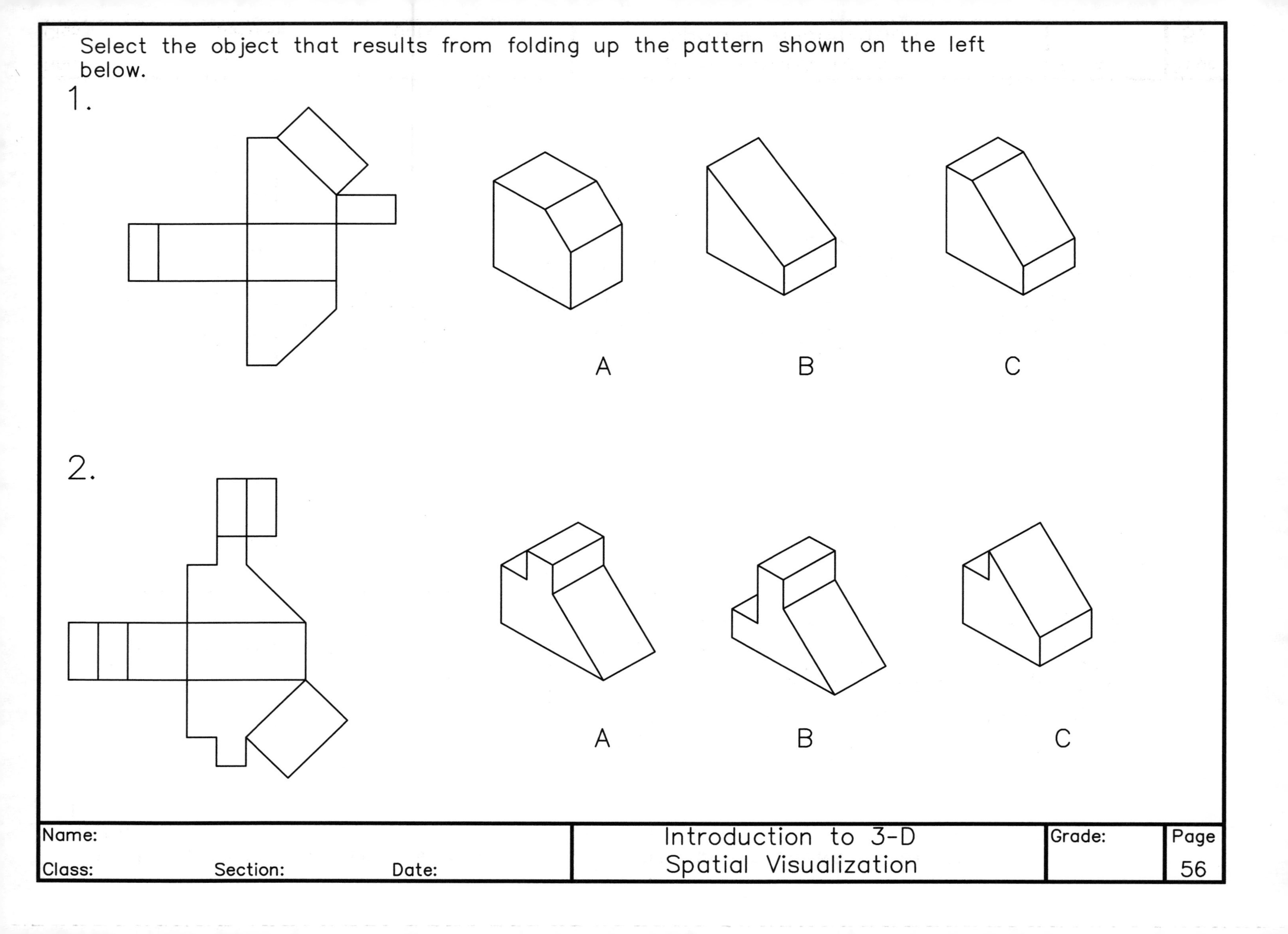

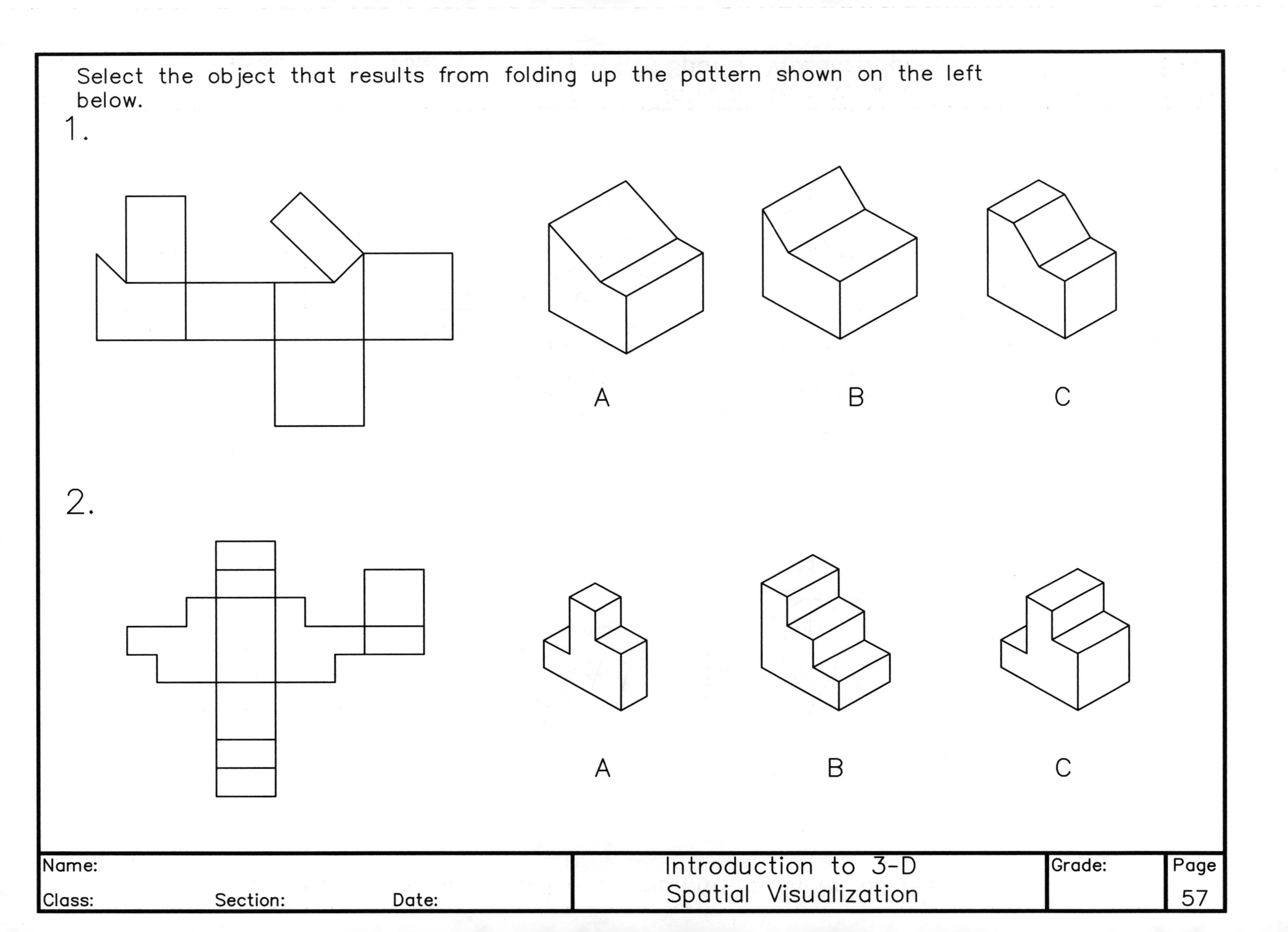
Select the object that results from folding up the pattern shown on the left below.
1.
A
B
C
2.
A
B
C

Select the object that results from folding up the pattern shown on the left below.

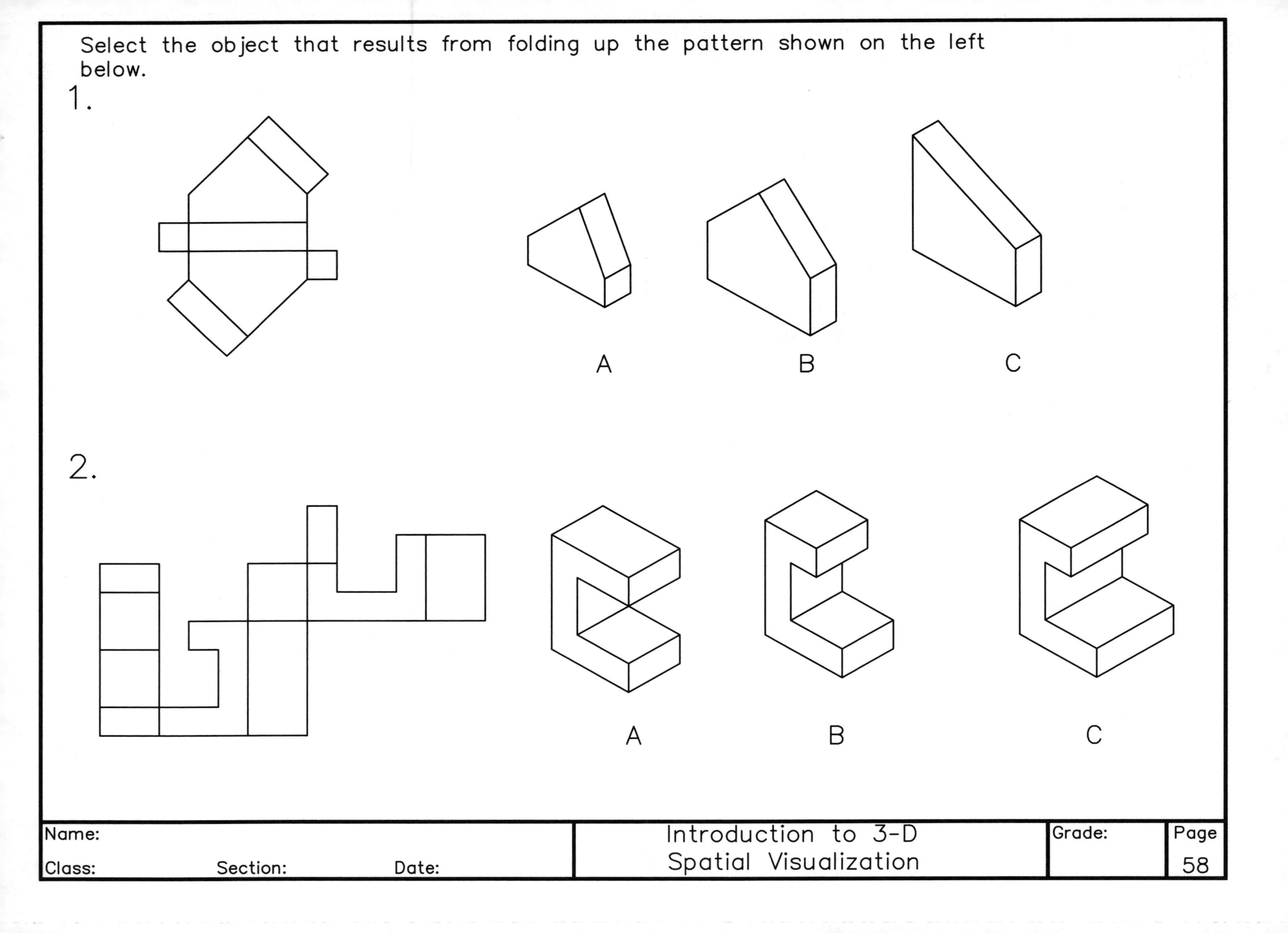

Select the object that results from folding up the pattern shown on the left below.

1.

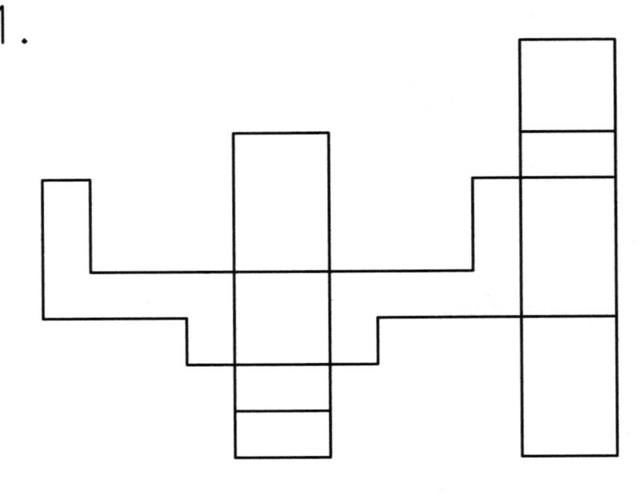

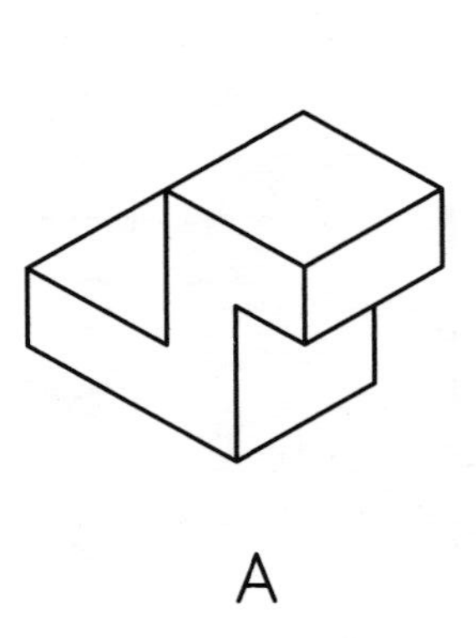

A

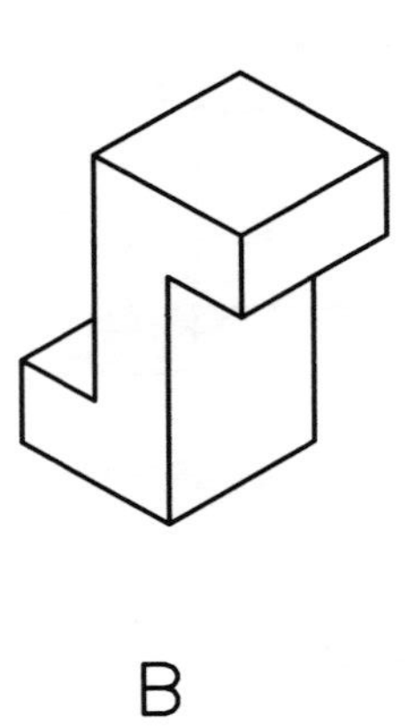

B

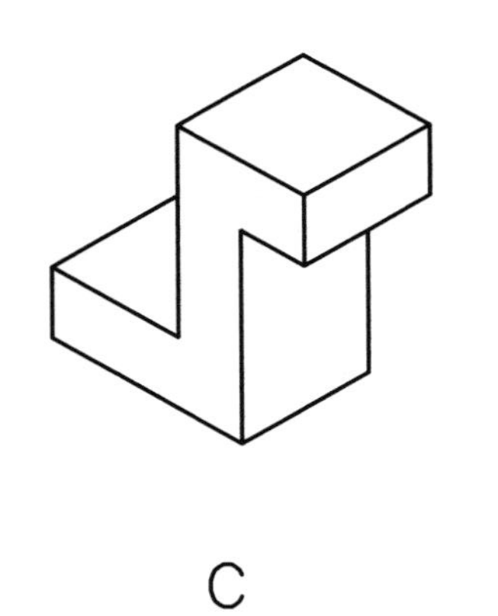

C

2.

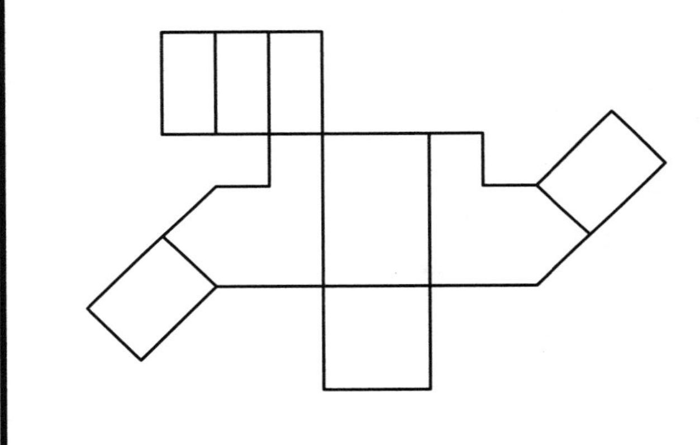

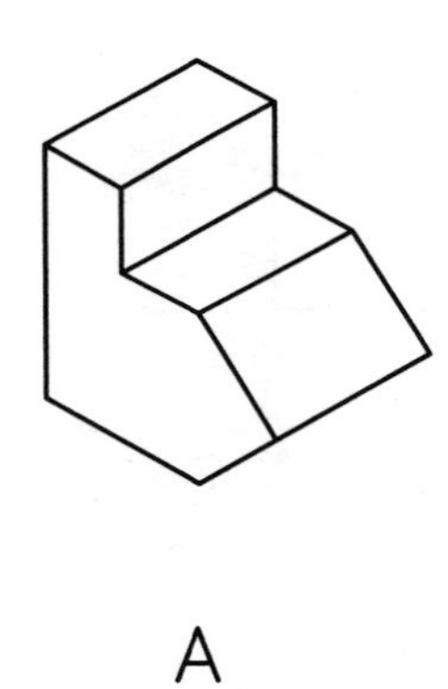

A

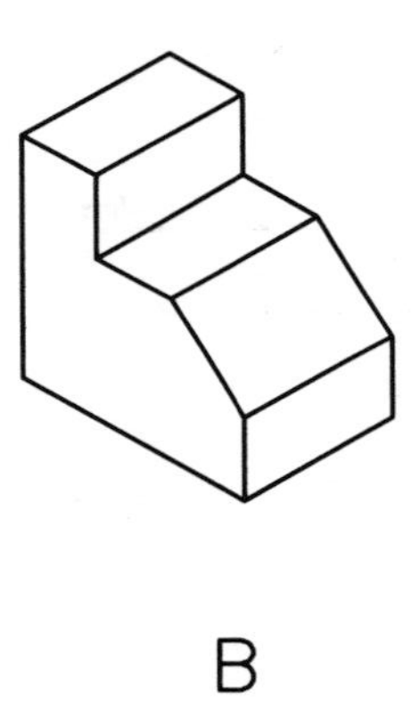

B

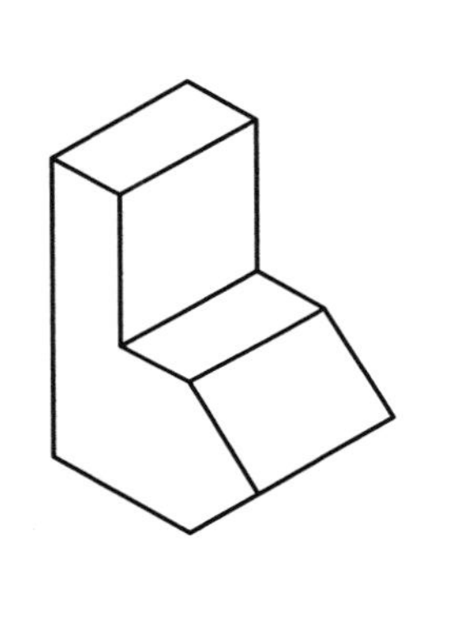

C

For the objects shown in isometric below, select the pattern from the choices given that could be folded up to obtain the object.

1.

A

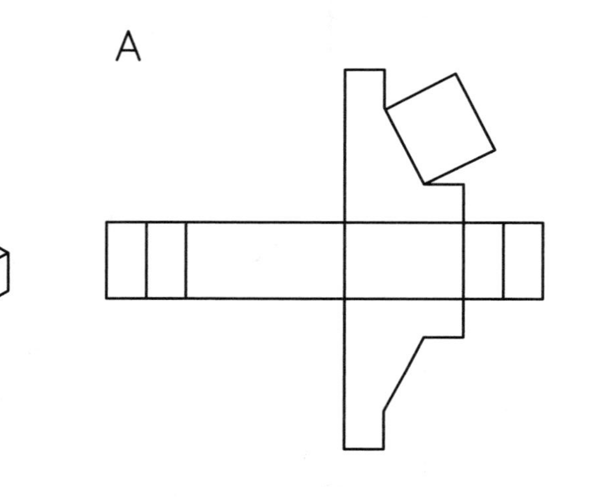

B

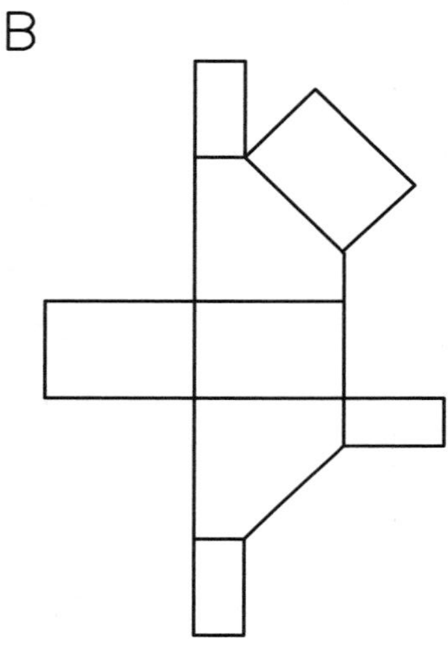

C

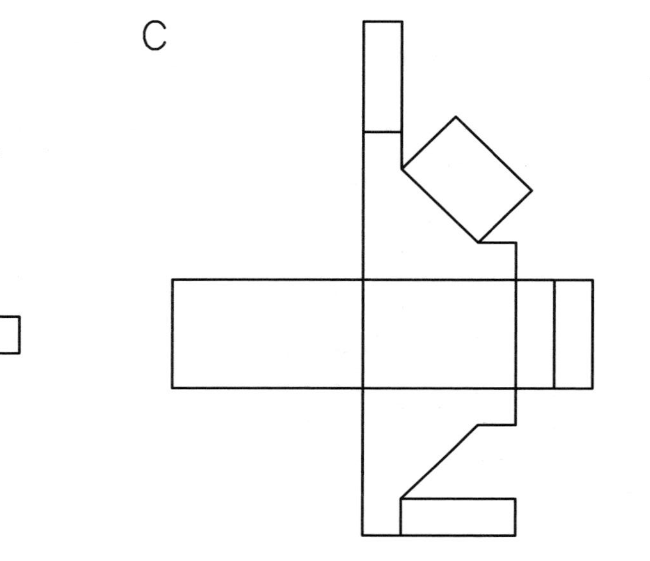

2.

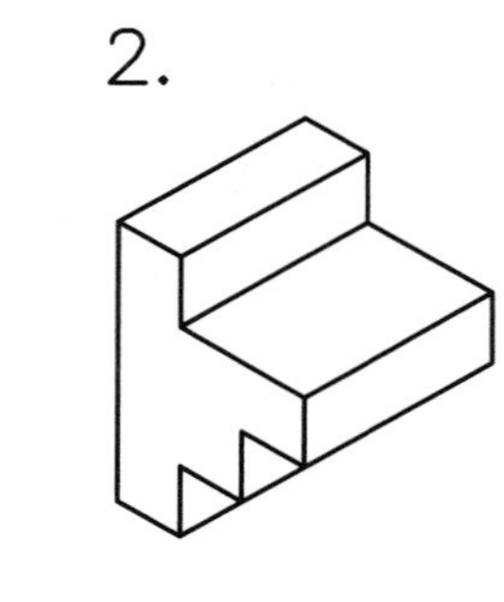

A

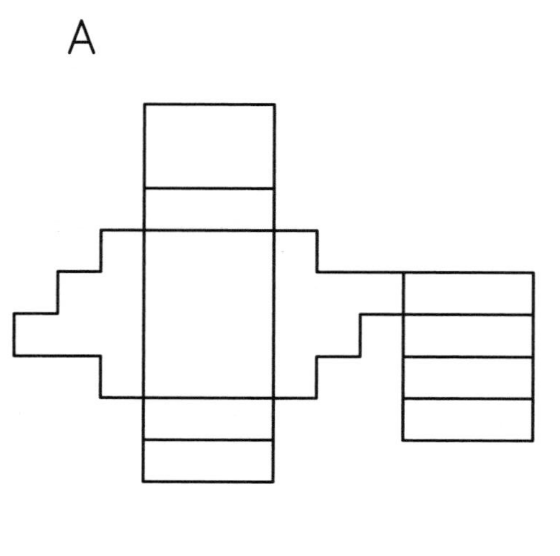

B

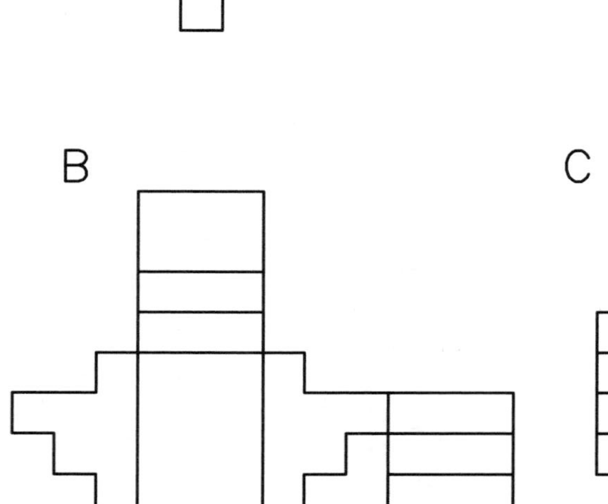

C

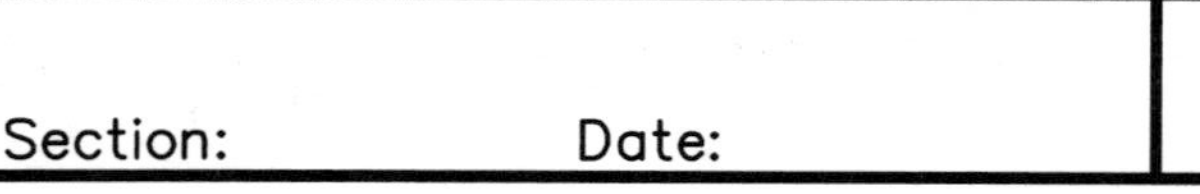

| Name: Class: Section: Date: | Introduction to 3-D Spatial Visualization | Grade: | Page 60 |
|---|---|---|---|

For the objects shown in isometric below, select the pattern from the choices given that could be folded up to obtain the object.

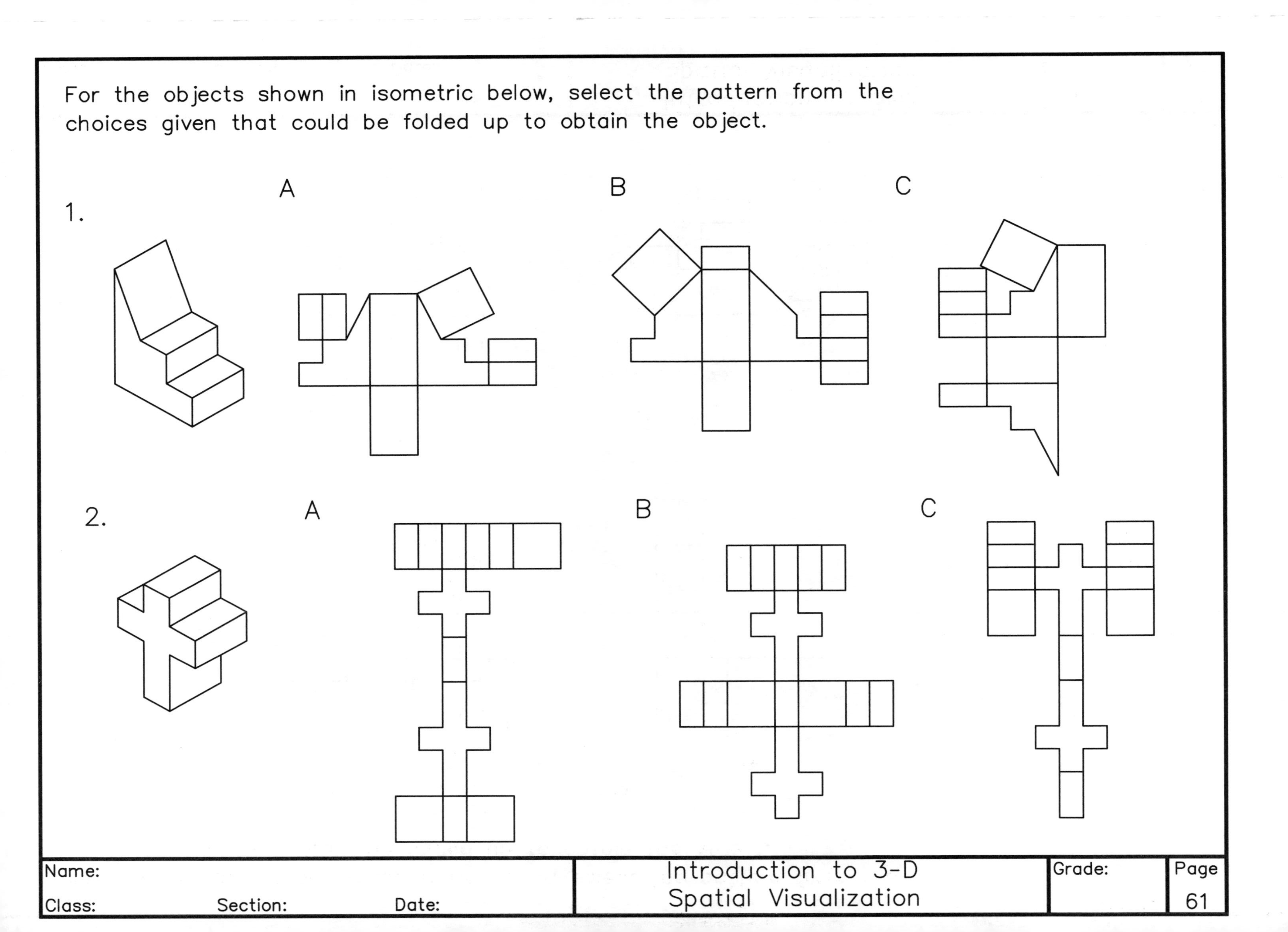

For the objects shown in isometric below, select the pattern from the choices given that could be folded up to obtain the object.

1.

A 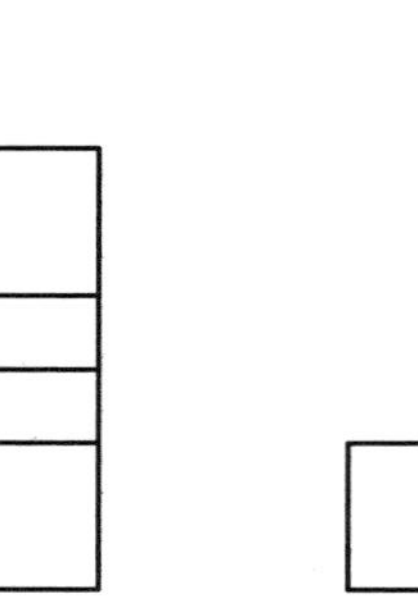
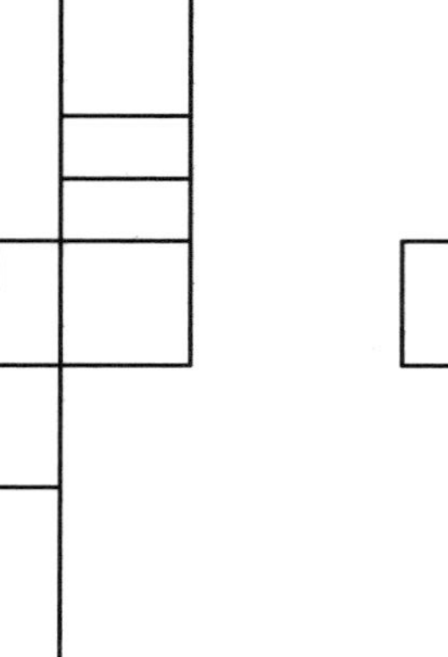

B 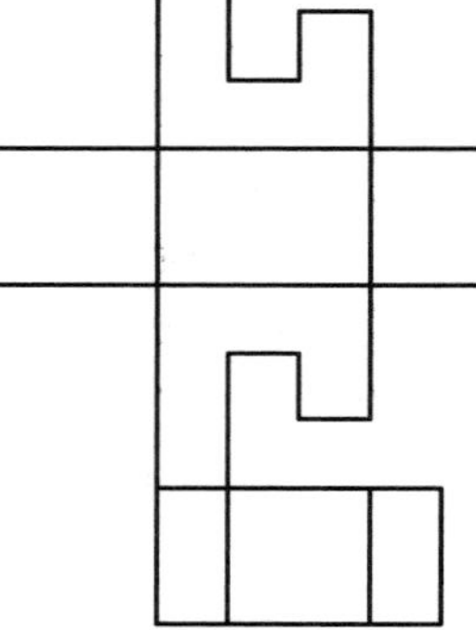

C 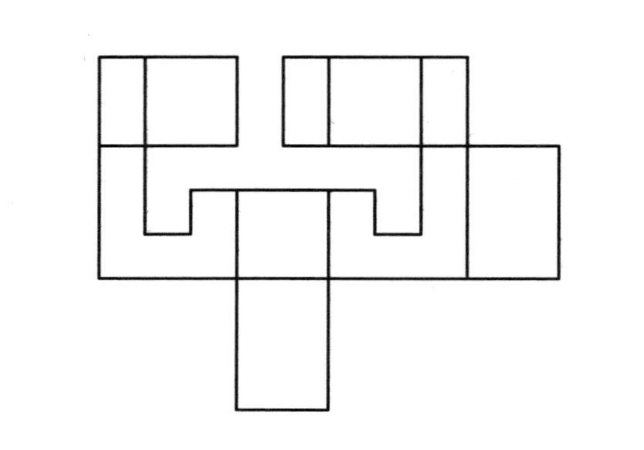

2.

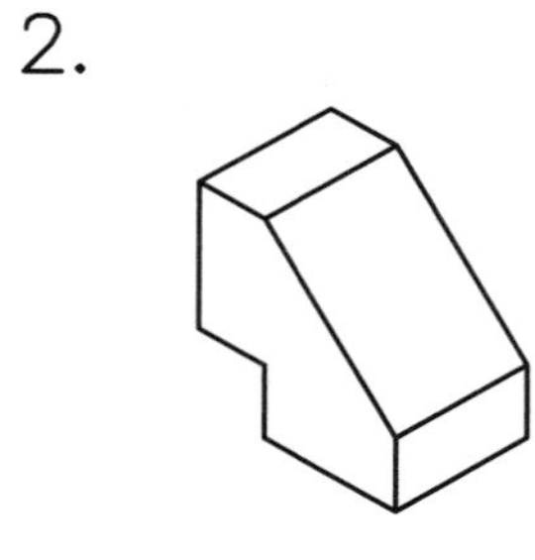

A 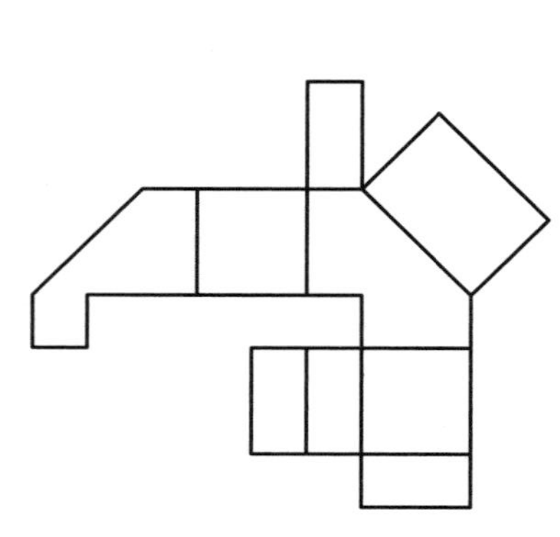

B 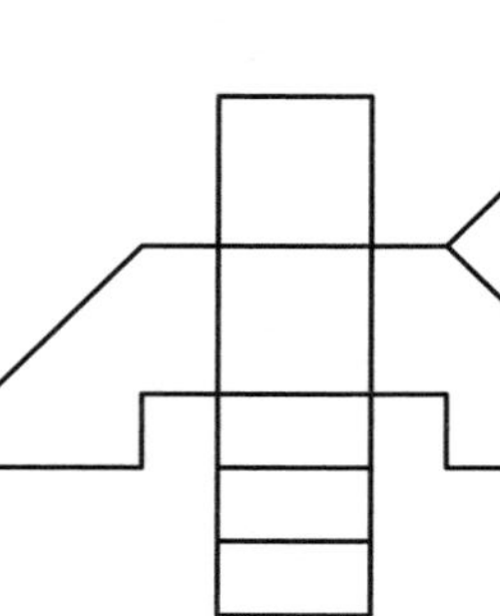

C 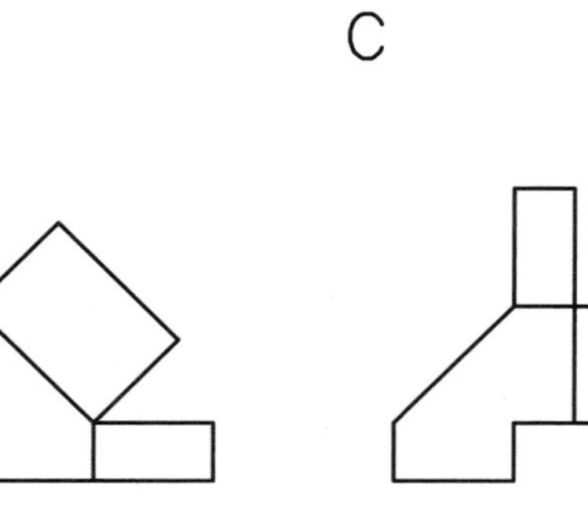

For the objects shown in isometric below, select the pattern from the choices given that could be folded up to obtain the object.

1.

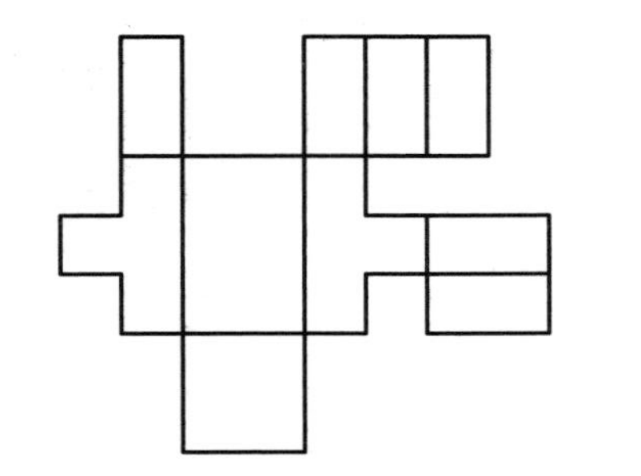

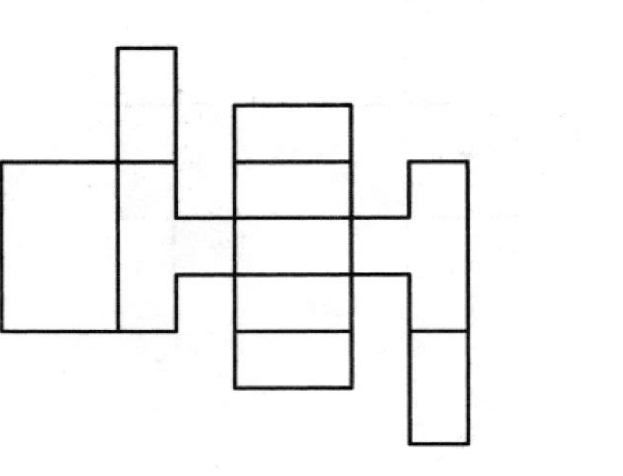

A B C

2.

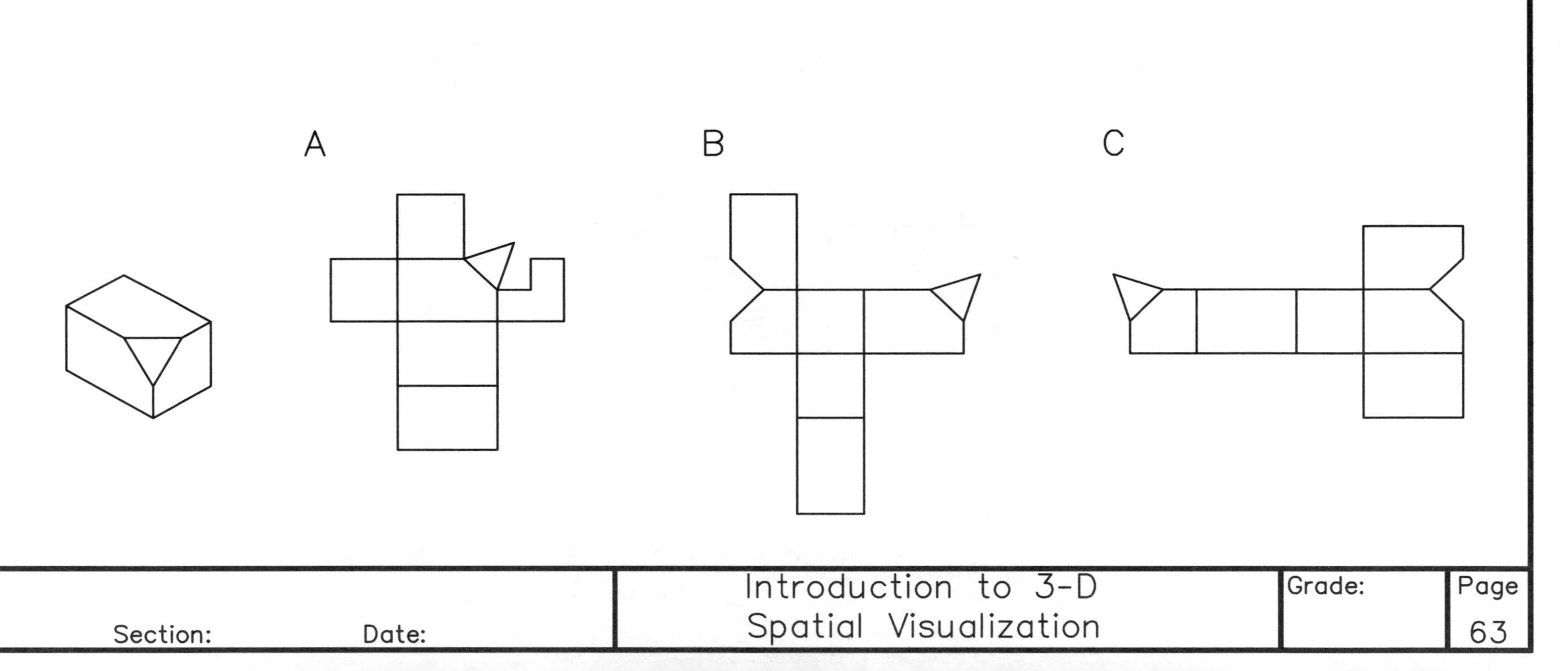

A B C

For the objects shown in isometric below, select the pattern from the choices given that could be folded up to obtain the object.

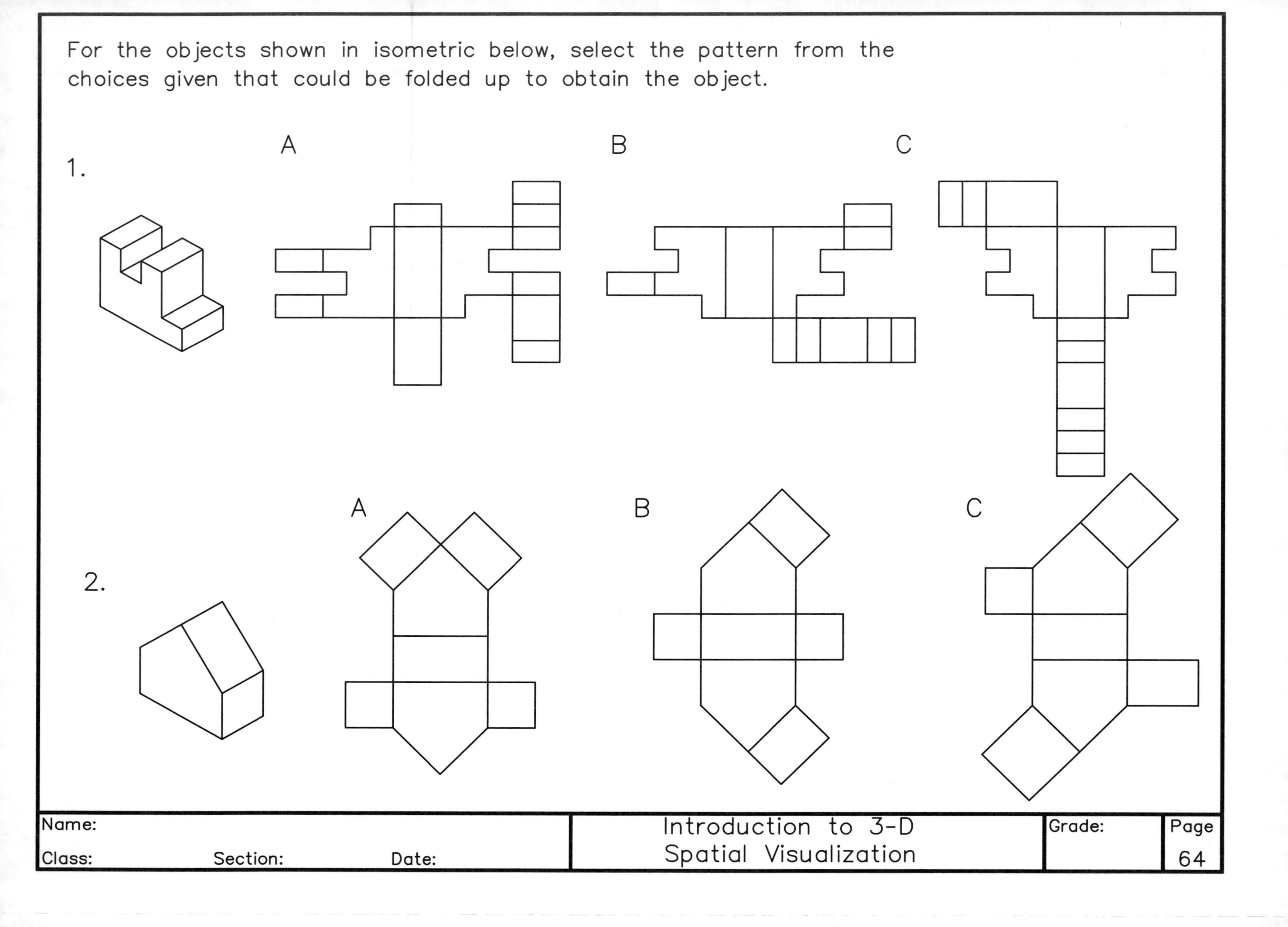

The rectangular prisms shown in isometric below have markings on each of their four long sides. Choose the panel for the flat pattern so that when folded up, the correct object is obtained.

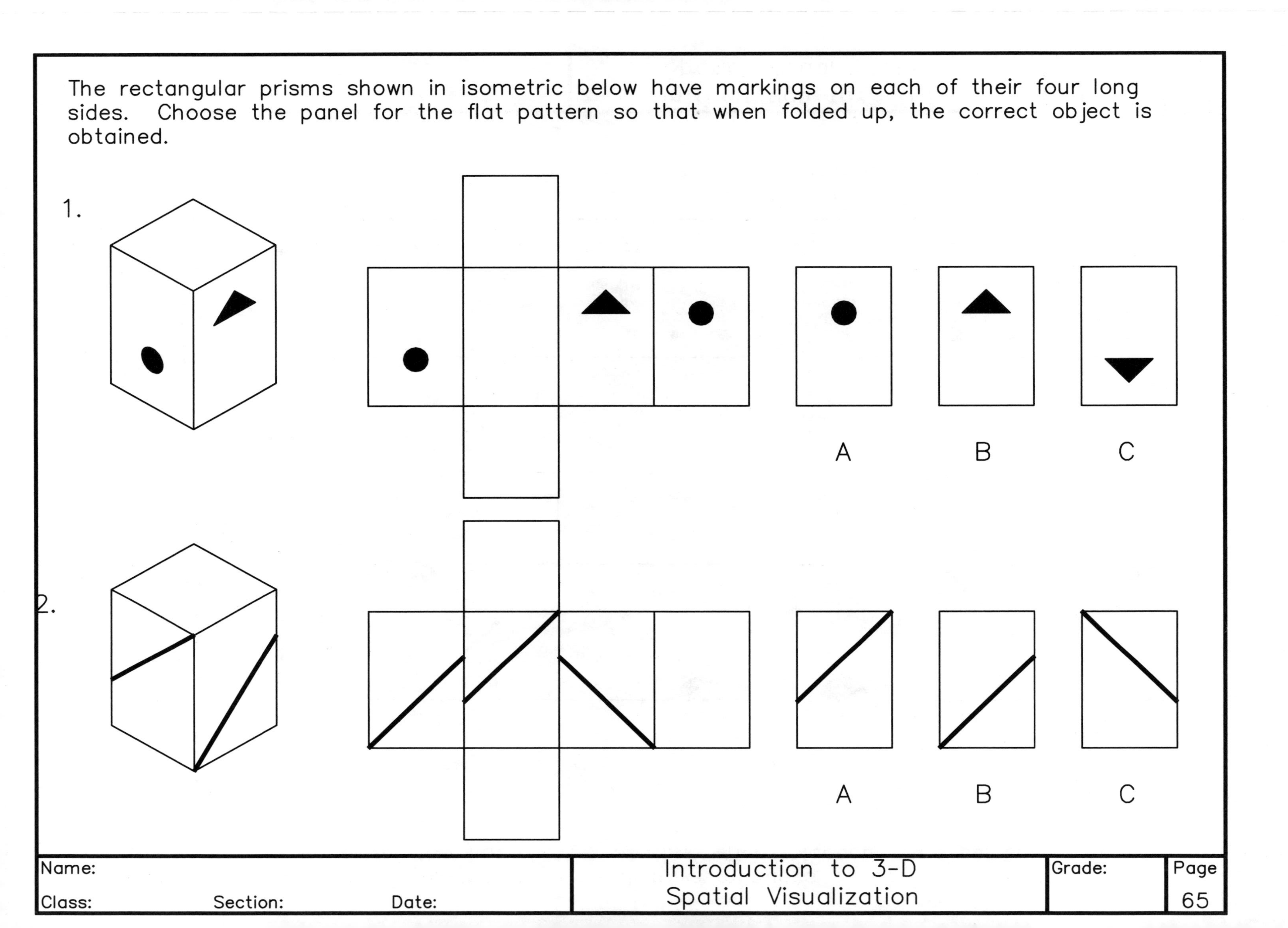

The rectangular prisms shown in isometric below have markings on each of their four long sides. Choose the panel for the flat pattern so that when folded up, the correct object is obtained.

1.

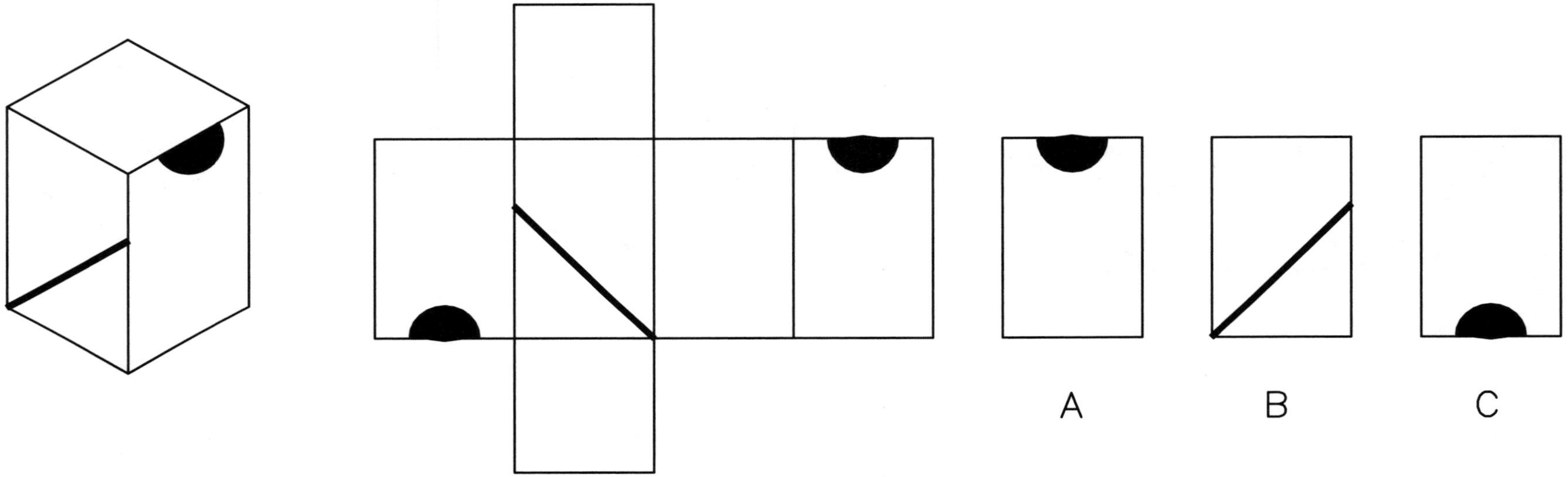

A B C

2.

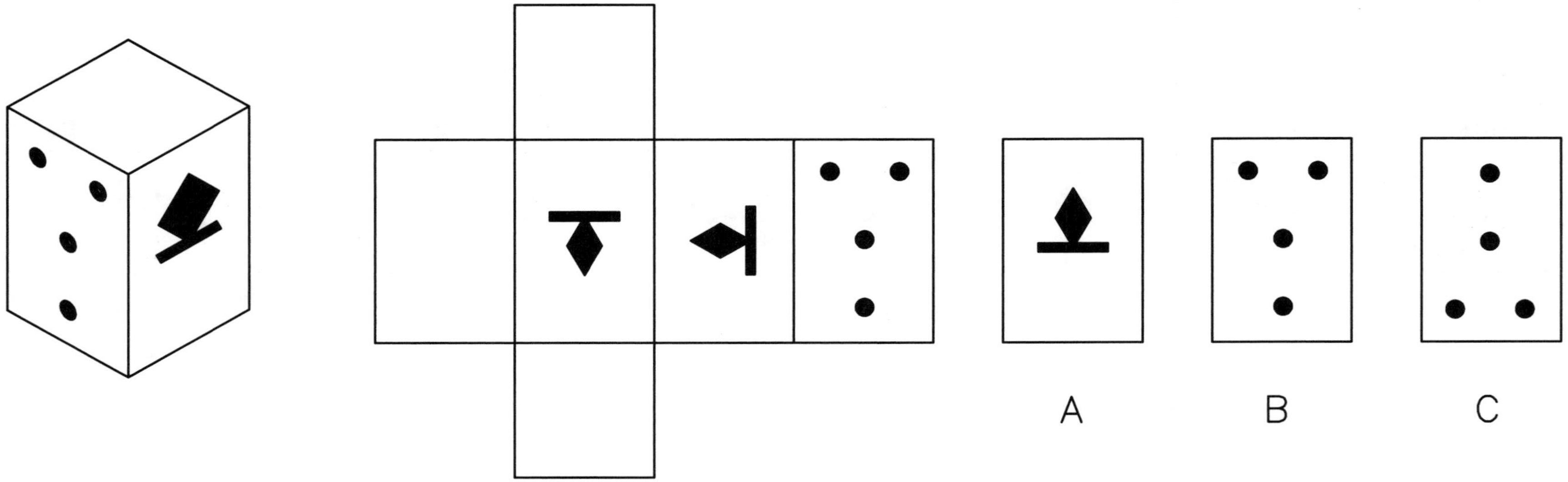

A B C

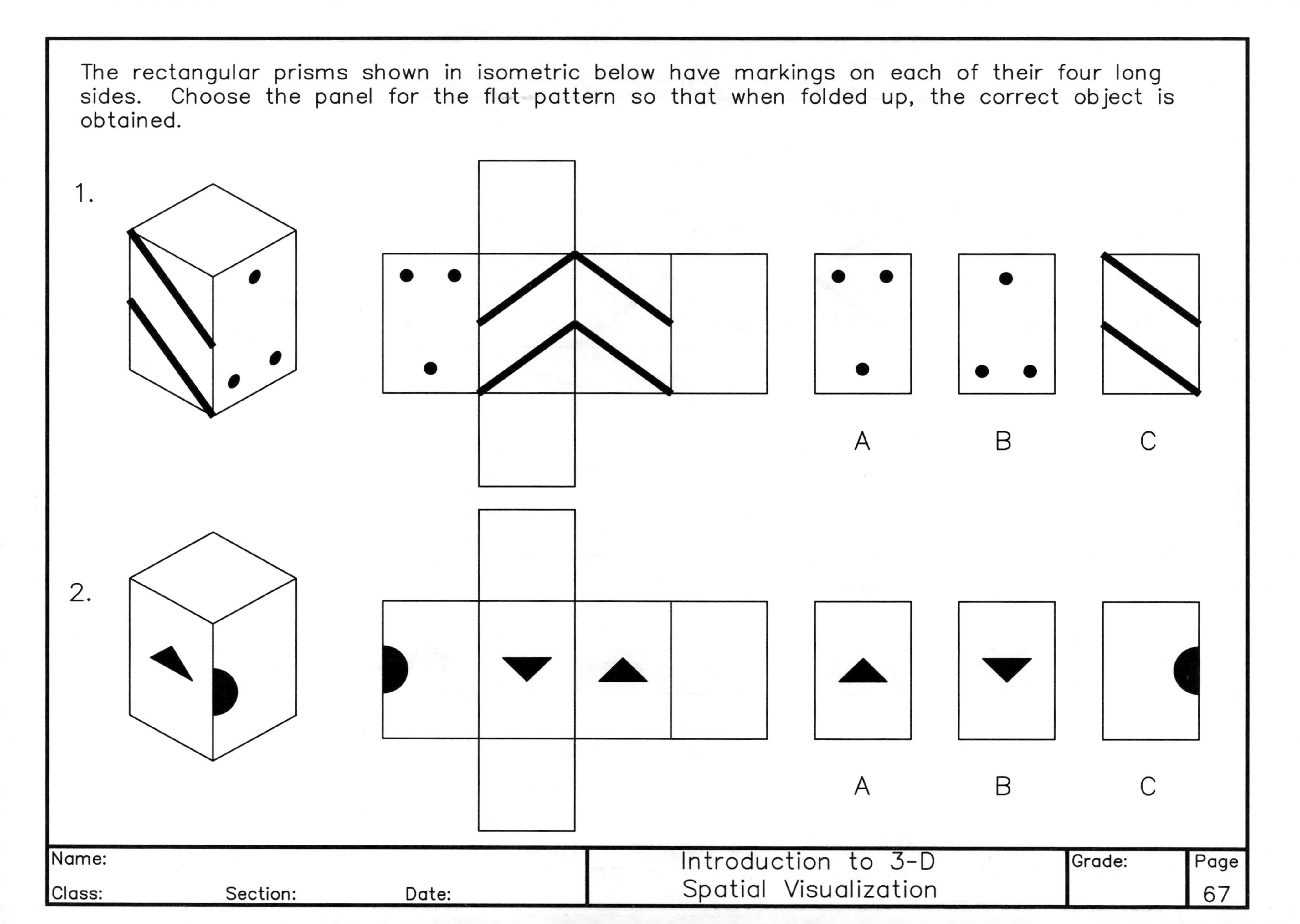
The rectangular prisms shown in isometric below have markings on each of their four long sides. Choose the panel for the flat pattern so that when folded up, the correct object is obtained.
1.
A
B
C
2.
A
B
C
Name:
Class:
Section:
Date:
Introduction to 3-D
Spatial Visualization
Grade:
Page
67

The rectangular prisms shown in isometric below have markings on each of their four long sides. Choose the panel for the flat pattern so that when folded up, the correct object is obtained.

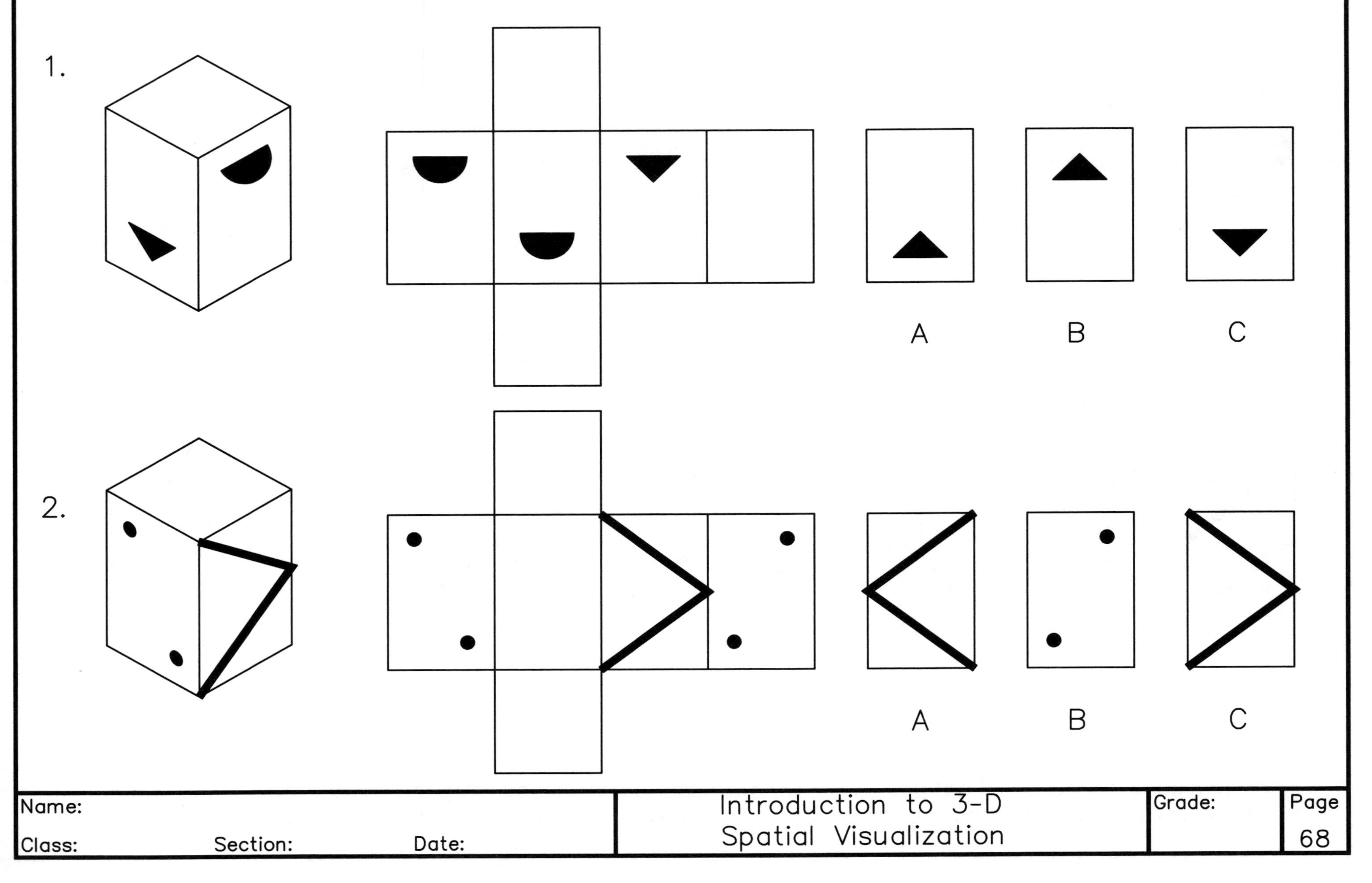

The rectangular prisms shown in isometric below have markings on each of their four long sides. Choose the panel for the flat pattern so that when folded up, the correct object is obtained.

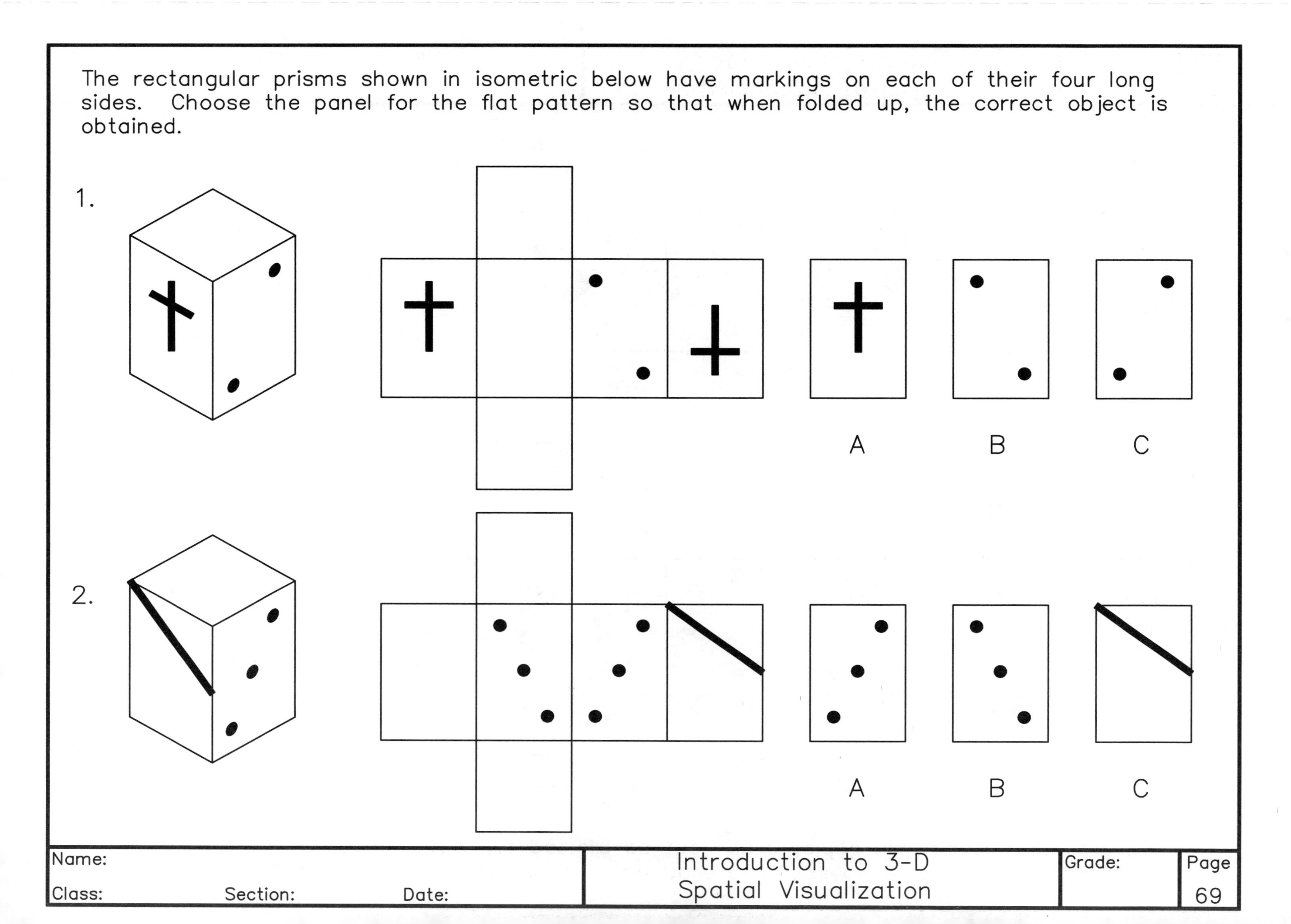

The rectangular prisms shown in isometric below have markings on each of their four long sides. Choose the panel for the flat pattern so that when folded up, the correct object is obtained.

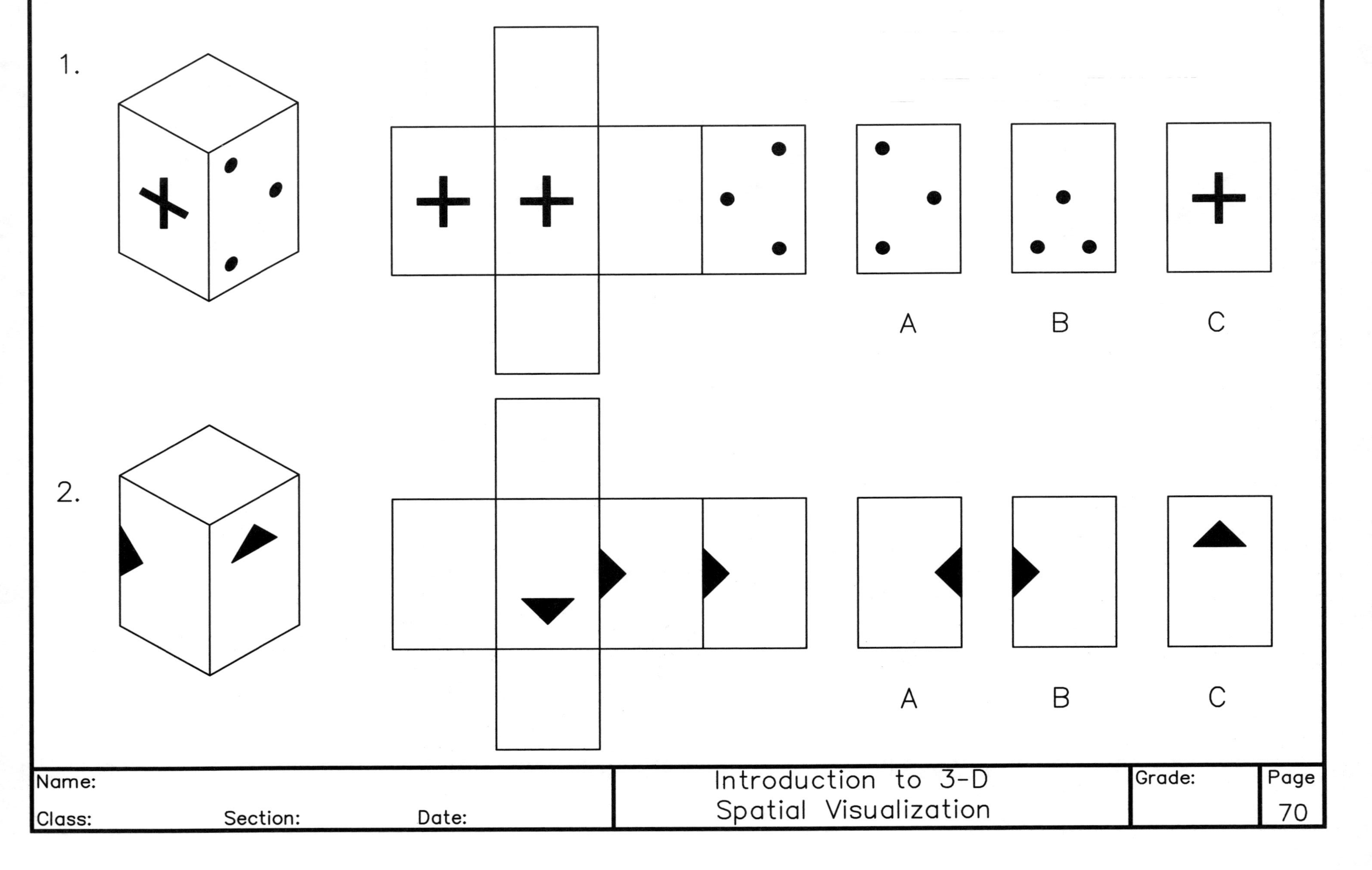

## Rotation of Objects about a Single Axis

A rotation of an object is a turning of it about a straight line. The line about which the object rotates is called the *axis of rotation.*

Original Object Position

An object can rotate either positively or negatively about an axis. If you look down the axis of rotation, a *positive* rotation is counterclockwise and a *negative* rotation is clockwise.

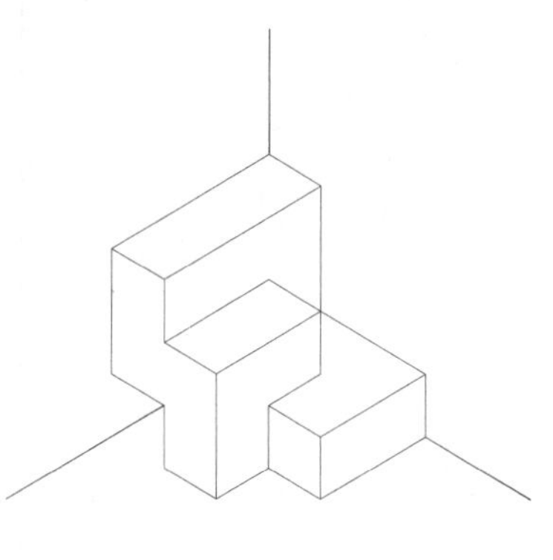

Original Object Position

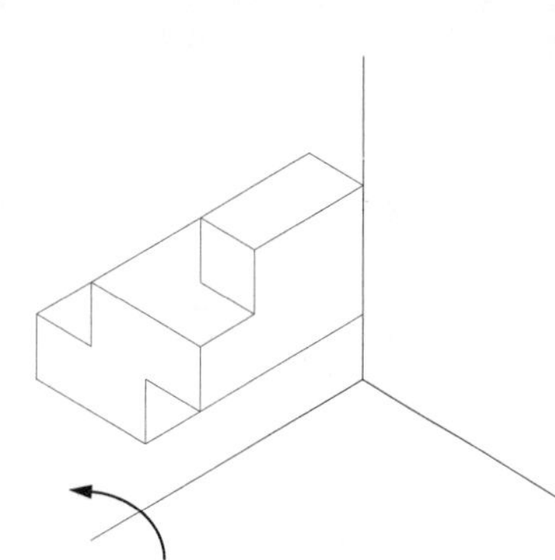

Positive Rotation

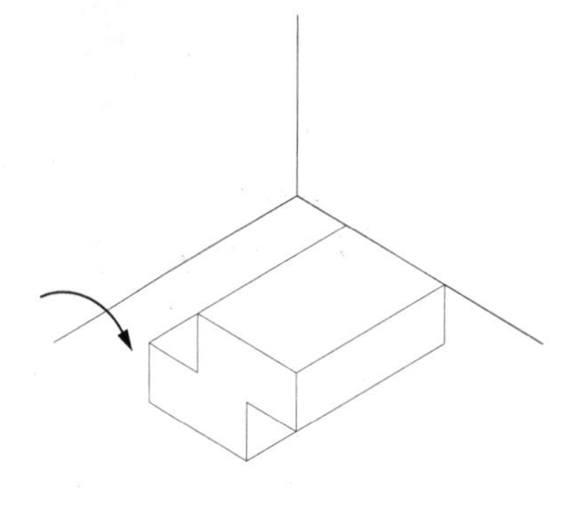

Negative Rotation

The direction of the rotation is determined by the right hand rule. For a positive rotation, if you point the thumb of your right hand along the positive direction of the axis of rotation, your fingers will curl in the direction of the rotation. For a negative rotation, if you point the thumb of your right hand along the negative axis of rotation, your fingers will curl in the direction of the rotation.

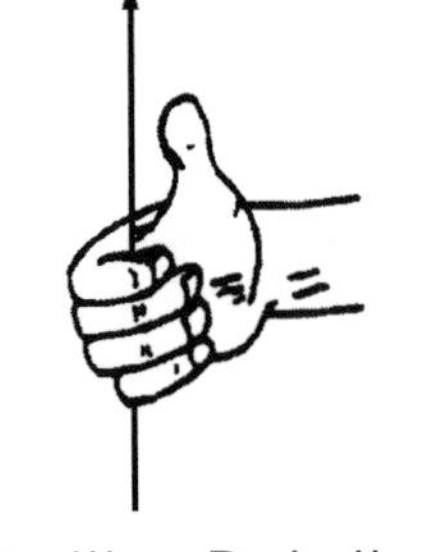

Positive Rotation

Negative Rotation

Object rotations can be represented by the following arrow coding scheme: A single arrow represents a 90 degree rotation about an axis. An arrow to the right indicates a positive rotation and an arrow to the left indicates a negative rotation. The axis about which the object rotates is given at the right end of the arrow.

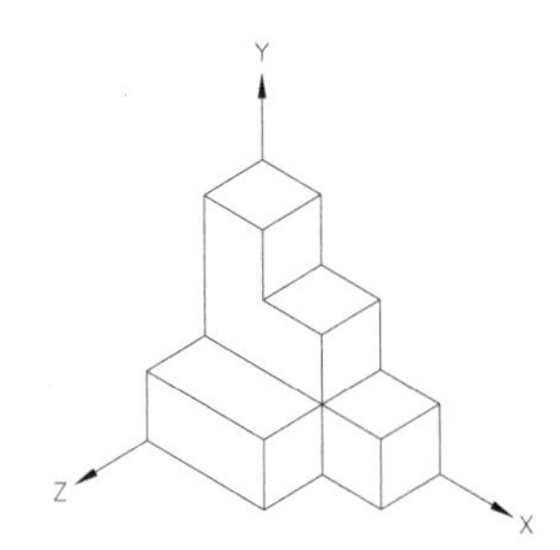


Original Object Position

→ X

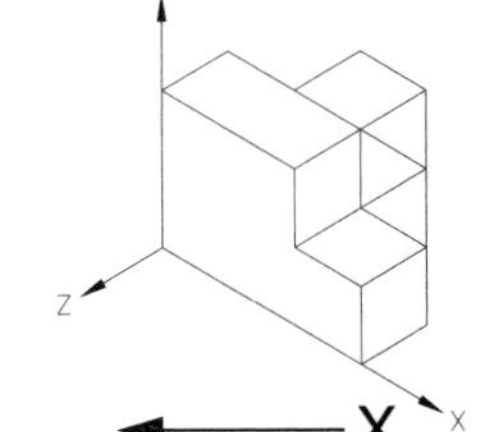


← X

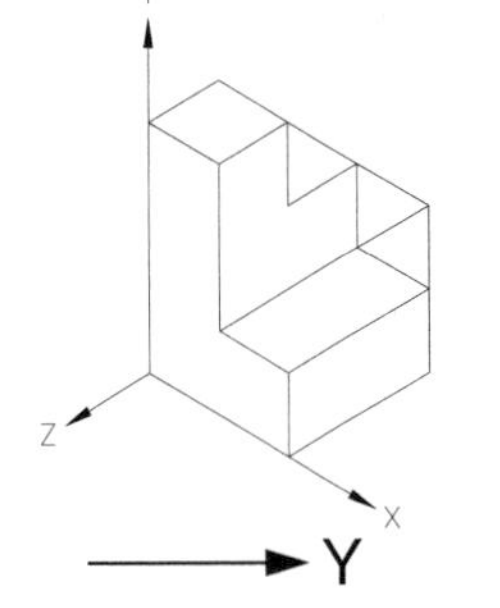


→ Y

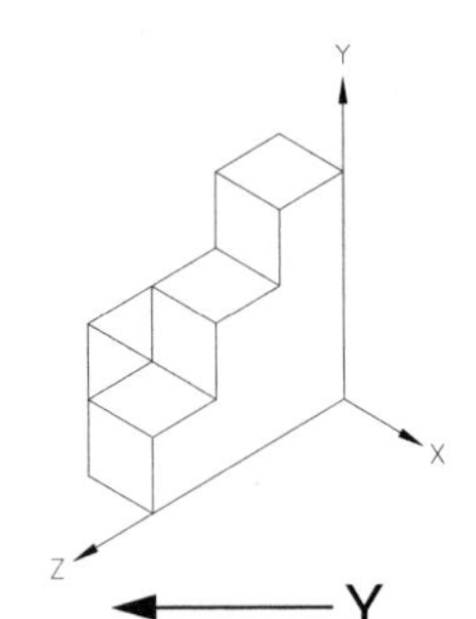


← Y

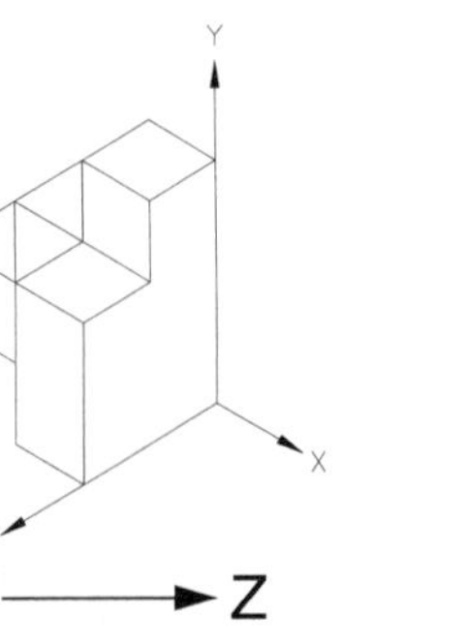


→ Z

← Z

An object can be rotated about an axis multiple times. For each increment of 90 degrees of rotation, a new arrow is included in the arrow coding scheme.

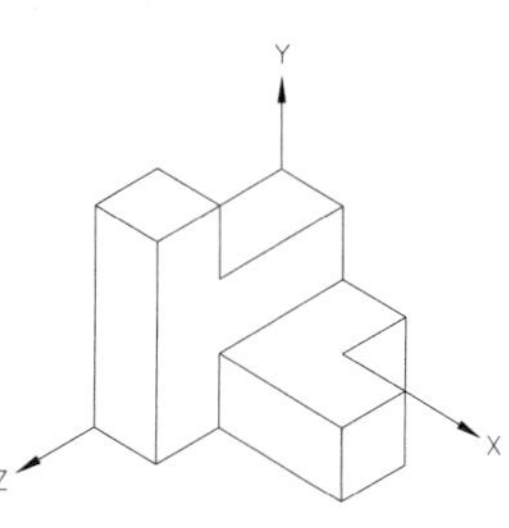


Original Object Position

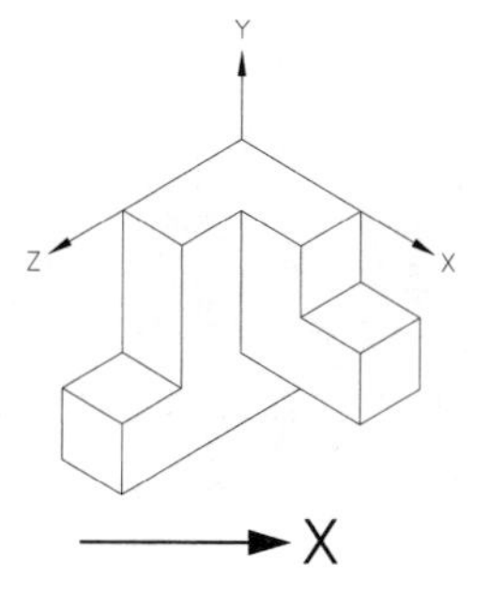


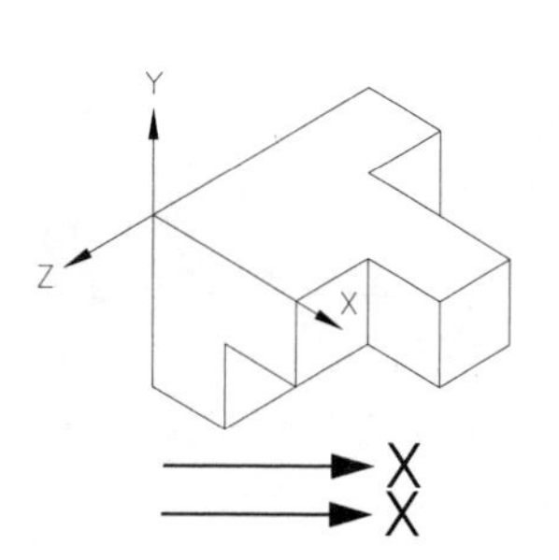


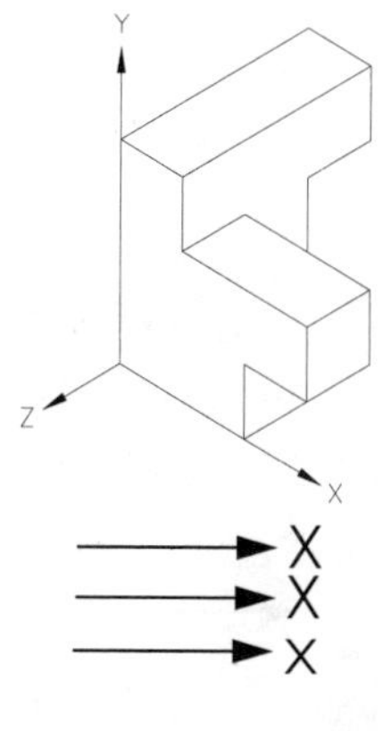


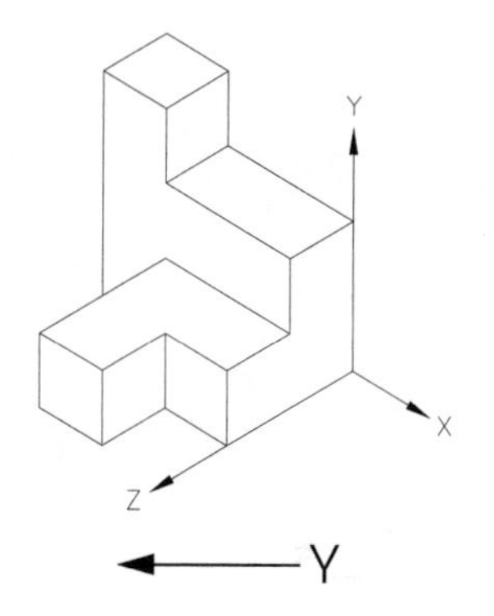


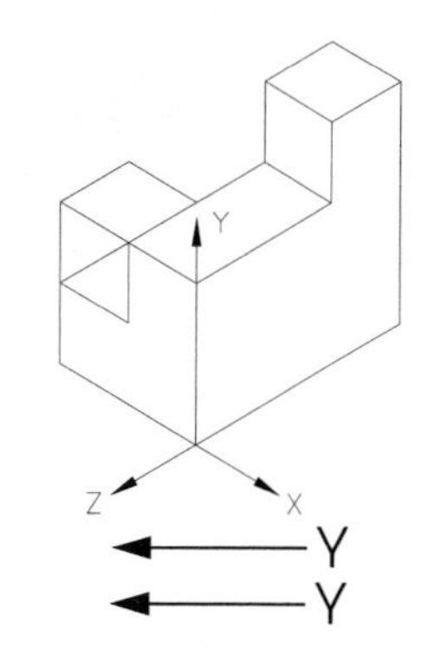


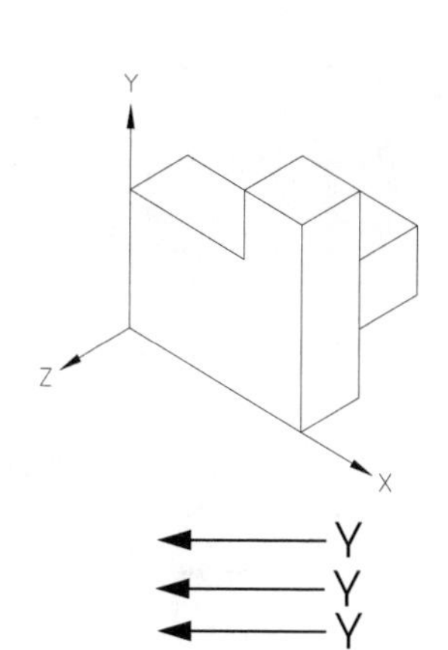

Different combinations of rotations can produce the same result. Thus, one arrow coding scheme can often be replaced by another one because they are *equivalent*.

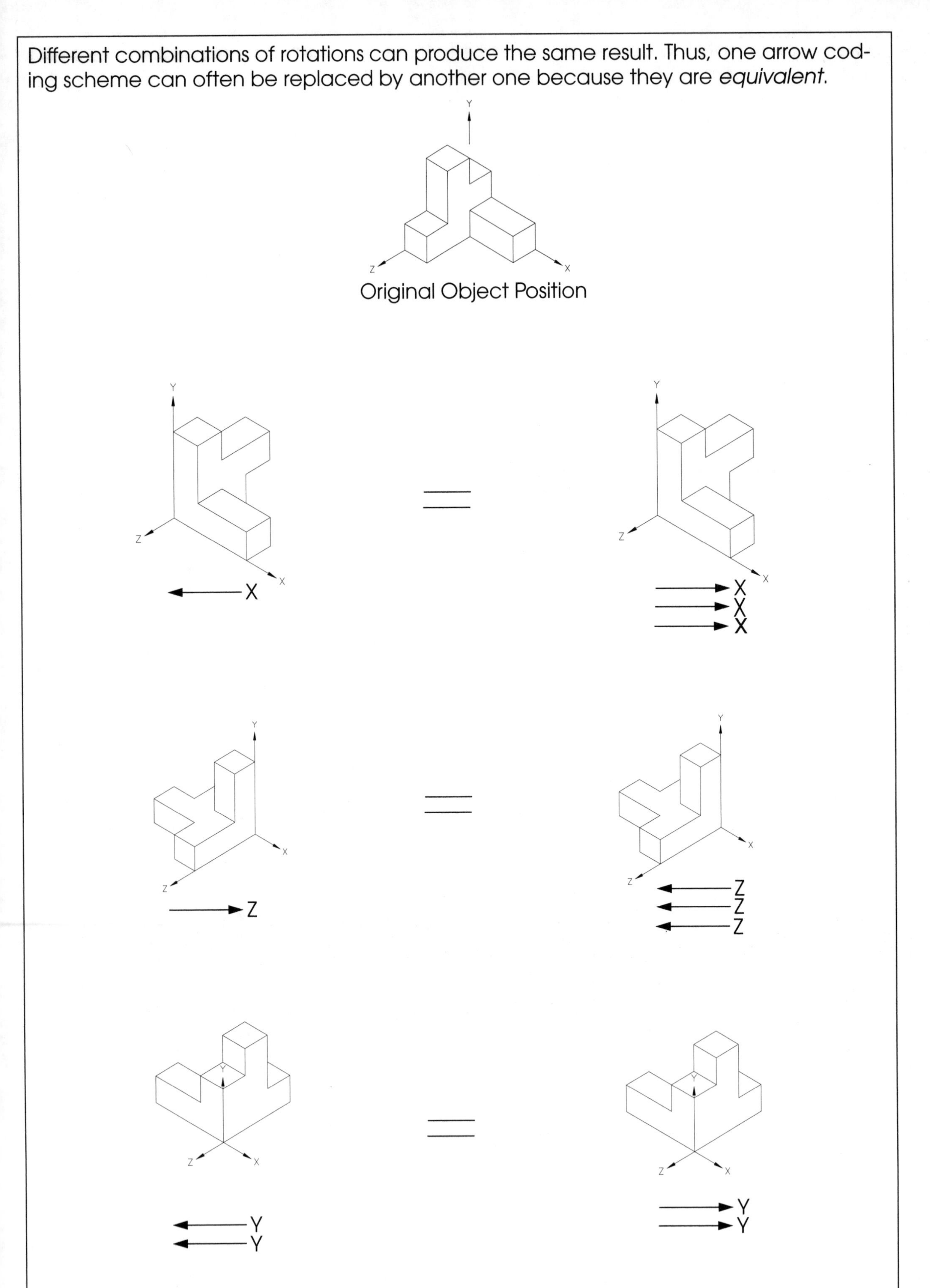

For the objects shown below, sketch the object in the space provided after rotating it about the axis by the indicated amount.

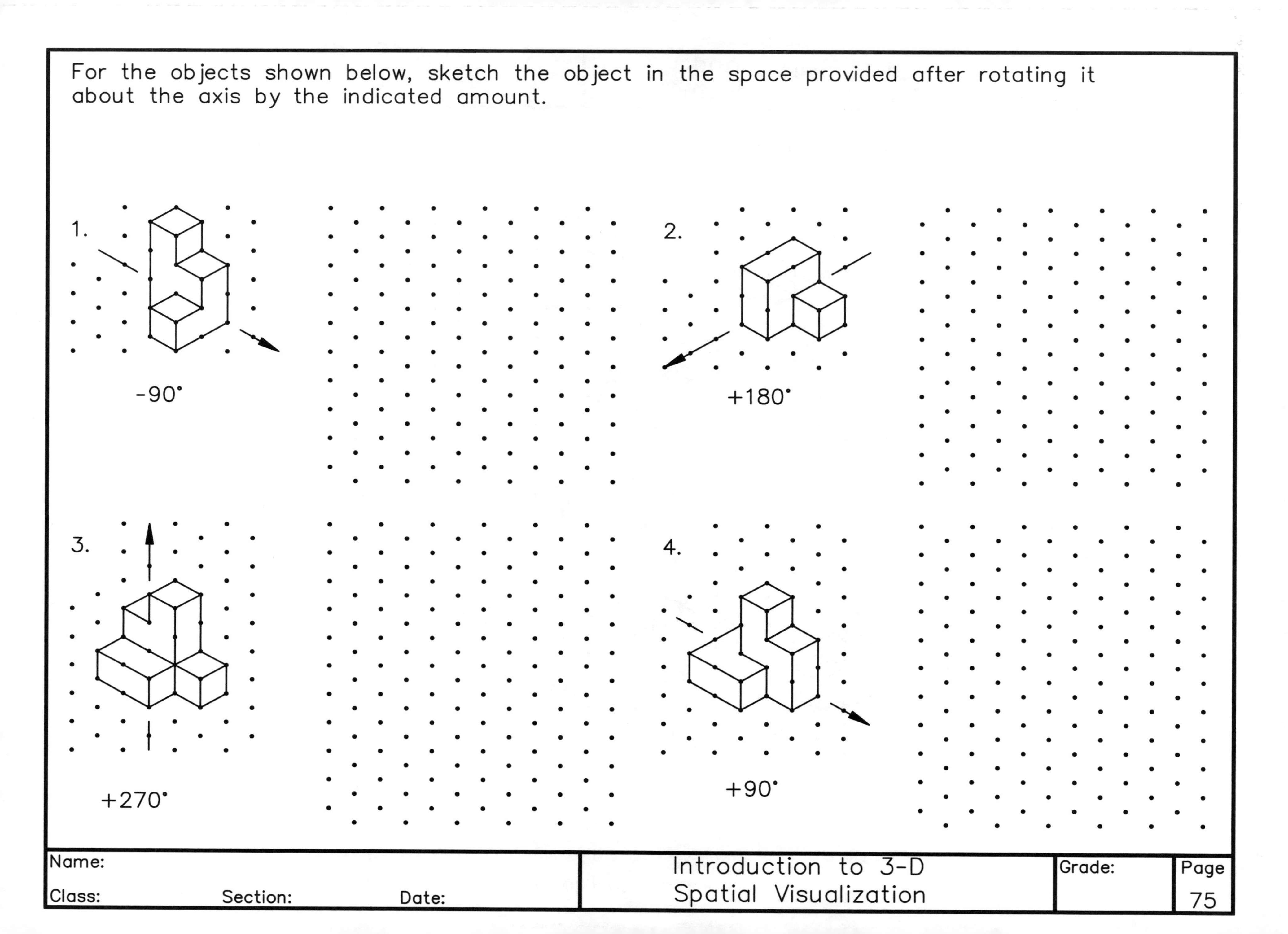


| Name: | Introduction to 3-D Spatial Visualization | Grade: | Page |
|---|---|---|---|
| Class: Section: Date: | | | 75 |

For the objects shown below, sketch the object in the space provided after rotating it about the axis by the indicated amount.

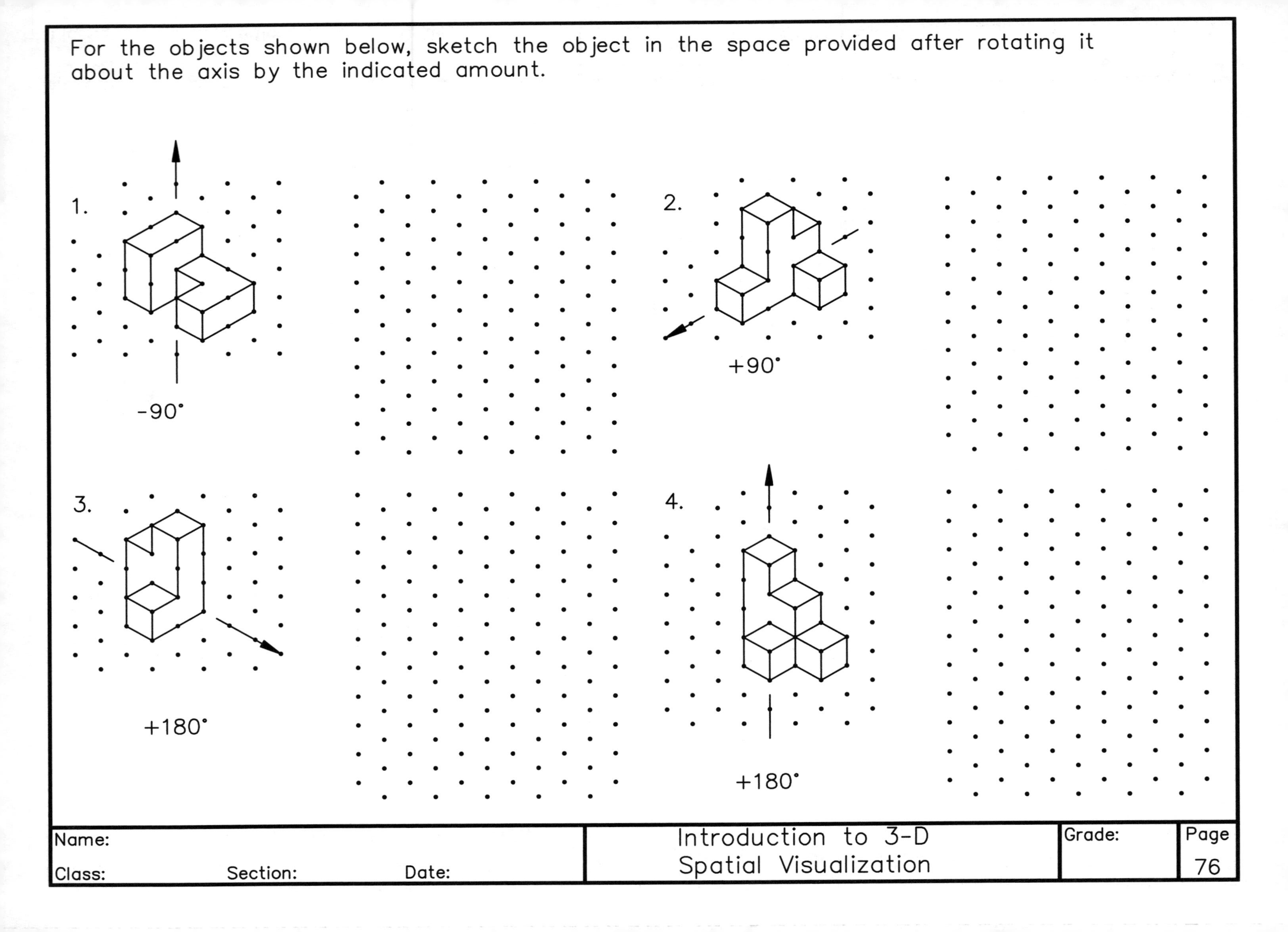

For the objects shown below, sketch the object in the space provided after rotating it about the axis by the indicated amount.

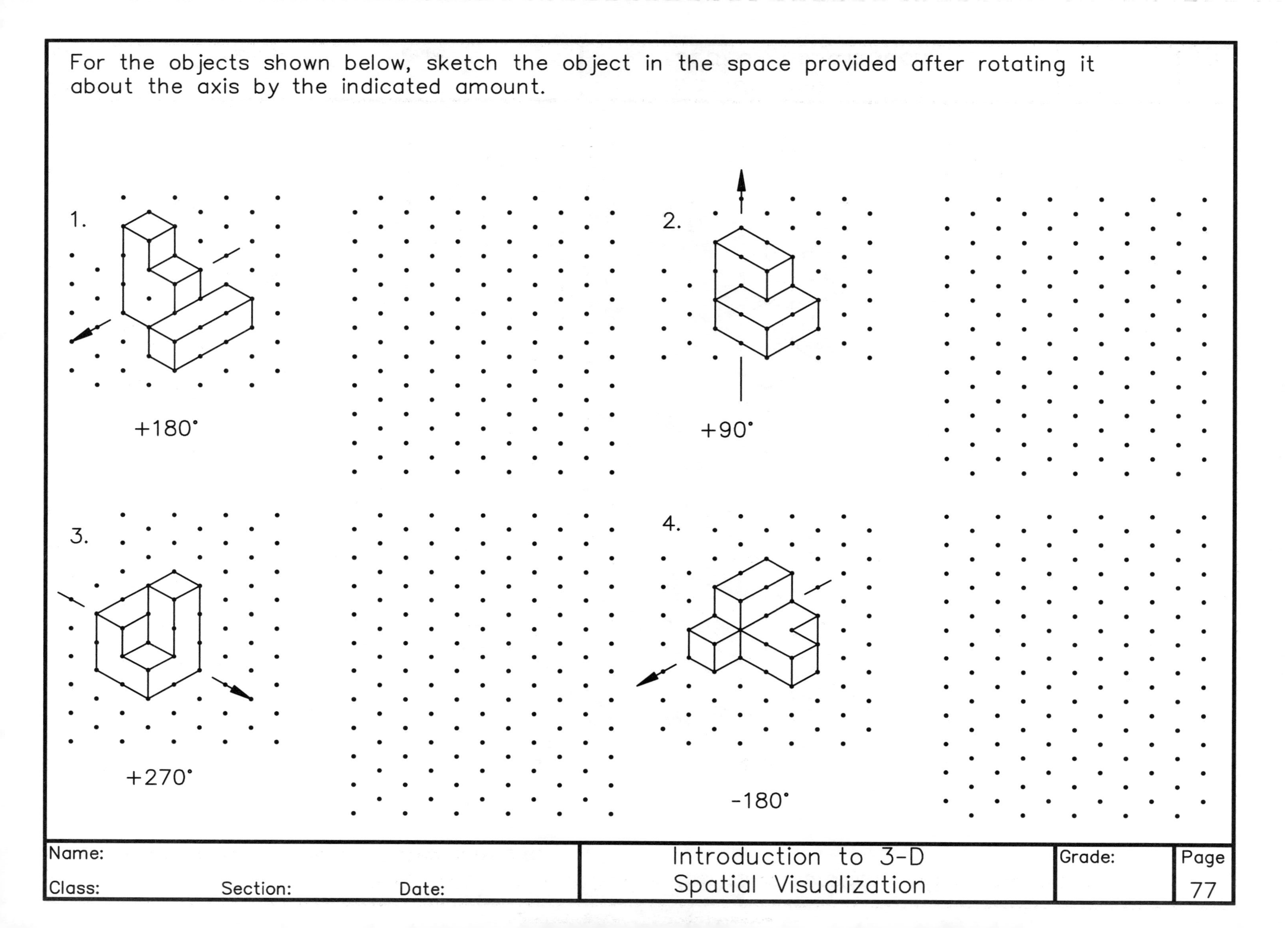


| Name: | | | Introduction to 3-D Spatial Visualization | Grade: | Page |
|---|---|---|---|---|---|
| Class: | Section: | Date: | | | 77 |

For the objects shown below, sketch the object in the space provided after rotating it about the axis by the indicated amount.

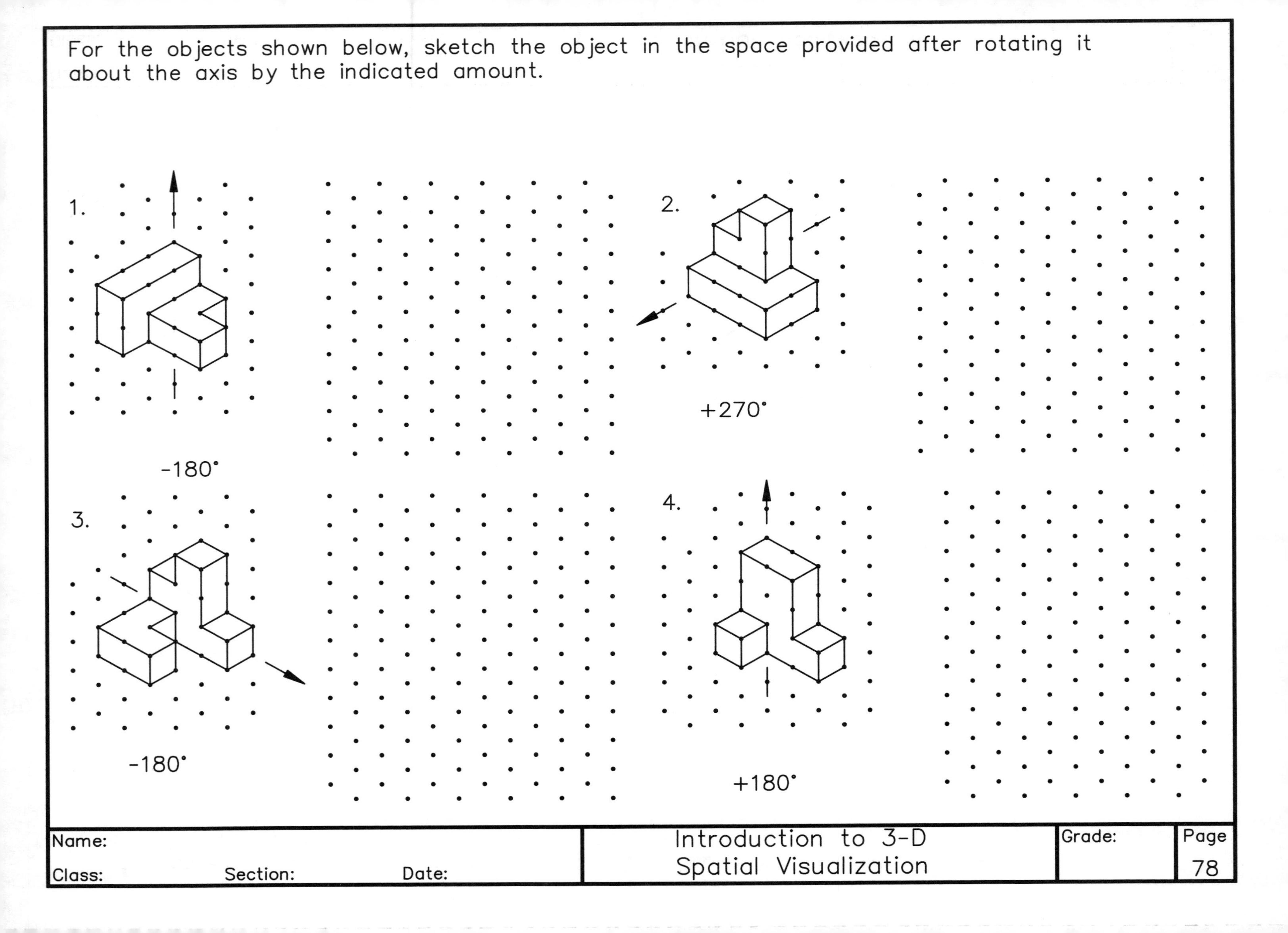

The objects shown below have been rotated <u>positively</u> about the given axis. In the space provided, indicate the amount of rotation (either 90°, 180°, or 270°).

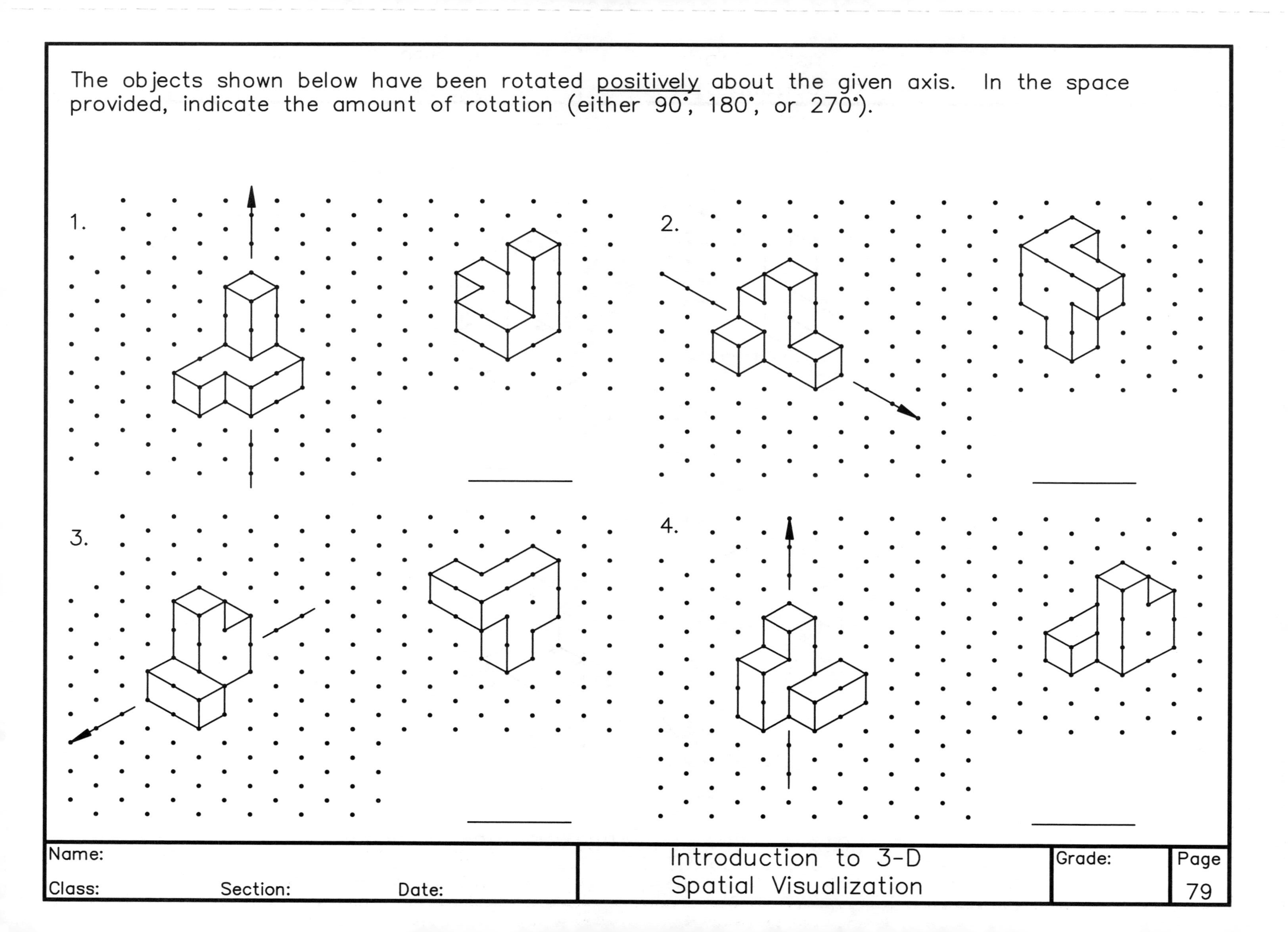

The objects shown below have been rotated <u>positively</u> about the given axis. In the space provided, indicate the amount of rotation (either 90°, 180°, or 270°).

1. __________

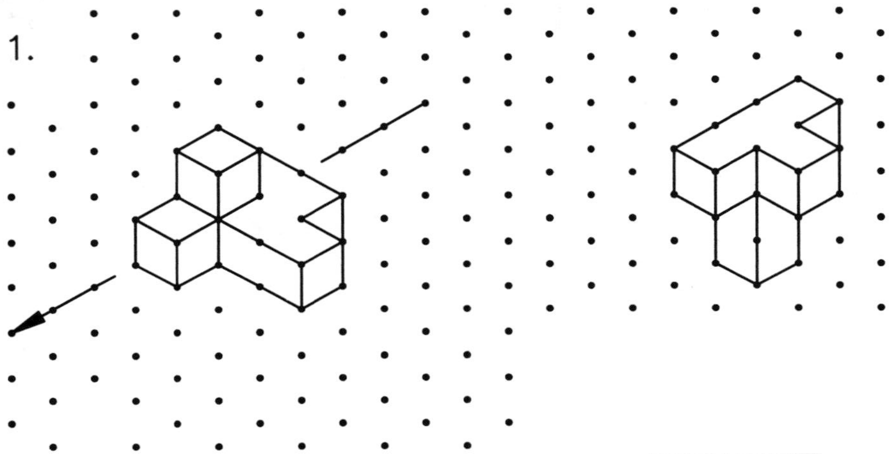

2. __________

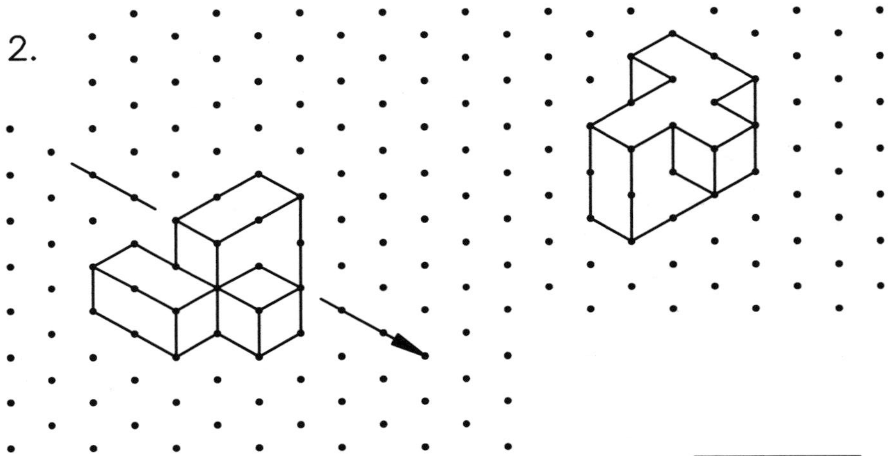

3. __________

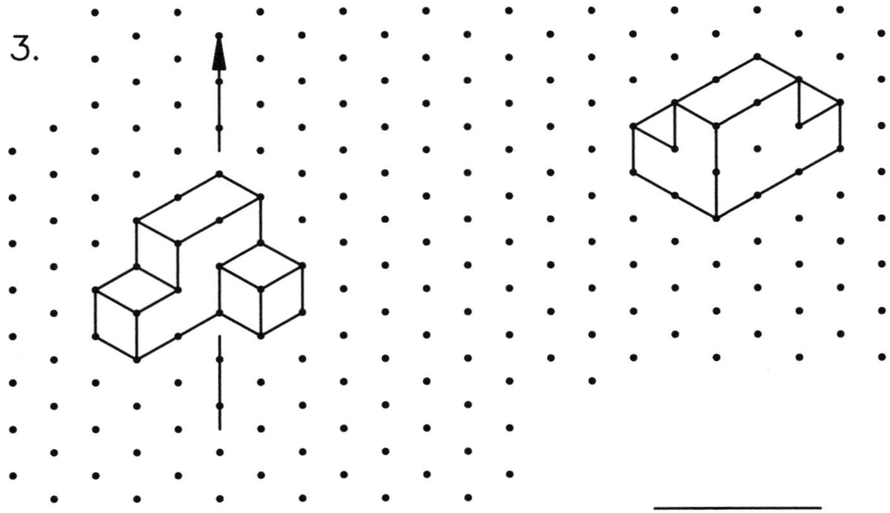

4. __________

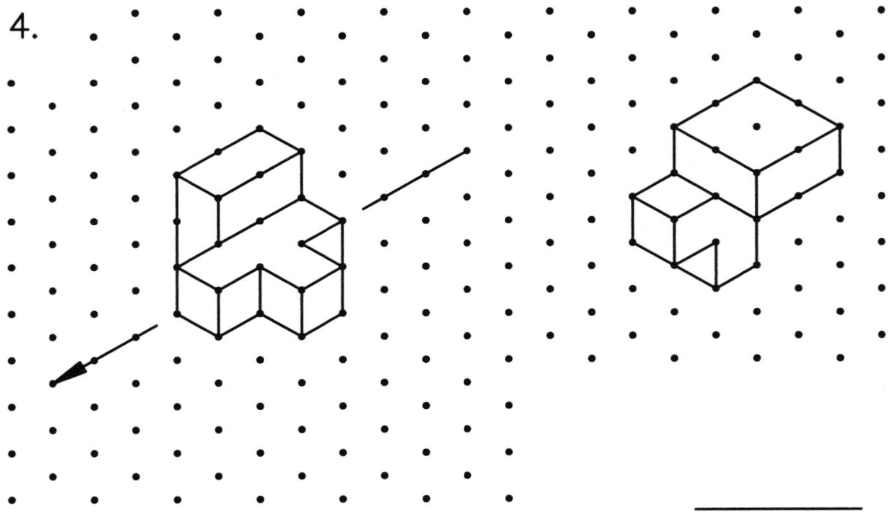

| Name: | Introduction to 3-D | Grade: | Page |
|---|---|---|---|
| Class: Section: Date: | Spatial Visualization | | 80 |

The objects shown below have been rotated <u>negatively</u> about the given axis. In the space provided, indicate the amount of rotation (either 90°, 180°, or 270°).

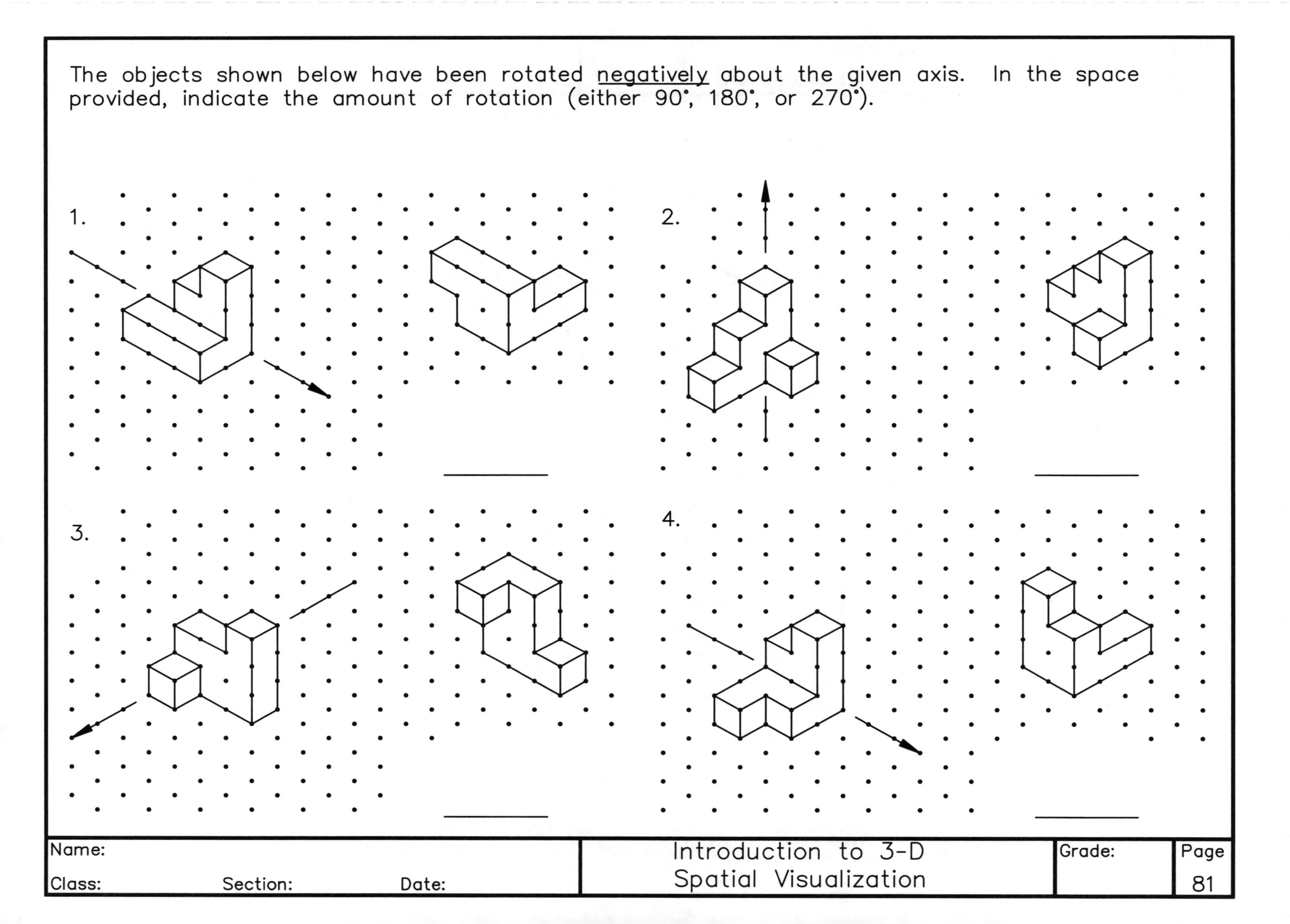


| Name: | | | Introduction to 3-D Spatial Visualization | Grade: | Page |
|---|---|---|---|---|---|
| Class: | Section: | Date: | | | 81 |

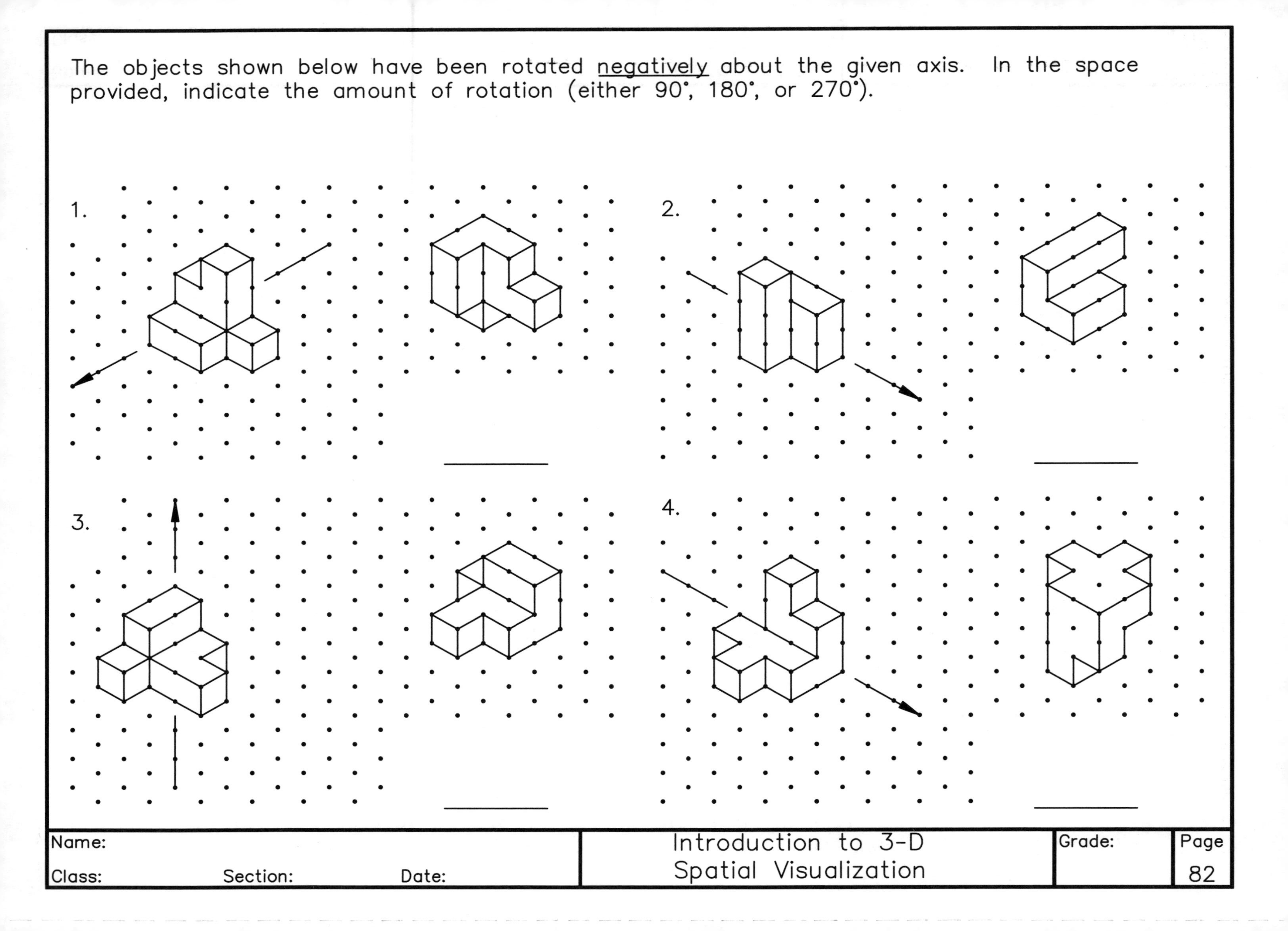
The objects shown below have been rotated negatively about the given axis. In the space provided, indicate the amount of rotation (either 90°, 180°, or 270°).
1. ______
2. ______
3. ______
4. ______
Name:
Class:
Section:
Date:
Introduction to 3-D Spatial Visualization
Grade:
Page 82

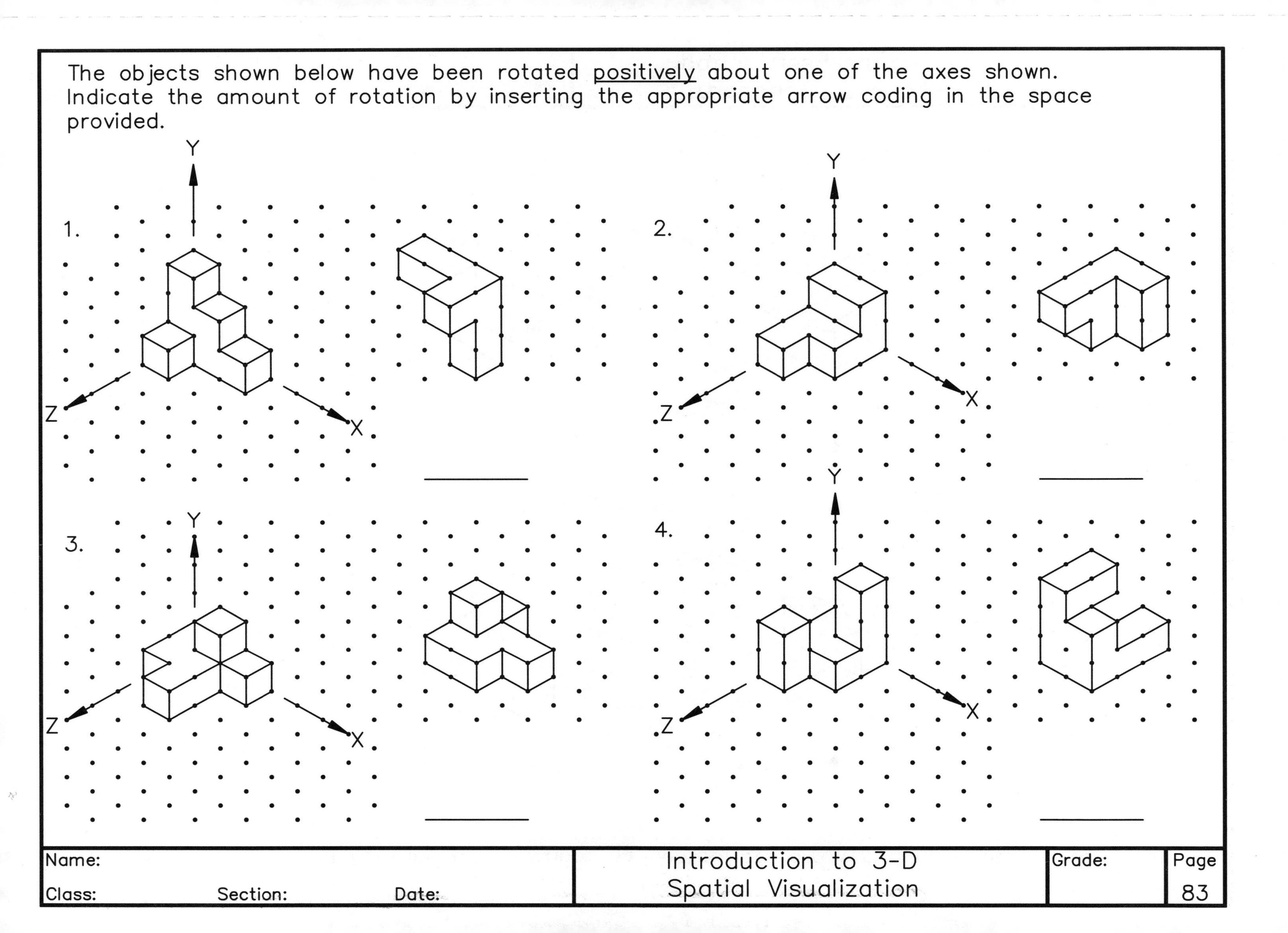

The objects shown below have been rotated positively about one of the axes shown. Indicate the amount of rotation by inserting the appropriate arrow coding in the space provided.
1.
Y
Z
X
2.
Y
Z
X
3.
Y
Z
X
4.
Y
Z
X
Name:
Class:
Section:
Date:
Introduction to 3-D
Spatial Visualization
Grade:
Page
83

The objects shown below have been rotated <u>positively</u> about one of the axes shown. Indicate the amount of rotation by inserting the appropriate arrow coding in the space provided.

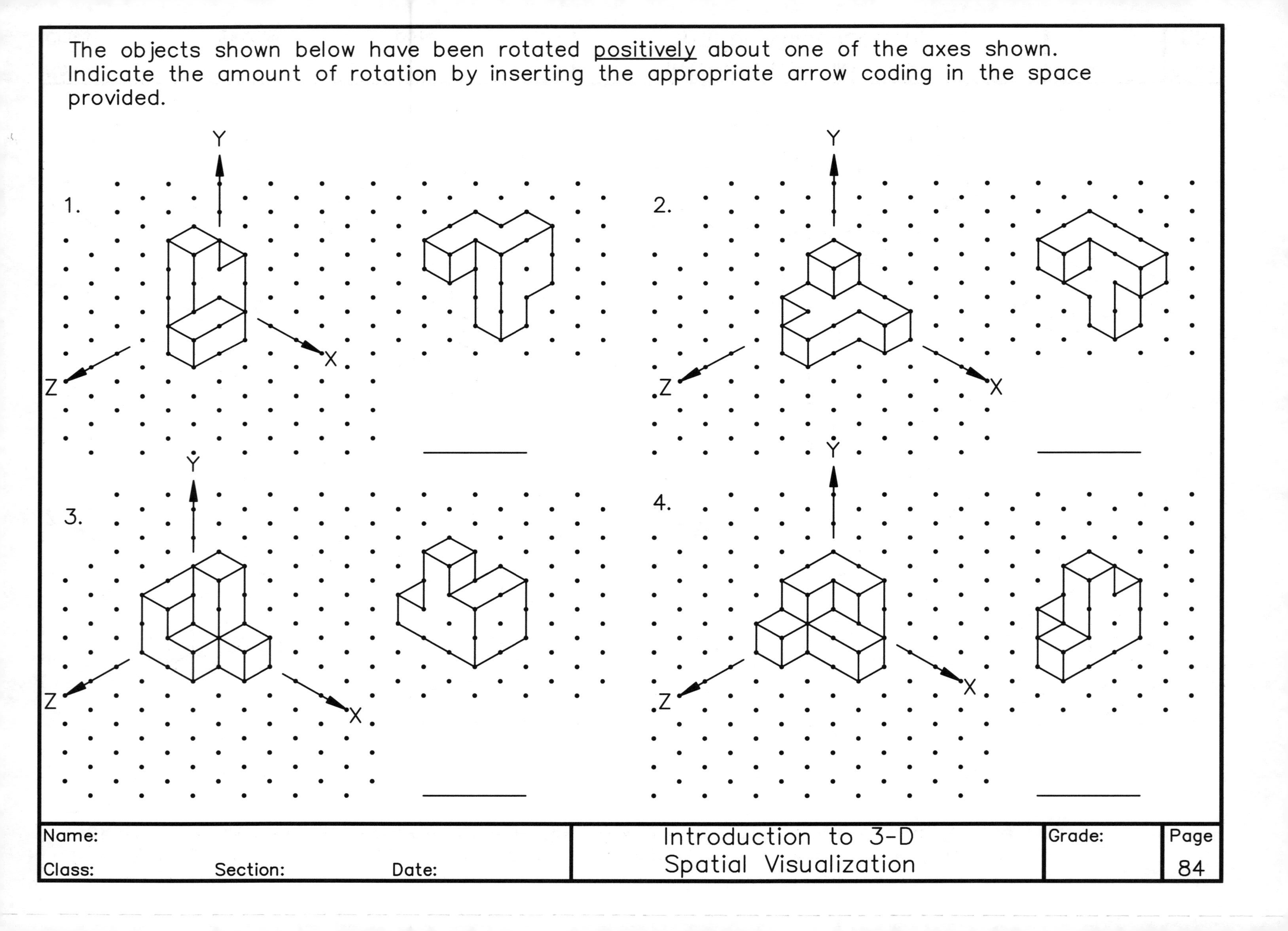

The objects shown below have been rotated <u>positively</u> about one of the axes shown. Indicate the amount of rotation by inserting the appropriate arrow coding in the space provided.

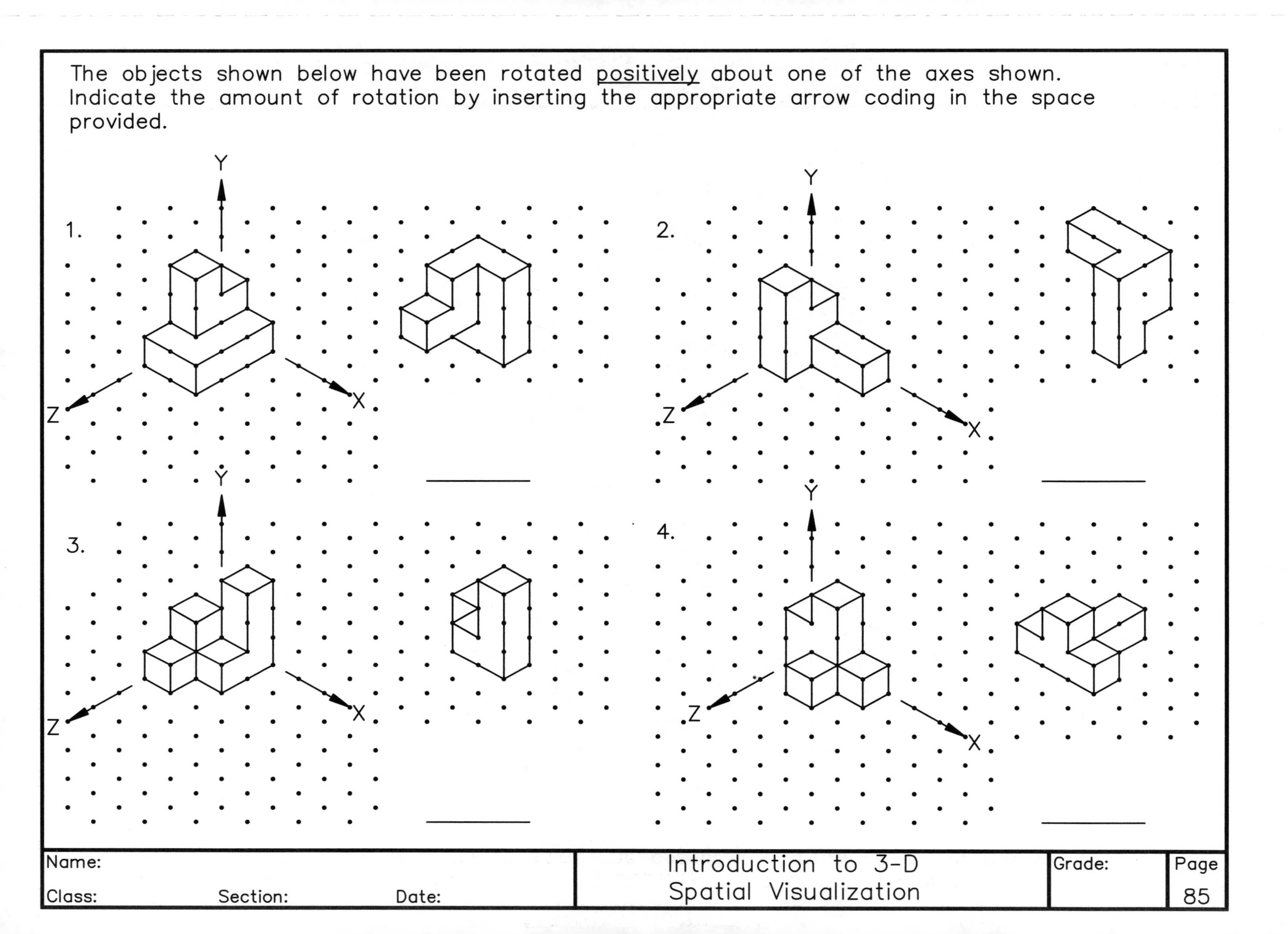

The objects shown below have been rotated <u>positively</u> about one of the axes shown. Indicate the amount of rotation by inserting the appropriate arrow coding in the space provided.

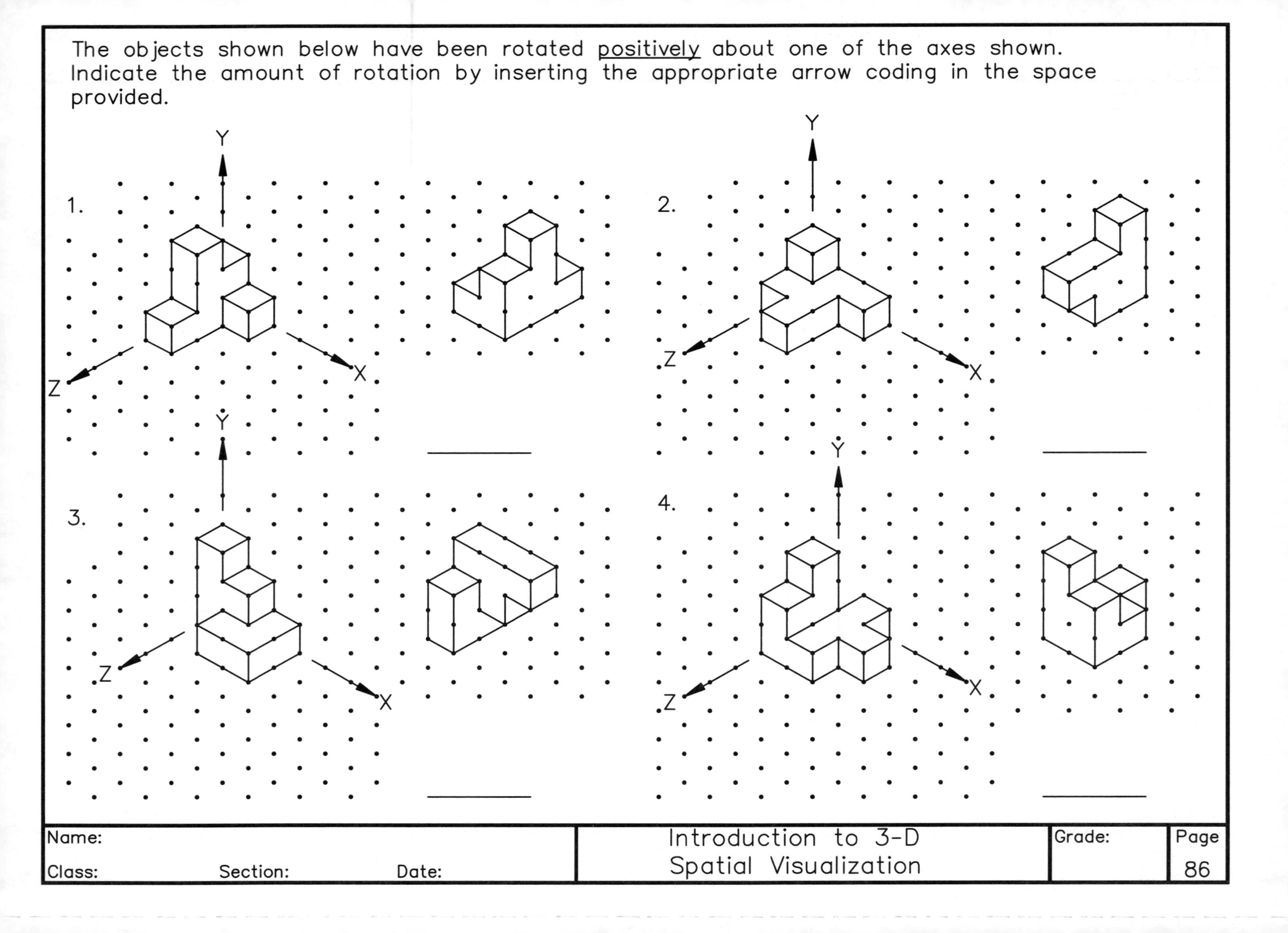

The objects shown below have experienced the rotations given by the arrows codings. In the space provided, indicate an equivalent arrow coding that would produce the same image.

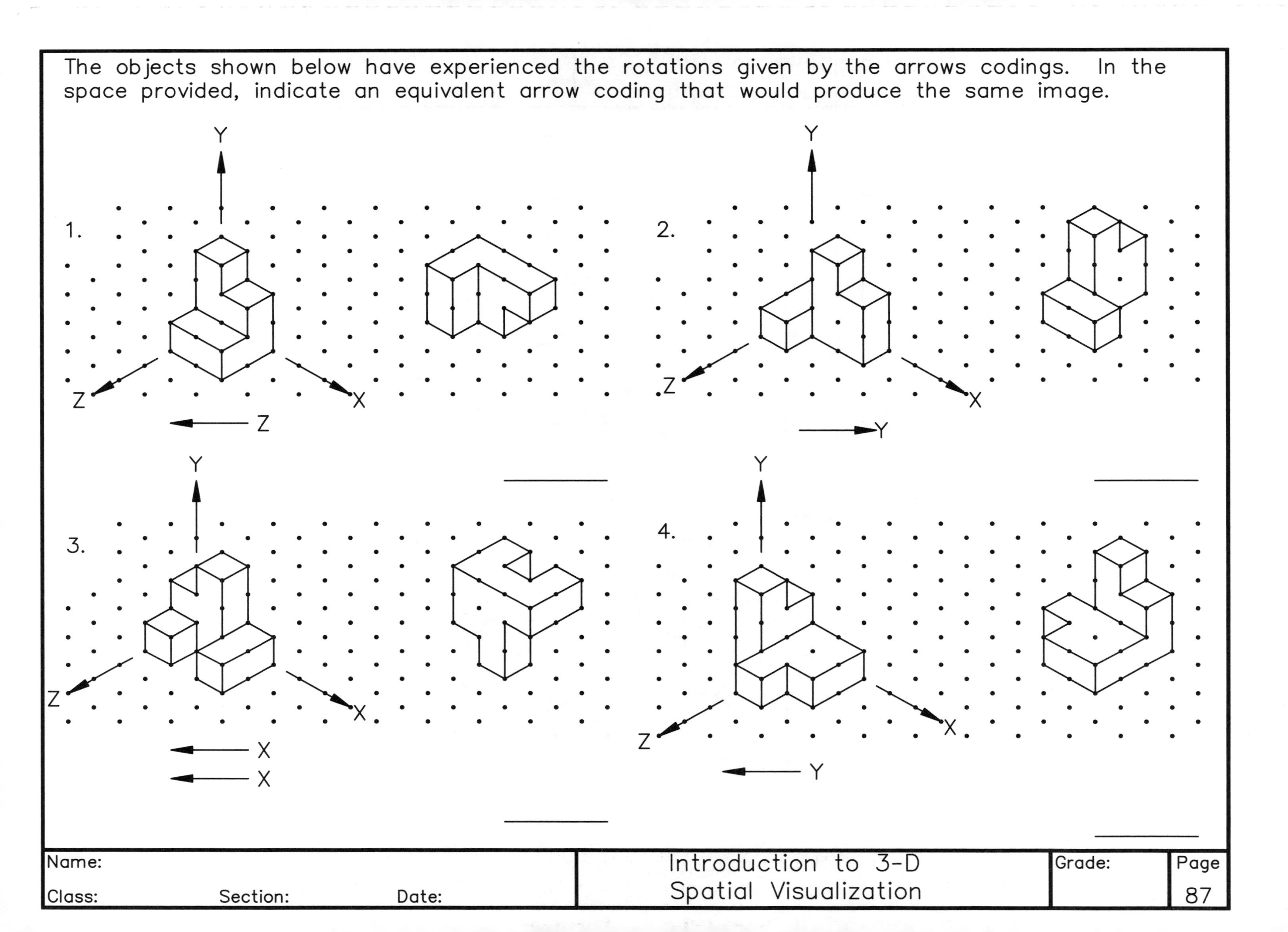

The objects shown below have experienced the rotations given by the arrows codings. In the space provided, indicate an equivalent arrow coding that would produce the same image.

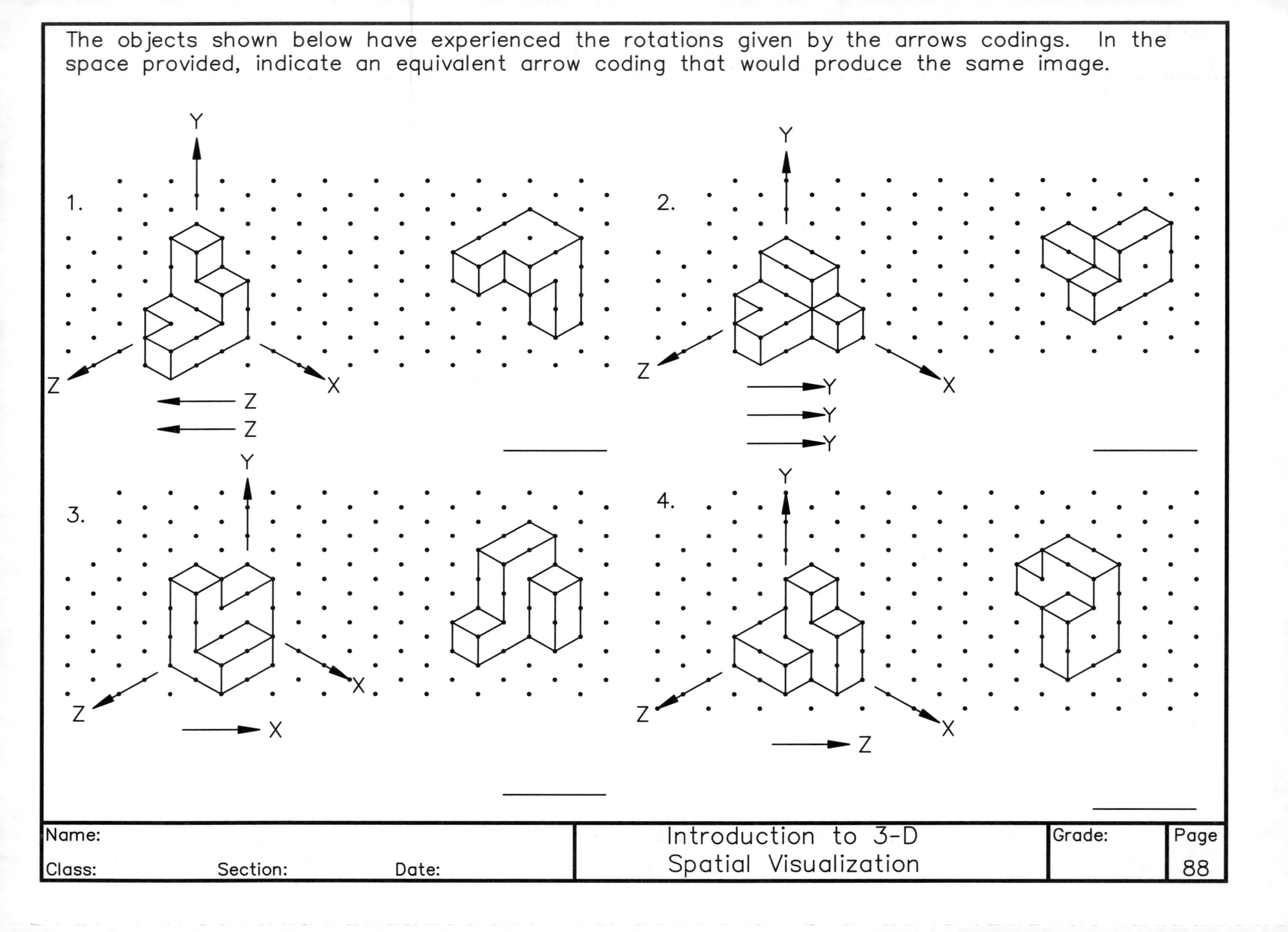

The objects shown below have experienced the rotations given by the arrows codings. In the space provided, indicate an equivalent arrow coding that would produce the same image.

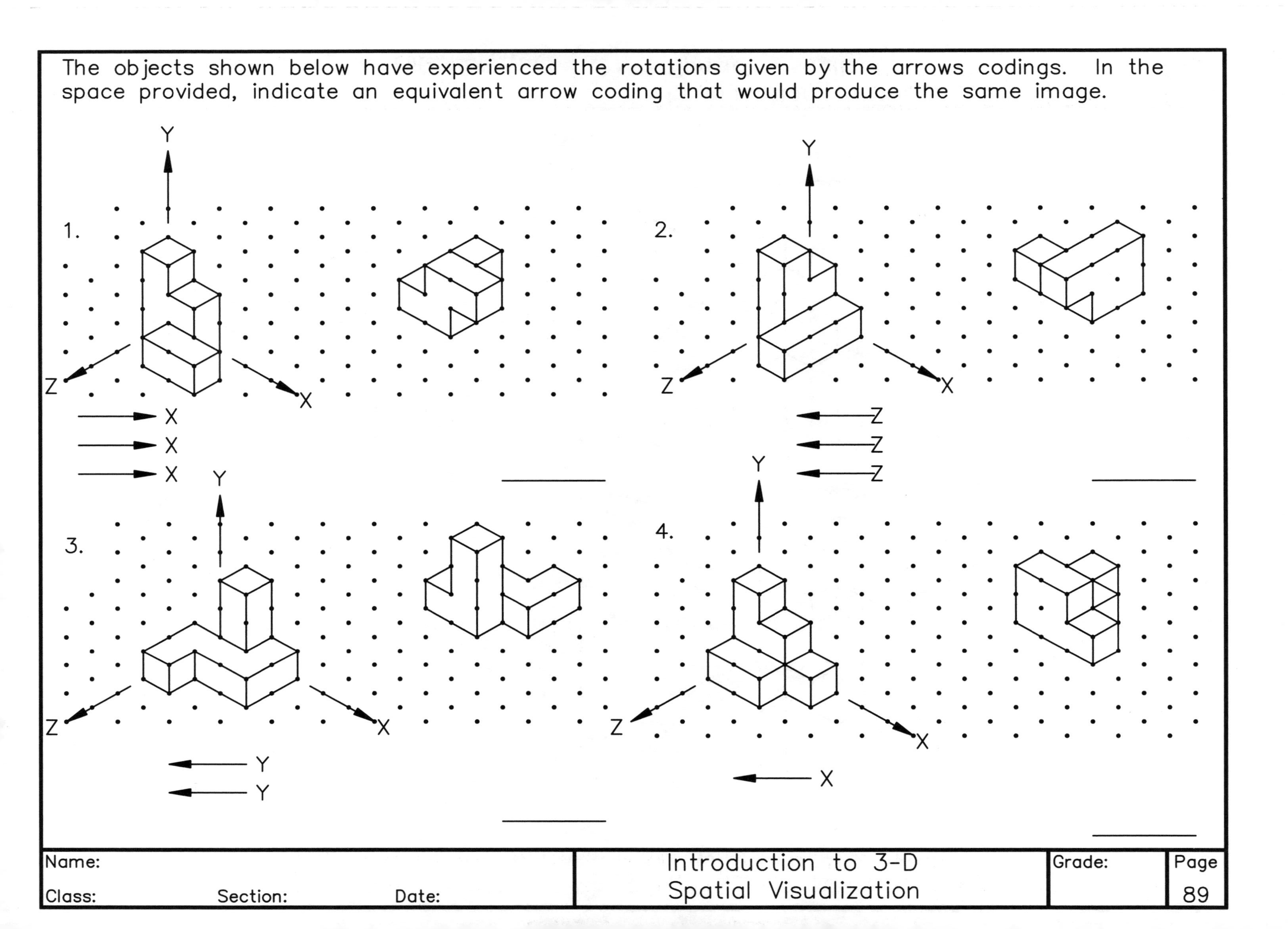


| Name: | Introduction to 3-D Spatial Visualization | Grade: | Page |
|---|---|---|---|
| Class: Section: Date: | | | 89 |

Rotate the objects shown below by the indicated amount and sketch the result in the space provided. You do not need to include the coordinate axes in your sketch.

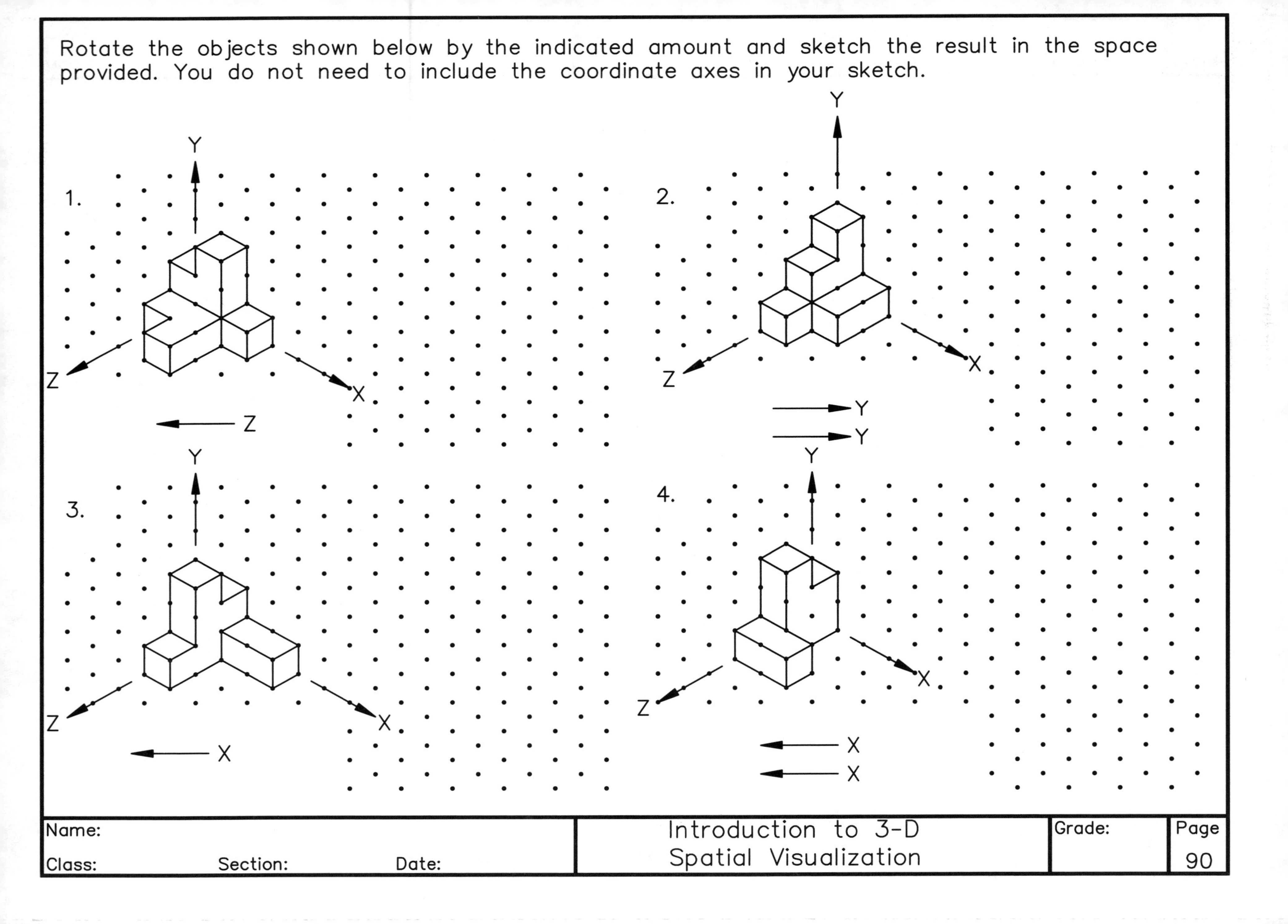


| Name: | Introduction to 3-D | Grade: | Page |
|---|---|---|---|
| Class: Section: Date: | Spatial Visualization | | 90 |

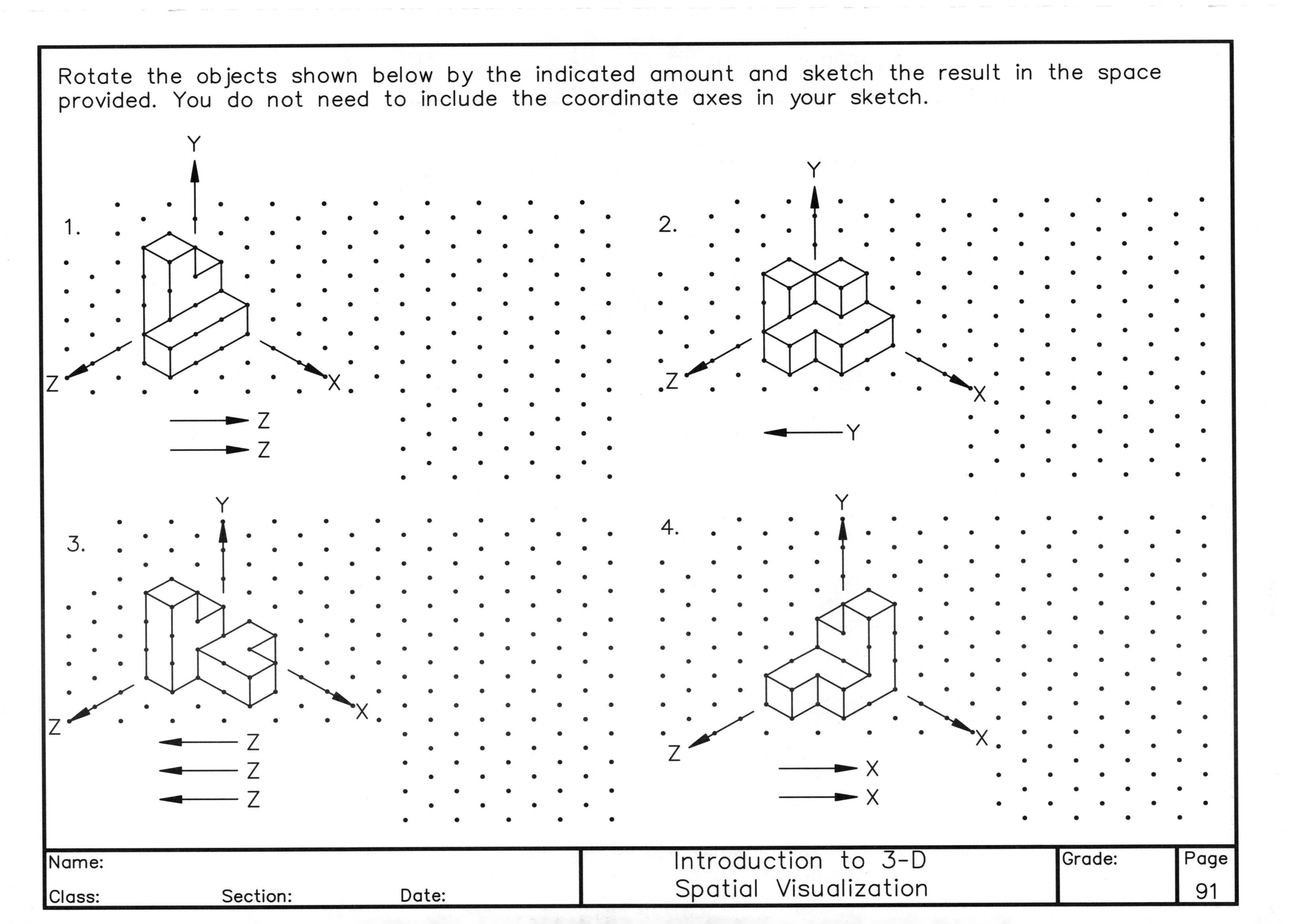
Rotate the objects shown below by the indicated amount and sketch the result in the space provided. You do not need to include the coordinate axes in your sketch.
1.
Y
Z
X
Z
Z
2.
Y
Z
X
Y
3.
Y
Z
X
Z
Z
Z
4.
Y
Z
X
X
X
Name:
Class:
Section:
Date:
Introduction to 3-D
Spatial Visualization
Grade:
Page
91

Rotate the objects shown below by the indicated amount and sketch the result in the space provided. You do not need to include the coordinate axes in your sketch.

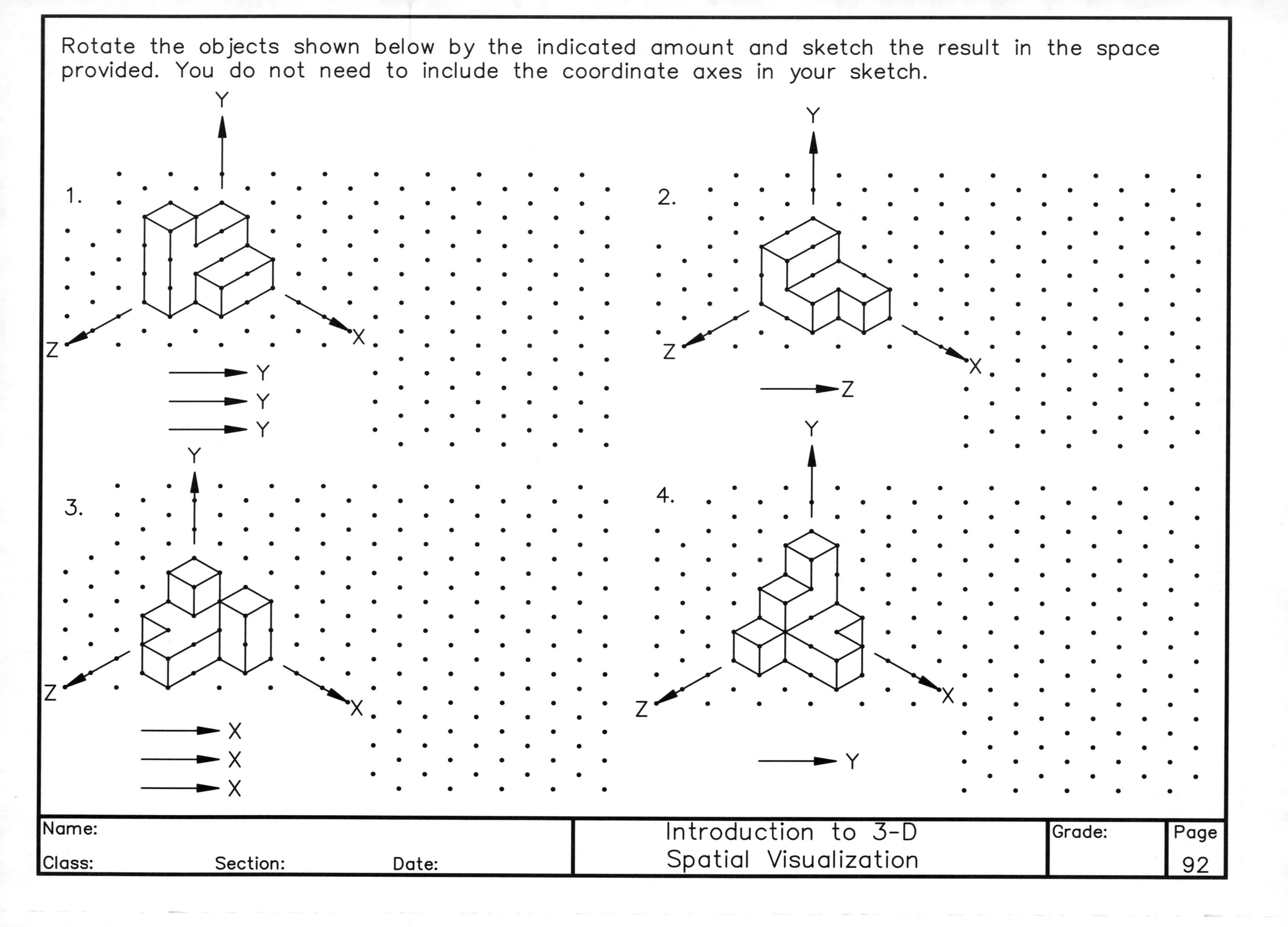


| Name: | | | Introduction to 3-D Spatial Visualization | Grade: | Page 92 |
|---|---|---|---|---|---|
| Class: | Section: | Date: | | | |

# Rotation of Objects about Two or More Axes

Objects can be rotated about two or more axes in the same manner that they can be rotated about a single axis.

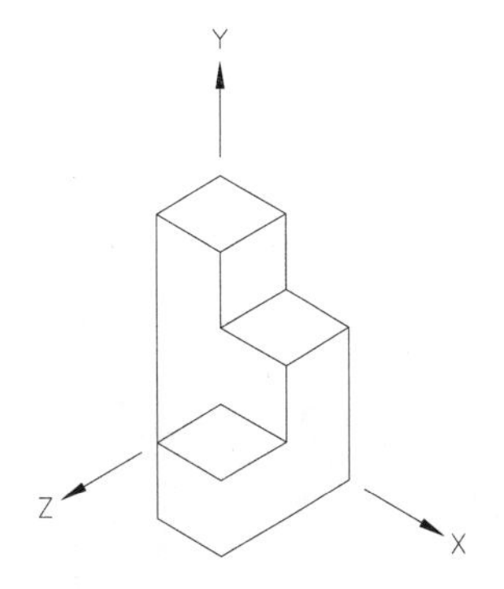


Original Object Position

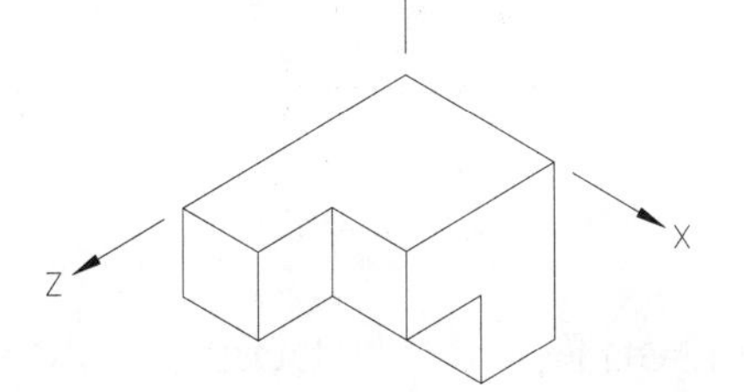


Object First Rotated about Positive X

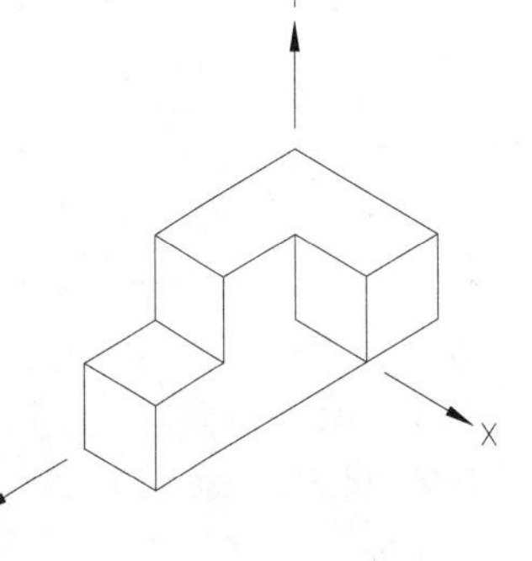

Then Rotated about Positive Z

→ X
→ Z

When rotation occurs about a single axis, an entire edge remains in its original position. When rotation occurs about two different axes, only a single pivot point remains in its original position.

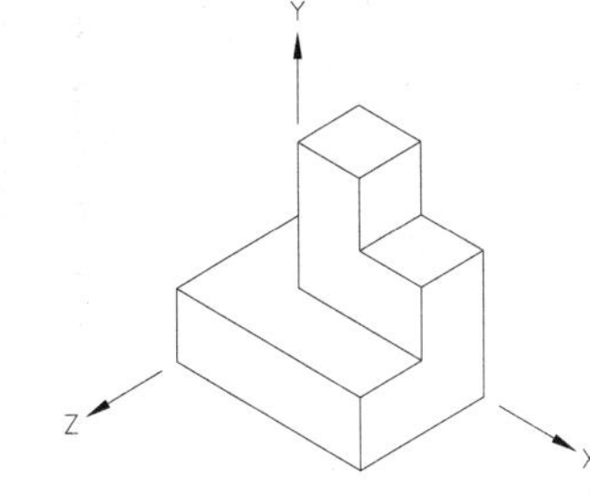


Original Object Position

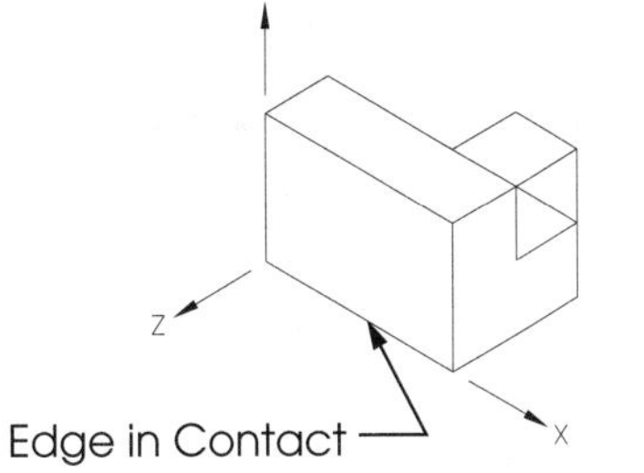


→ X
→ X
→ X

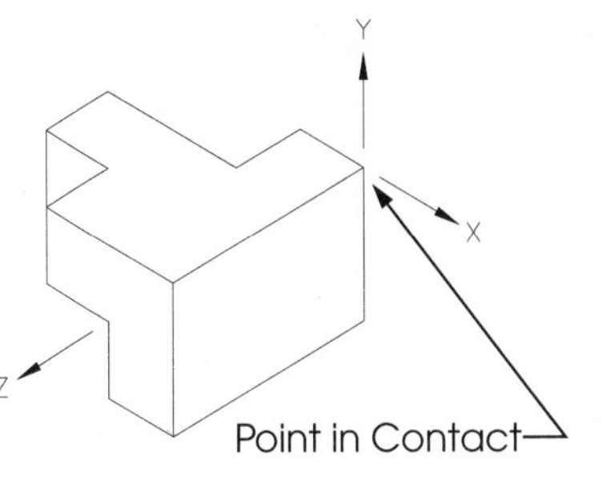


→ X
← Y

Objects can be rotated positively (counterclockwise) or negatively (clockwise) about each axis, just like when rotating about a single axis. (Remember the right hand rule!)

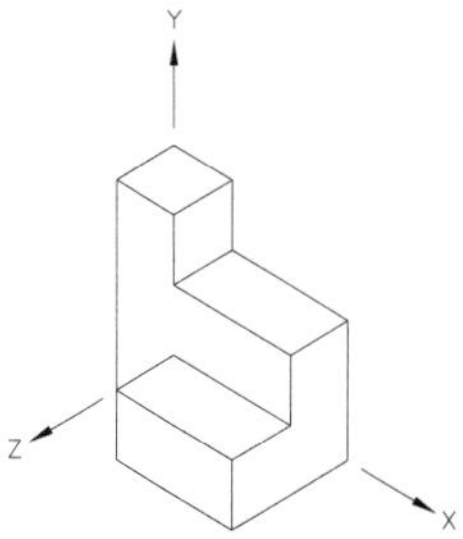

Original Object Position

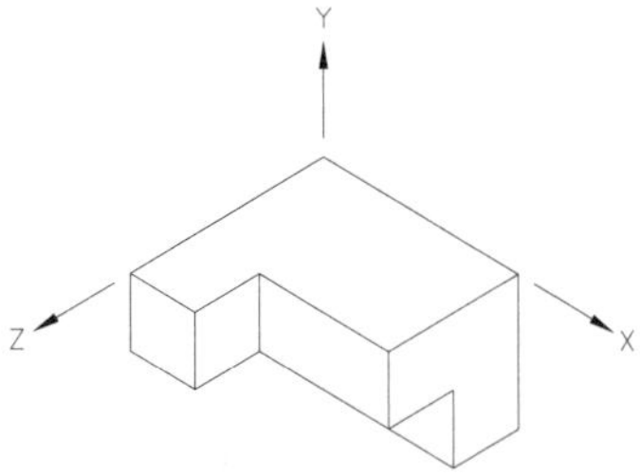

Object First Rotated about Positive X

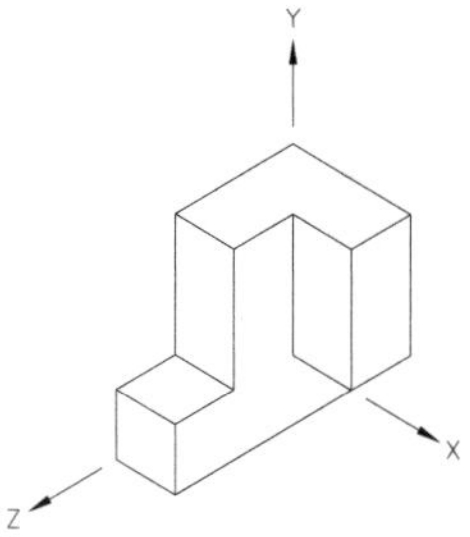

Then Rotated about Positive Z

→ X
→ Z

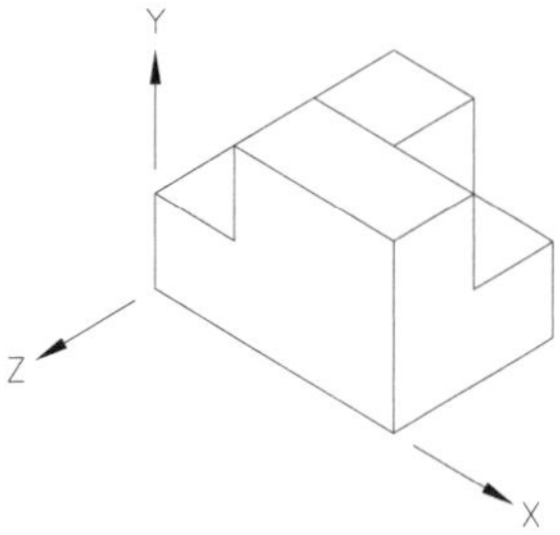

Object First Rotated about Negative X

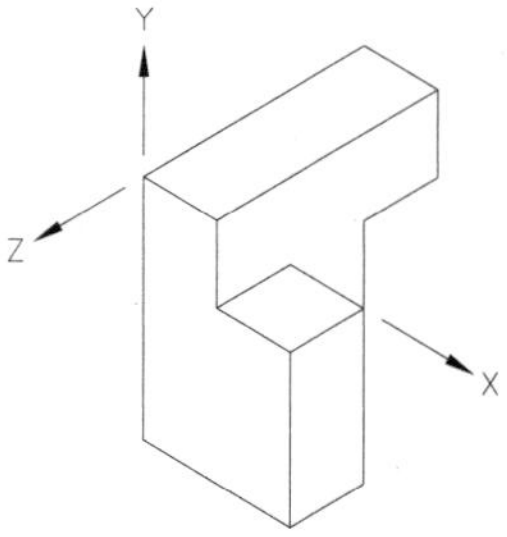

Then Rotated about Negative Z

← X
← Z

Rotations about two or more axes are not commutative. In other words, the order in which the rotations occur is important. If you switch the order of the rotations, you will not end up with the same result.

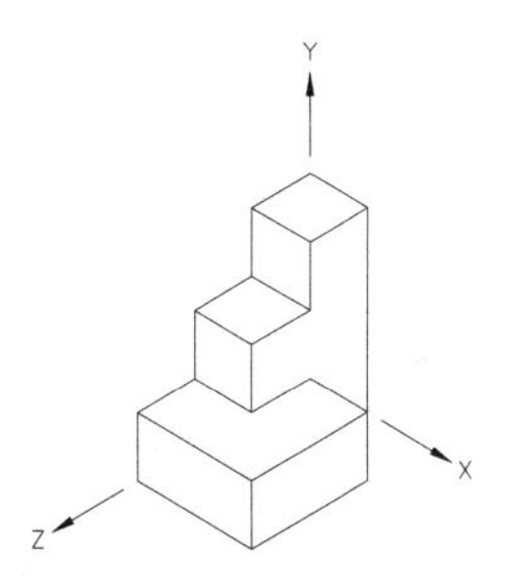


Original Object Position

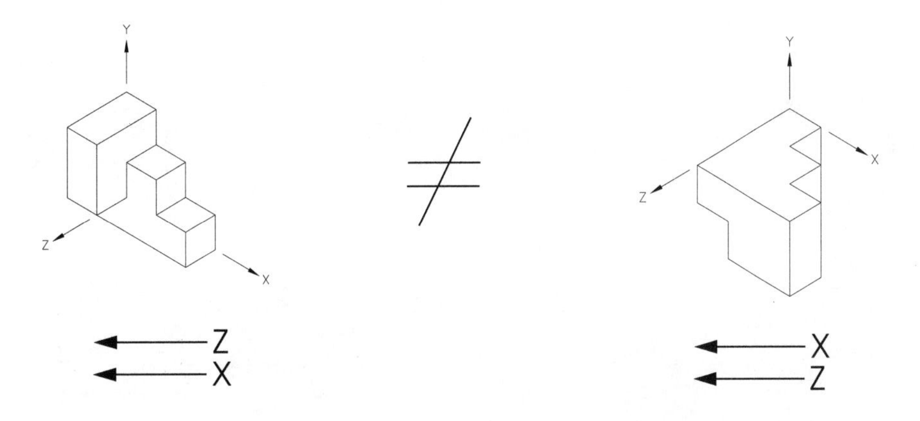


You can rotate an object in space as many times as you wish which can result in complex arrow codes.

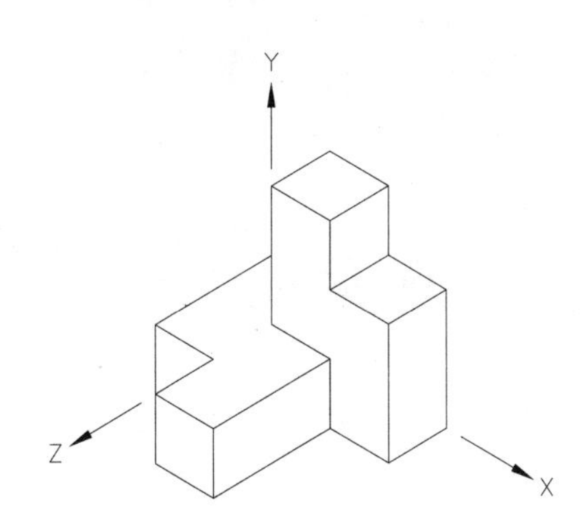


Original Object Position

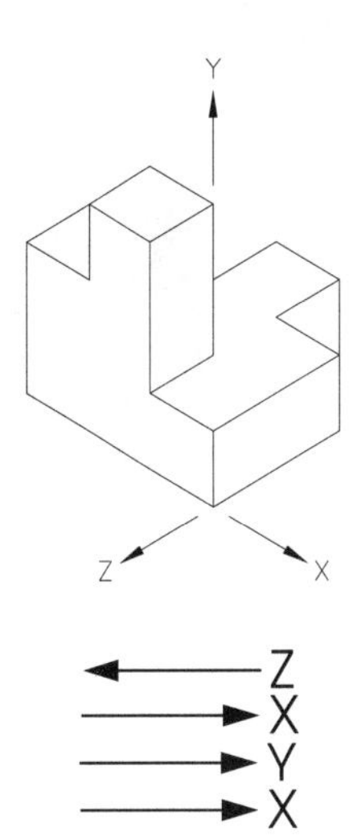

As with single axis rotations, different arrow codes can give you the same result. Two sets of arrow codes that result in the same final orientation of the object are said to be *equivalent*.

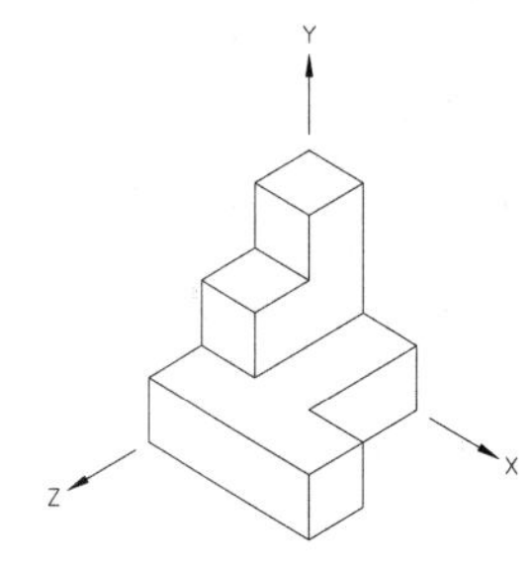

Original Object Position

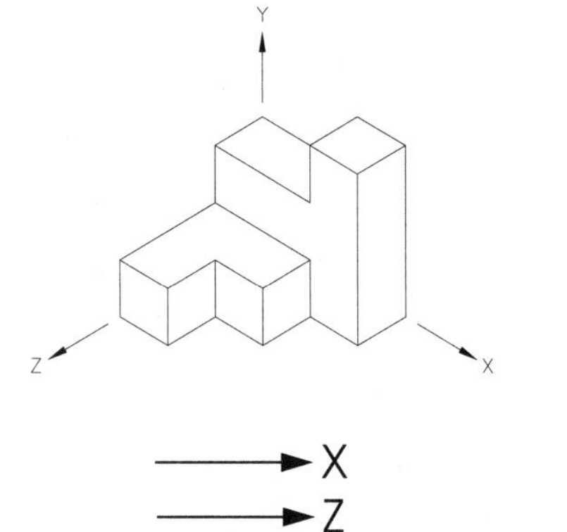


=

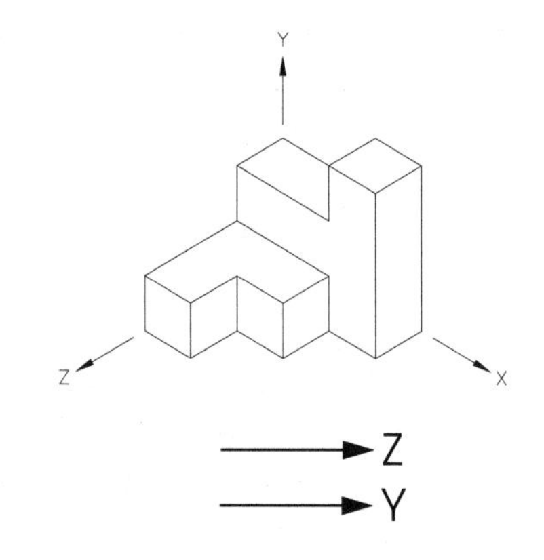


Complex arrow codes can sometimes be reduced to simpler ones.

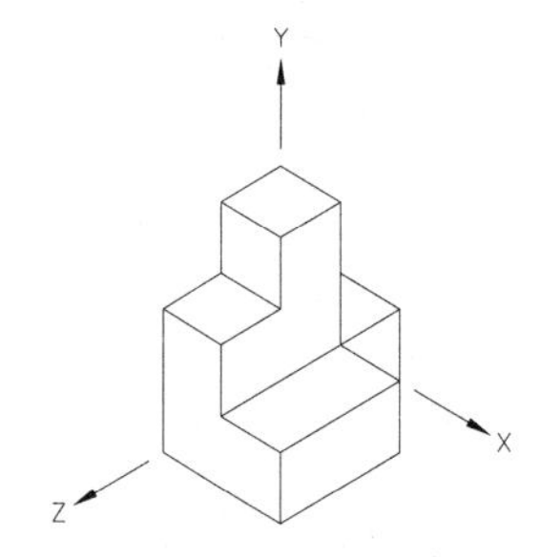

Original Object Position

Circle the letter corresponding to the view on the right that shows the result of rotating the object on the left by the indicated amount.

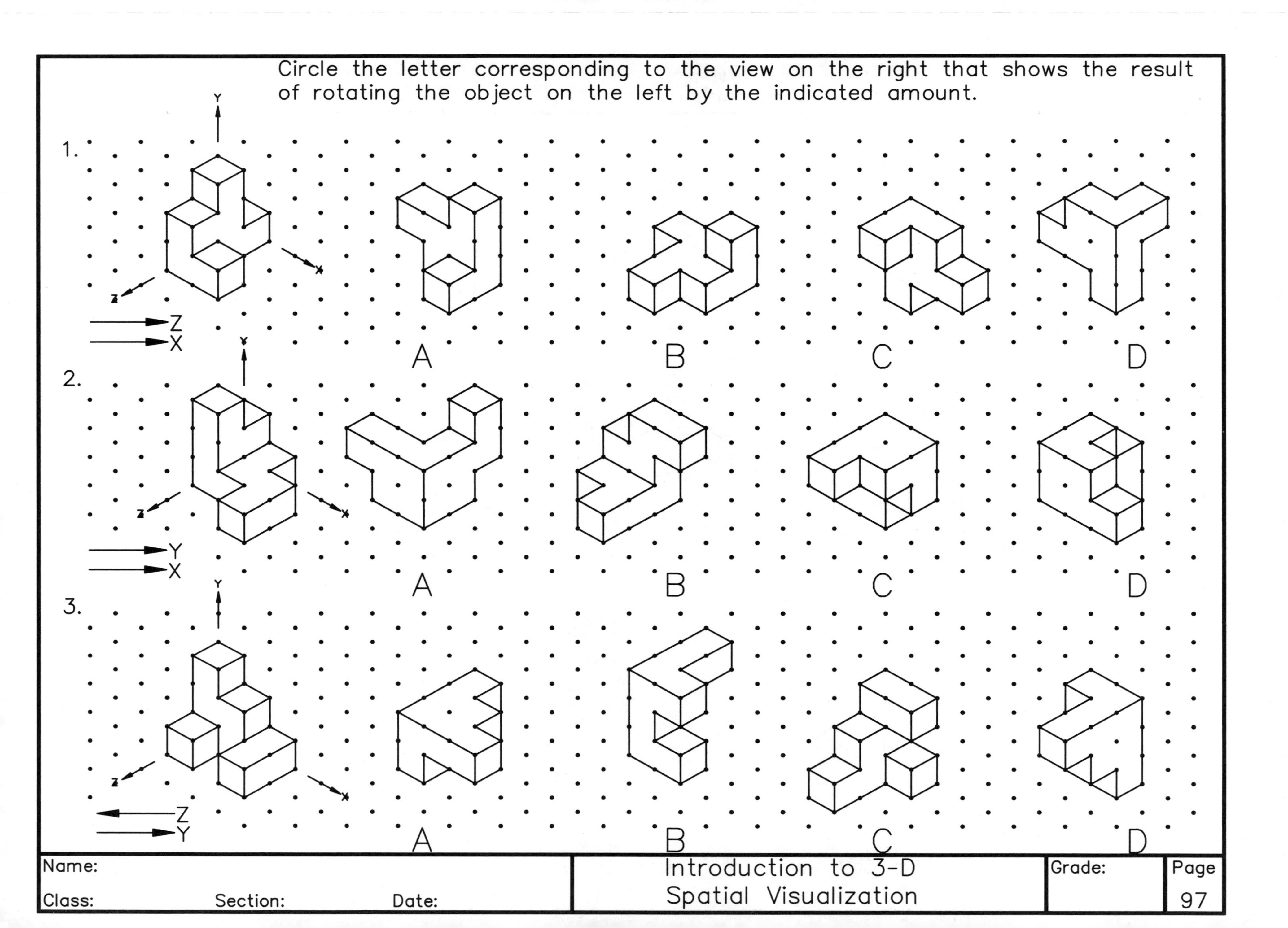


| Name: | Introduction to 3-D Spatial Visualization | Grade: | Page |
|---|---|---|---|
| Class: Section: Date: | | | 97 |

Circle the letter corresponding to the view on the right that shows the result of rotating the object on the left by the indicated amount.

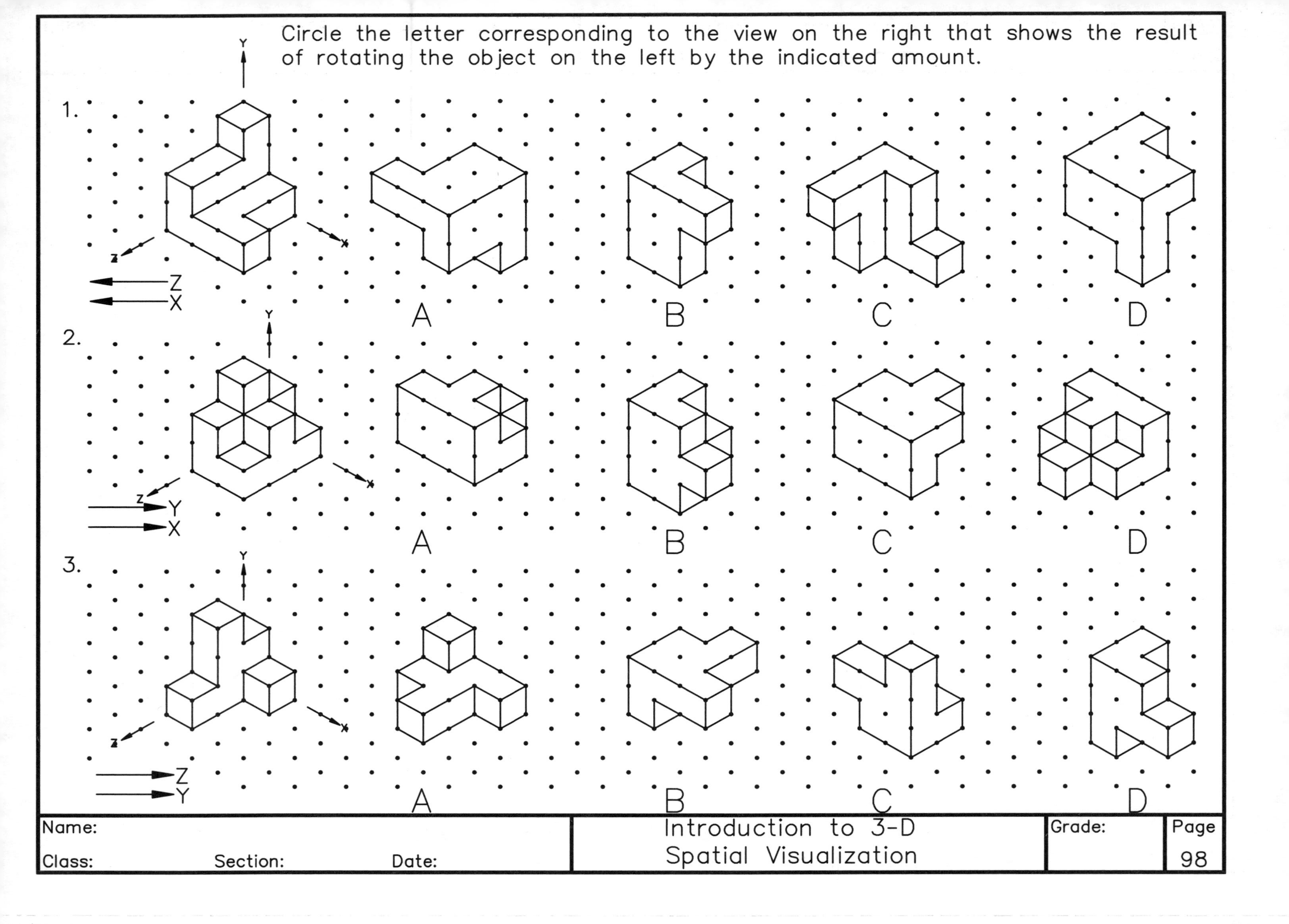


| Name: | Introduction to 3-D | Grade: | Page |
|---|---|---|---|
| Class: Section: Date: | Spatial Visualization | | 98 |

Circle the letter corresponding to the view on the right that shows the result of rotating the object on the left by the indicated amount.

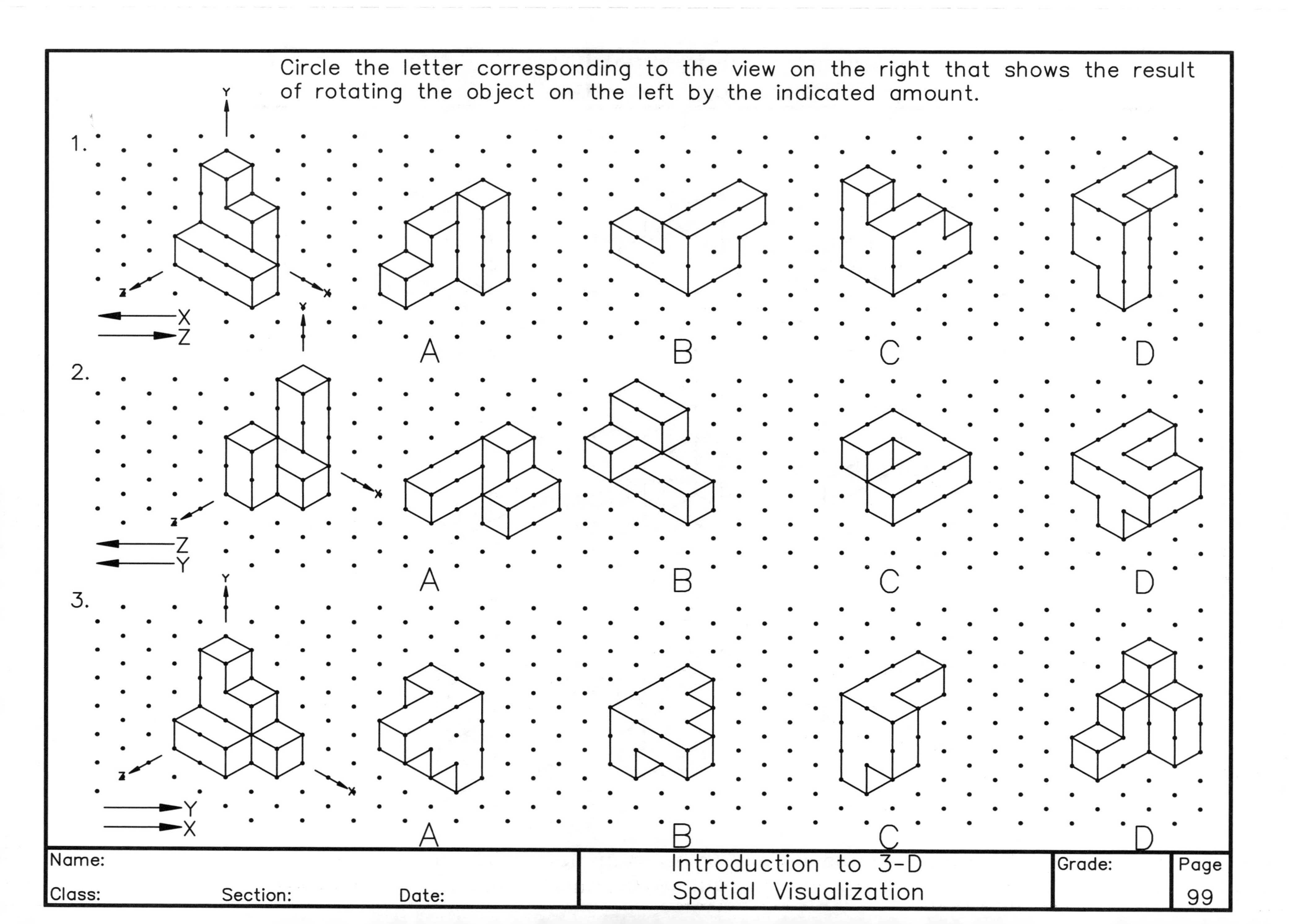

Circle the letter corresponding to the view on the right that shows the result of rotating the object on the left by the indicated amount.

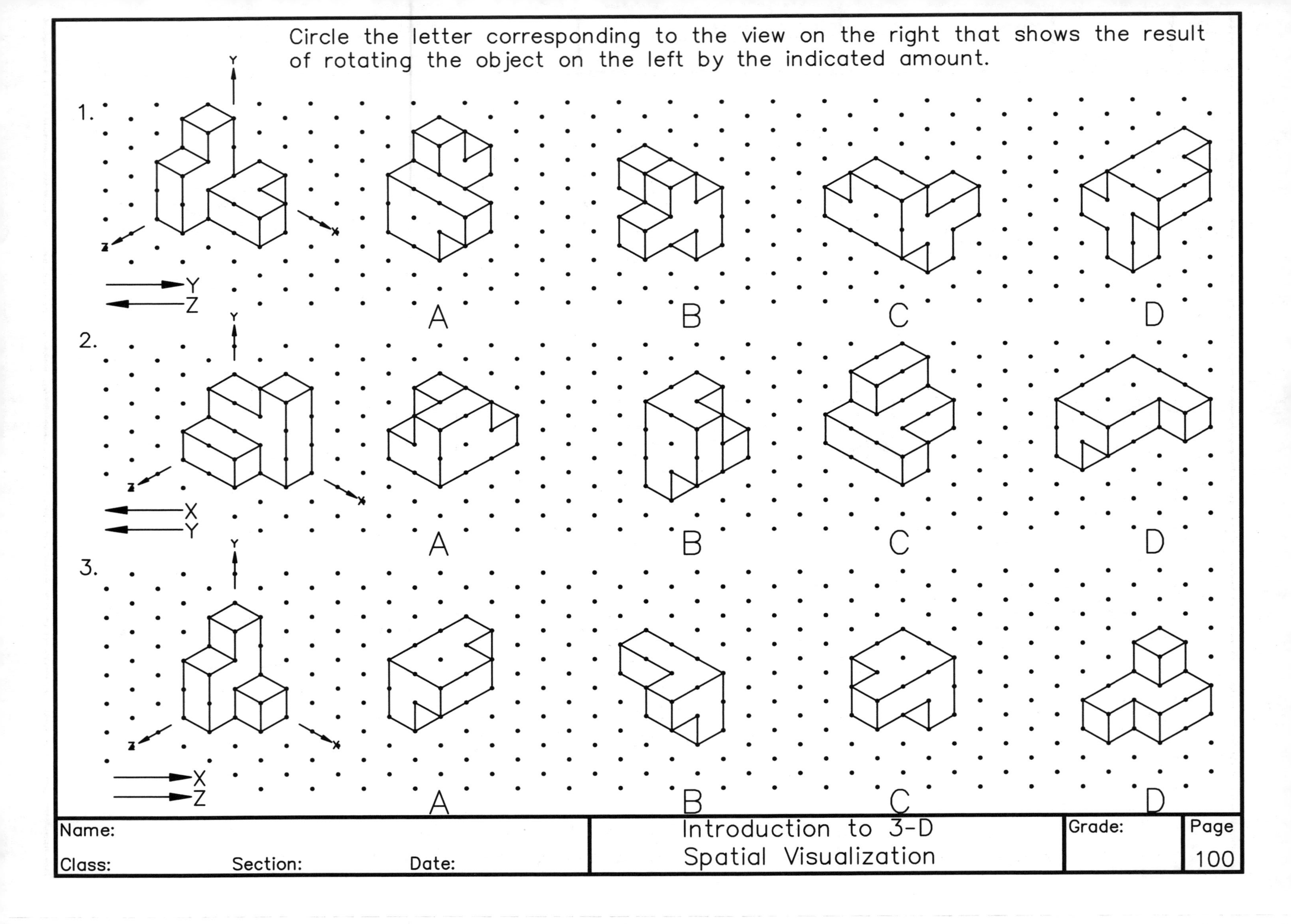


Name:

Class: Section: Date:

Introduction to 3-D Spatial Visualization

Grade:

Circle the letter corresponding to the view on the right that shows the result of rotating the object on the left by the indicated amount.

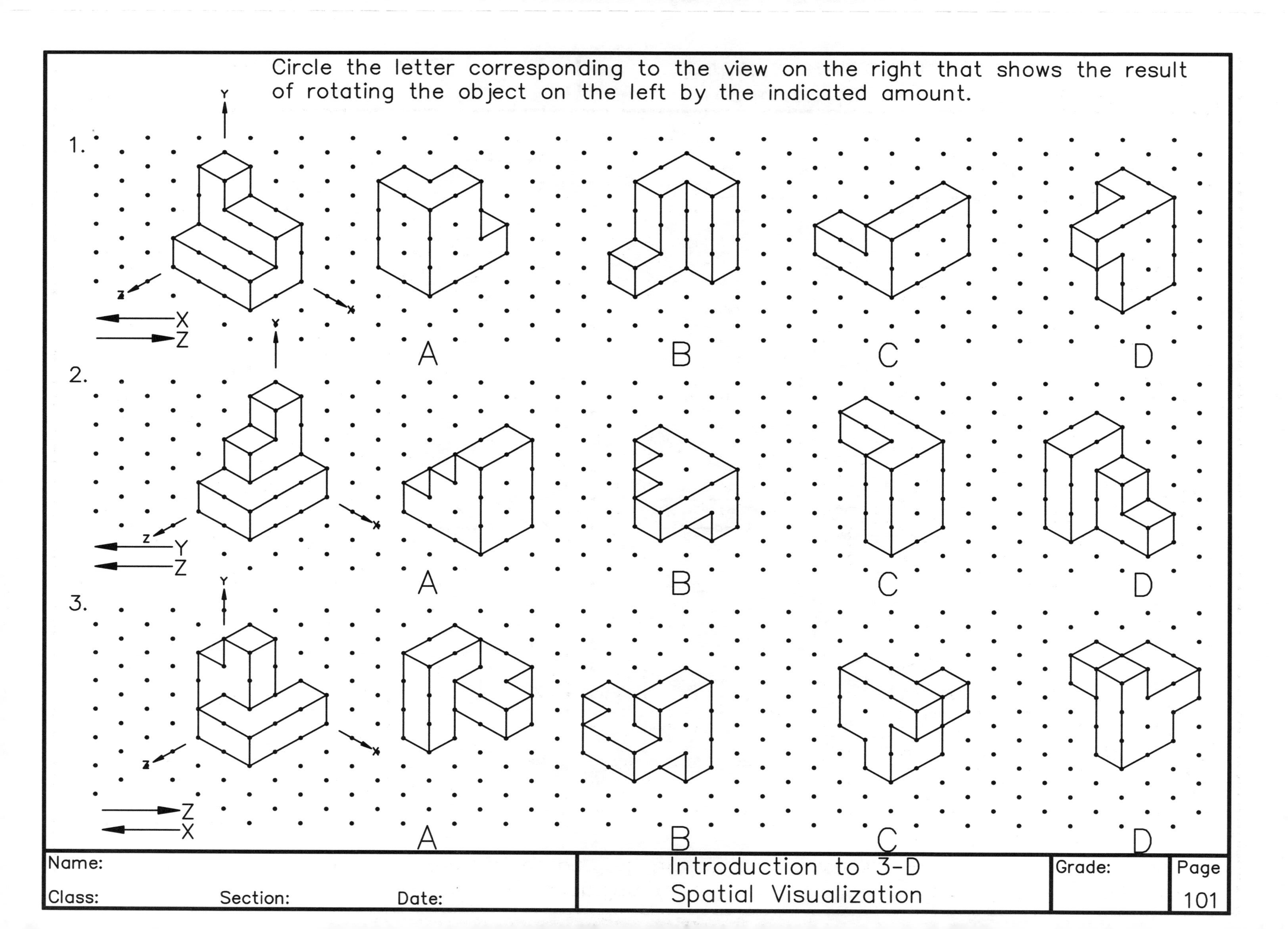

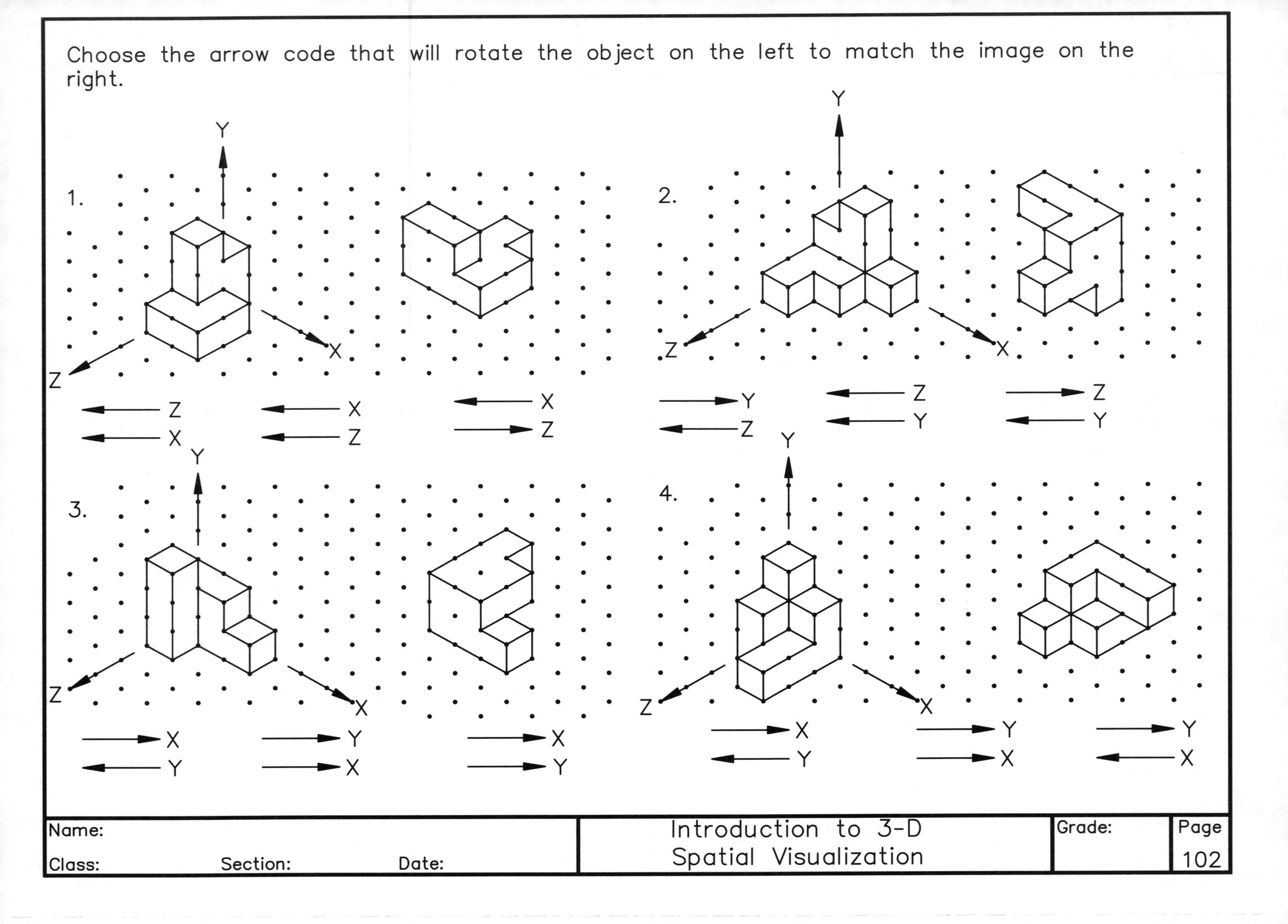
Choose the arrow code that will rotate the object on the left to match the image on the right.
1.
Y
Z
X
Z
X
X
Z
X
Z
2.
Y
Z
X
Y
Z
Z
Y
Z
Y
3.
Y
Z
X
X
Y
Y
X
X
Y
4.
Y
Z
X
X
Y
Y
X
Y
X
Name:
Class:
Section:
Date:
Introduction to 3-D
Spatial Visualization
Grade:
Page
102

Choose the arrow code that will rotate the object on the left to match the image on the right.

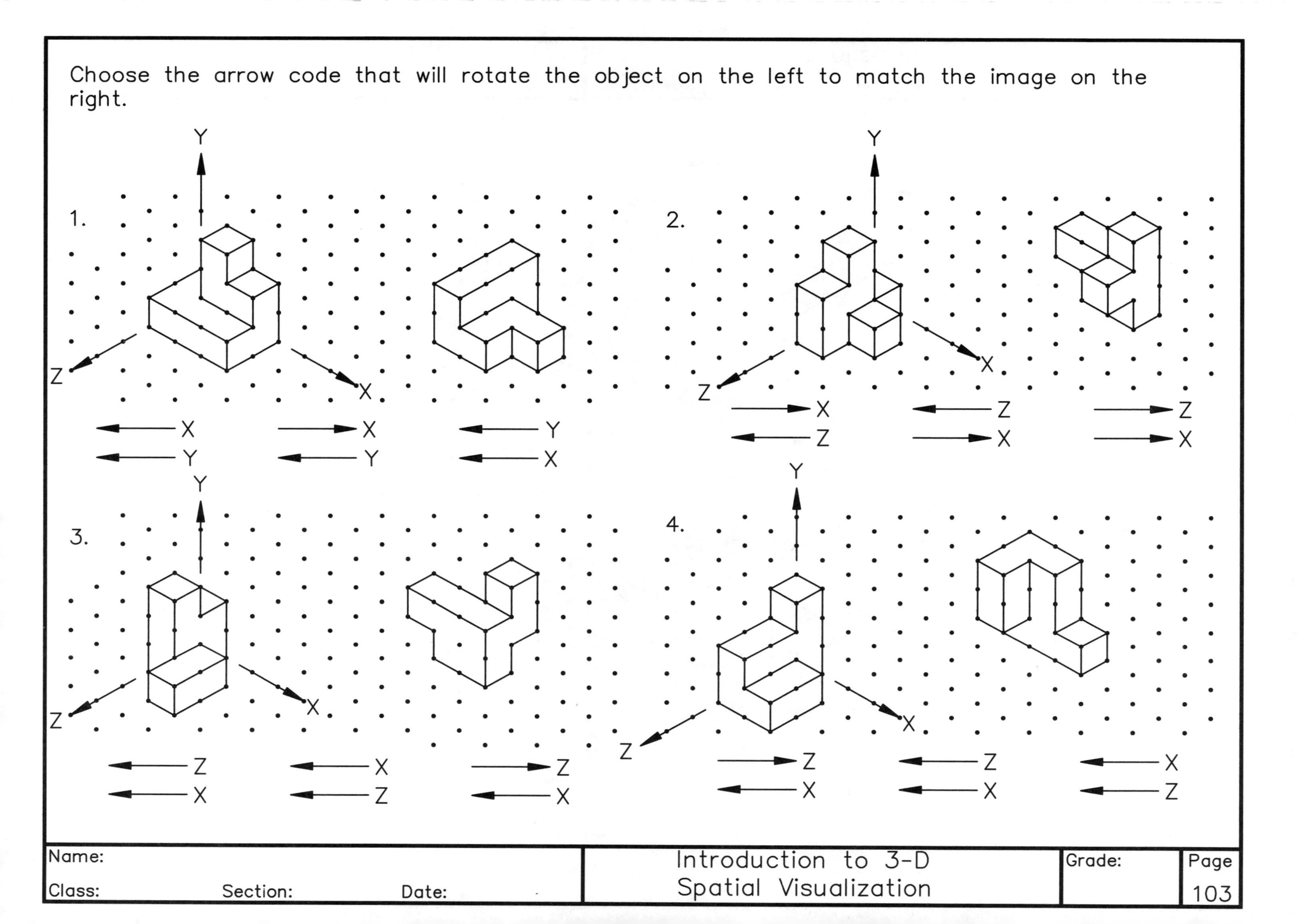

Choose the arrow code that will rotate the object on the left to match the image on the right.

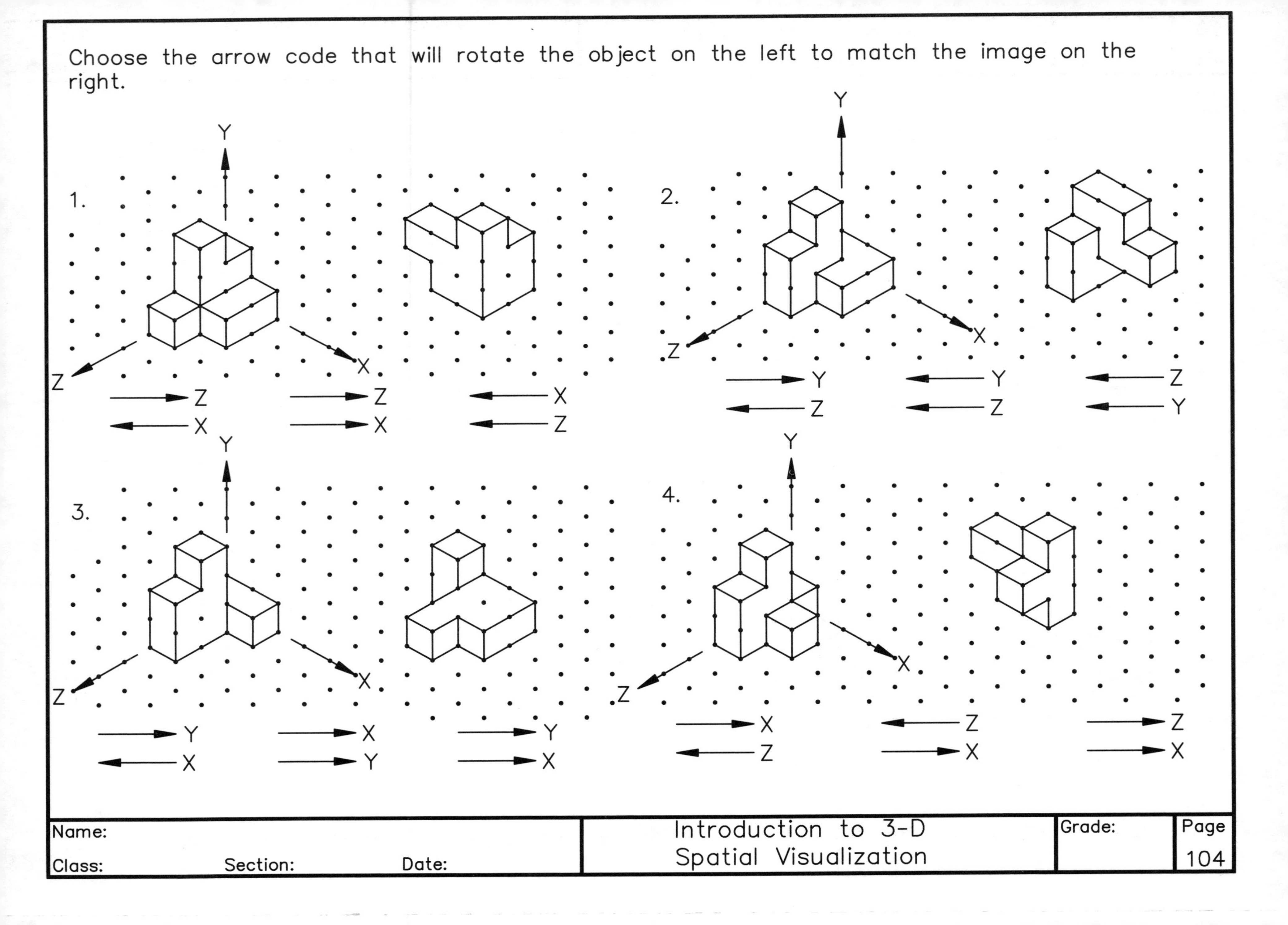

Choose the arrow code that will rotate the object on the left to match the image on the right.

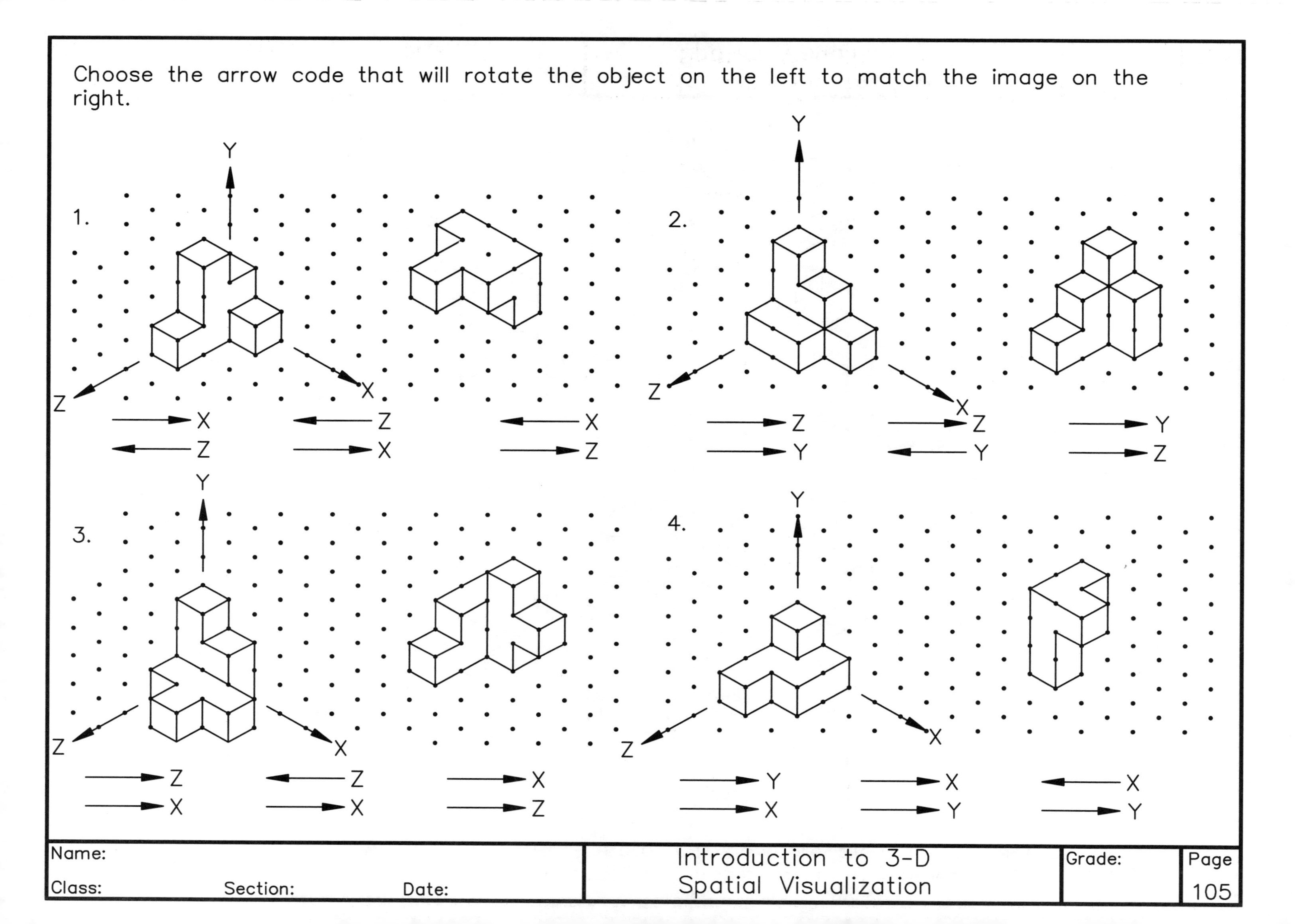

In the space provided, indicate the arrow code that would rotate the object on the left to obtain the view of it shown on the right. (There may be more than one correct response.)

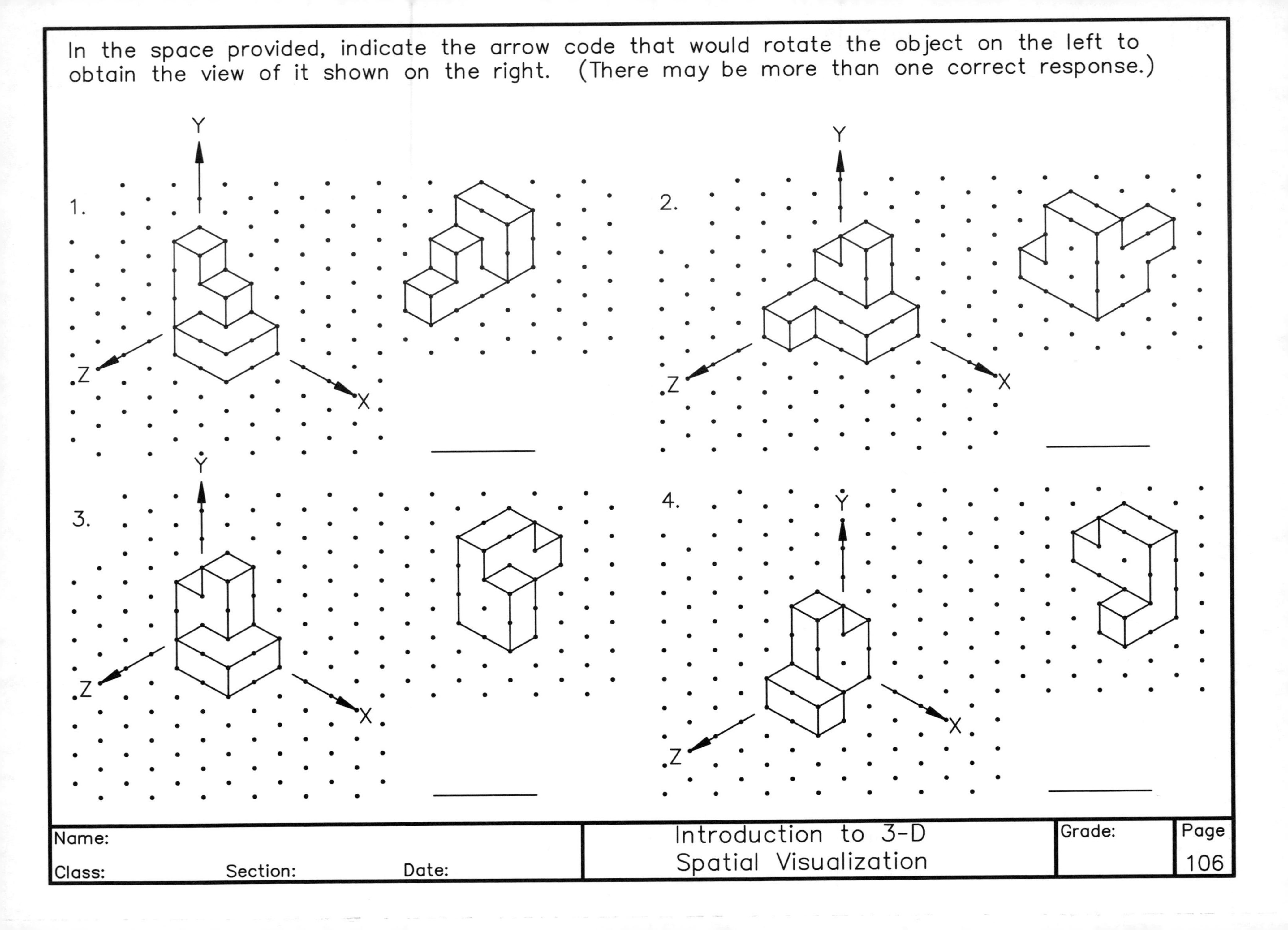

In the space provided, indicate the arrow code that would rotate the object on the left to obtain the view of it shown on the right. (There may be more than one correct response.)

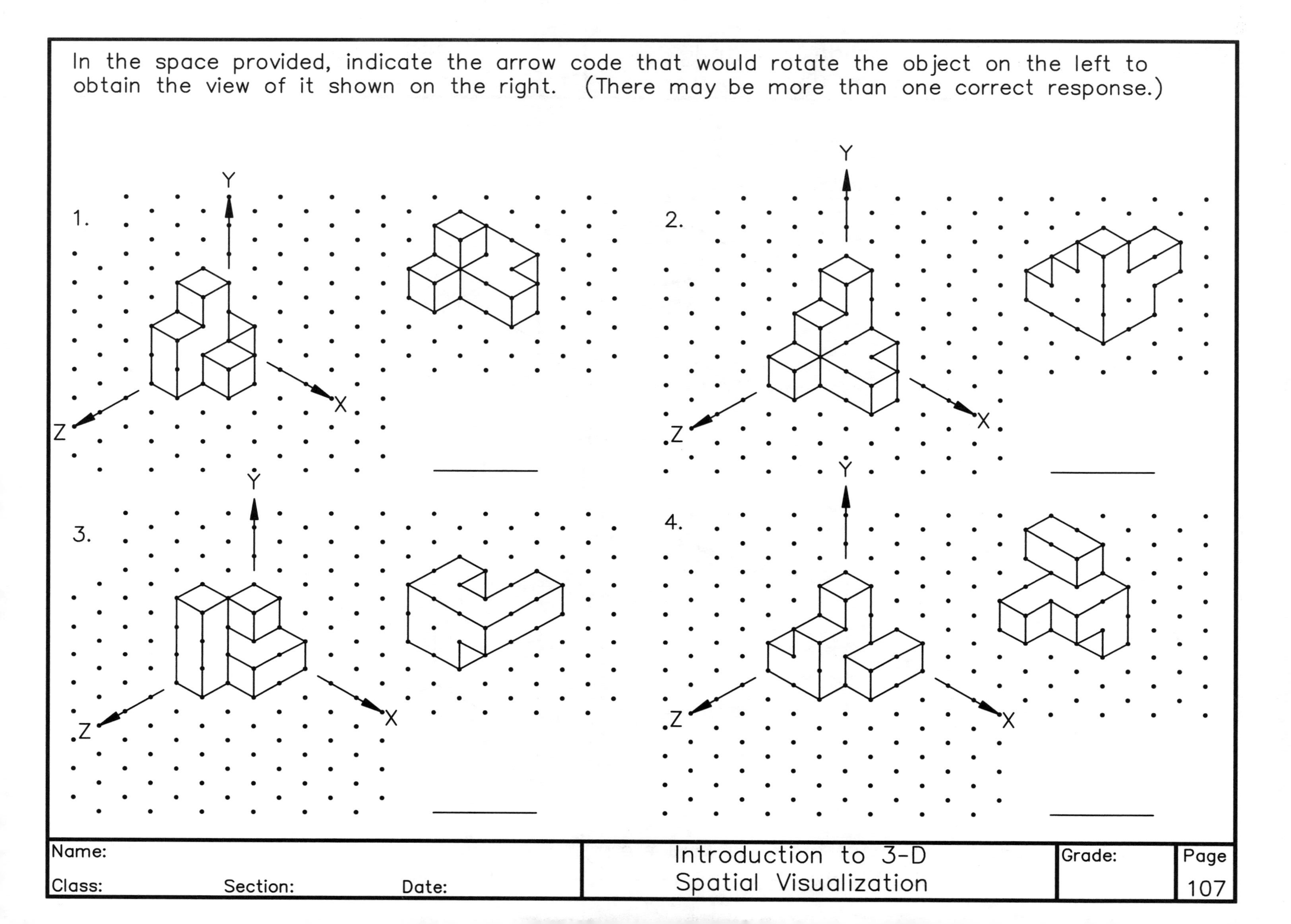


| Name: | Introduction to 3-D Spatial Visualization | Grade: | Page 107 |
|---|---|---|---|
| Class: Section: Date: | | | |

In the space provided, indicate the arrow code that would rotate the object on the left to obtain the view of it shown on the right. (There may be more than one correct response.)

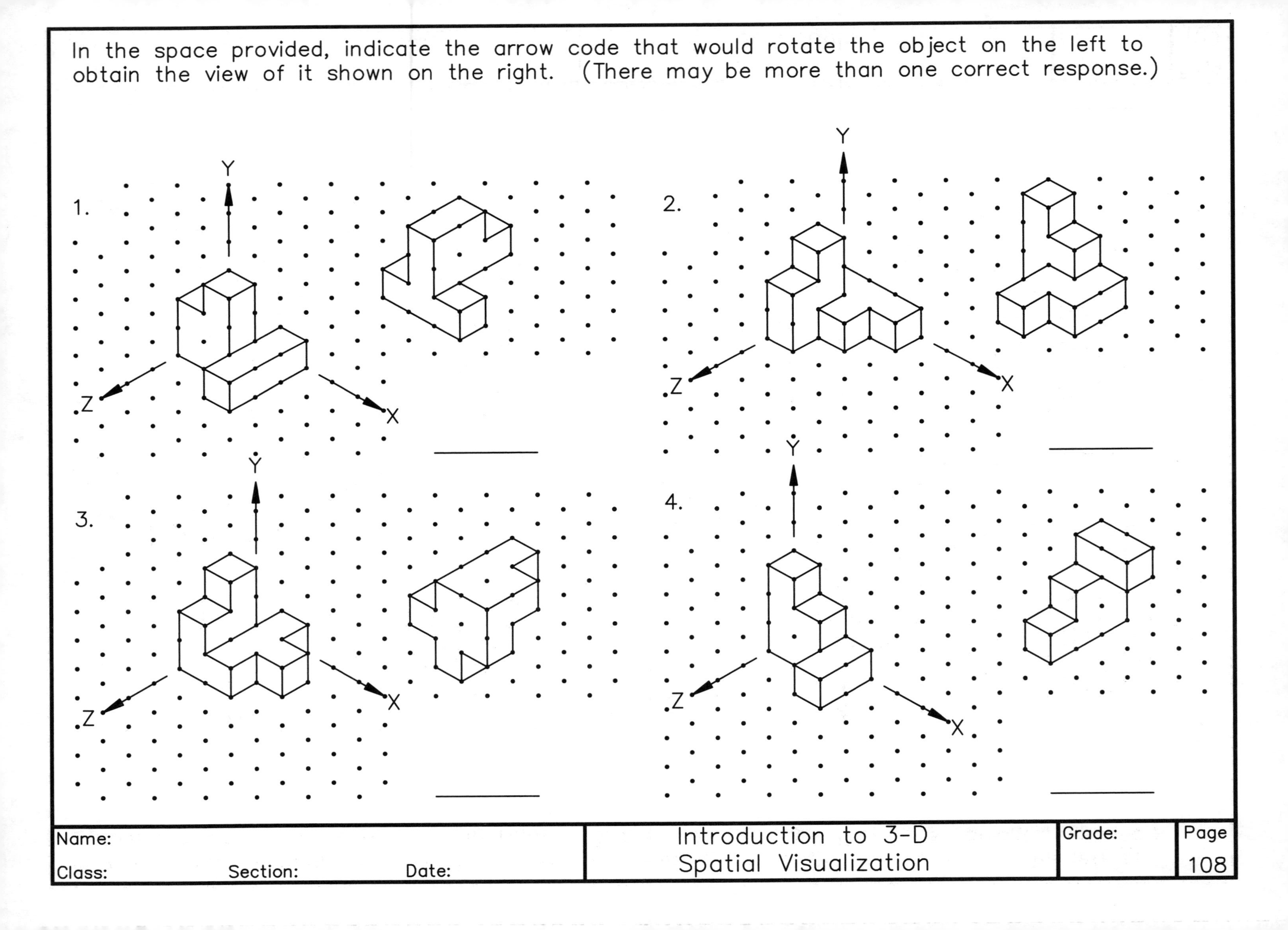

In the space provided, indicate the arrow code that would rotate the object on the left to obtain the view of it shown on the right. (There may be more than one correct response.)

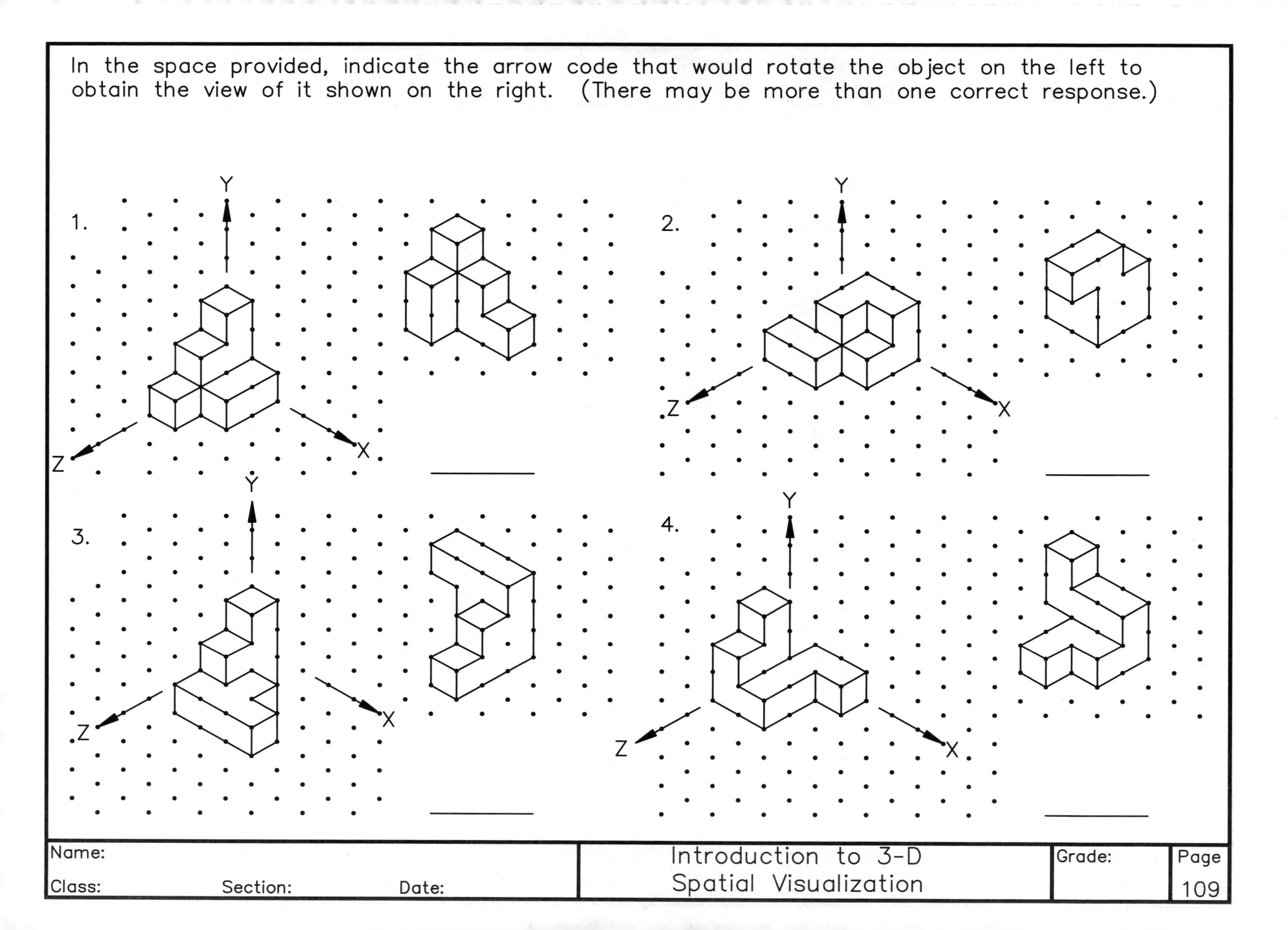

Rotate the objects shown below by the indicated amount. Sketch the result in the space provided. Make sure you perform the rotations in the given order.

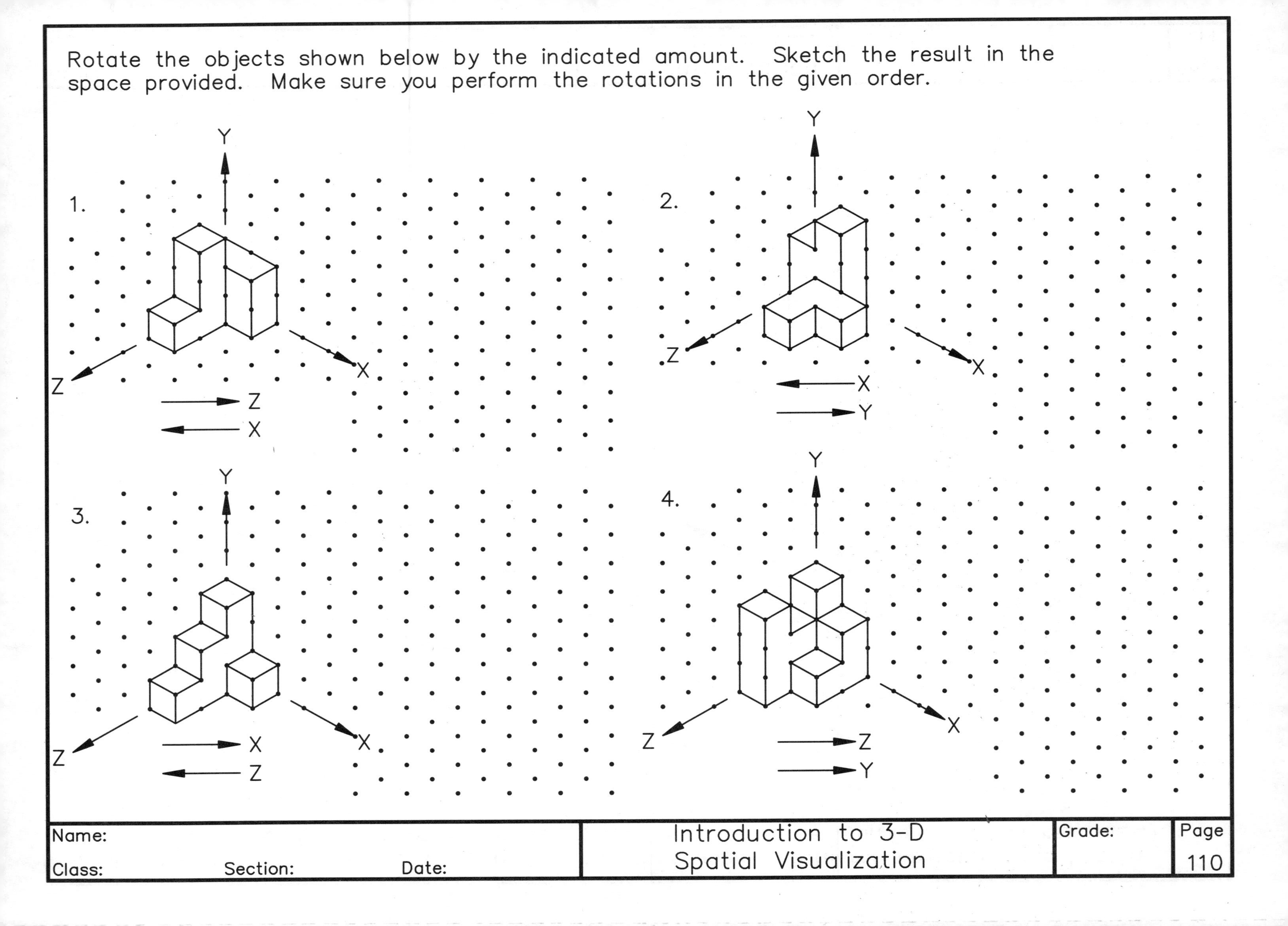

Rotate the objects shown below by the indicated amount. Sketch the result in the space provided. Make sure you perform the rotations in the given order.

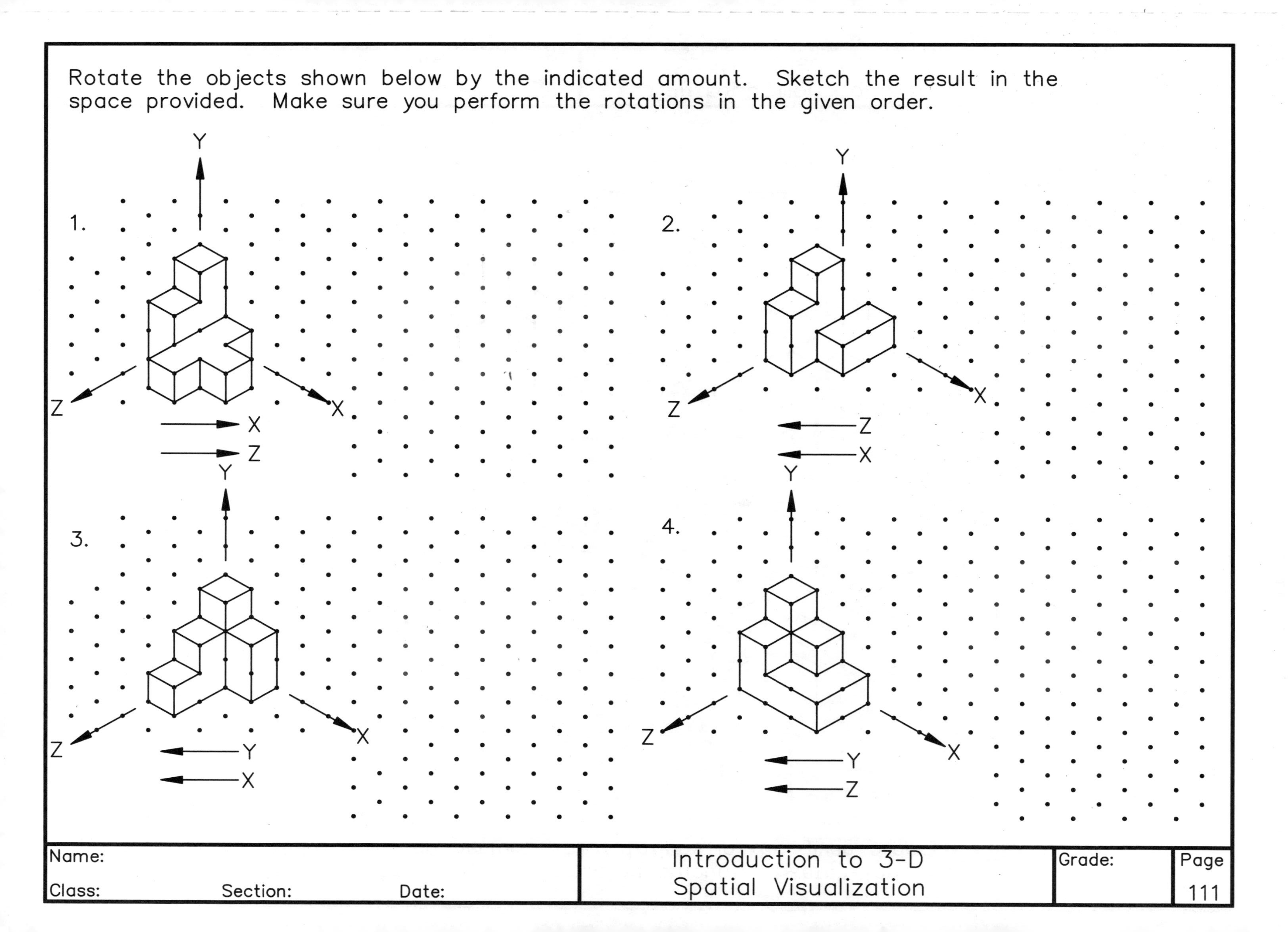

Rotate the objects shown below by the indicated amount. Sketch the result in the space provided. Make sure you perform the rotations in the given order.

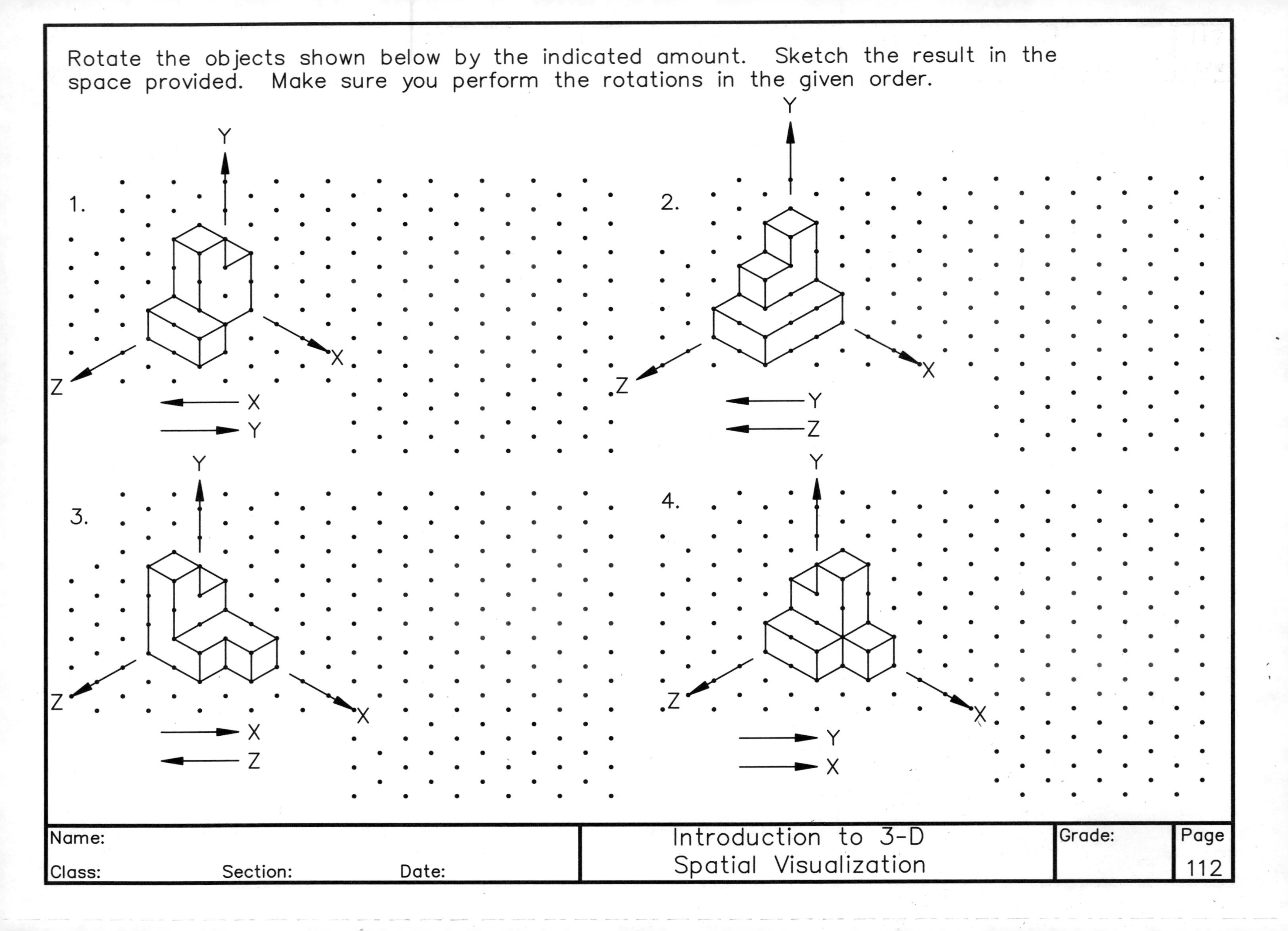


| Name: | Introduction to 3-D Spatial Visualization | Grade: | Page |
|---|---|---|---|
| Class: Section: Date: | | | 112 |

Rotate the objects shown below by the indicated amount. Sketch the result in the space provided. Make sure you perform the rotations in the given order.

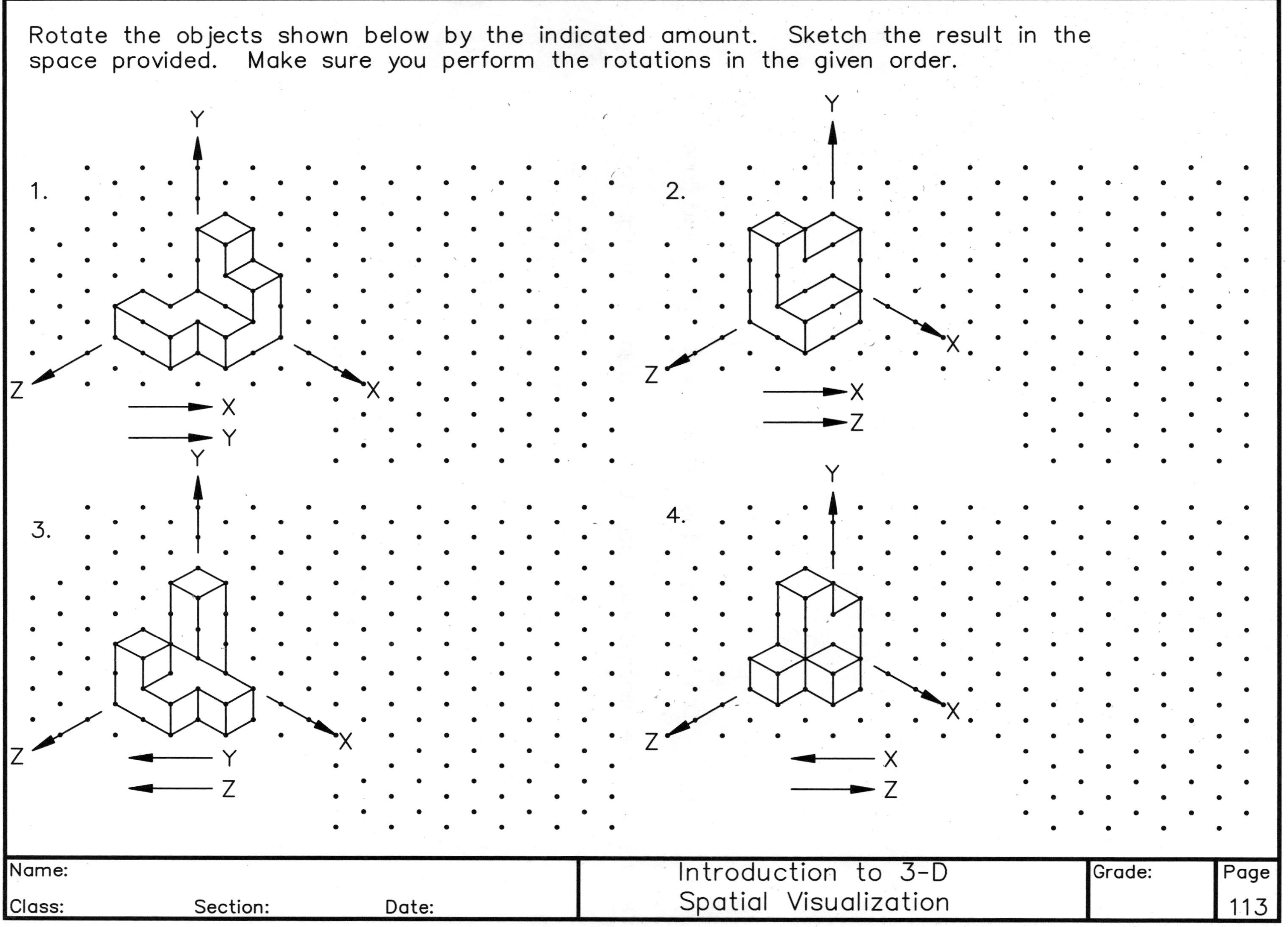

# Object Reflections and Symmetry

When an object is reflected across a plane, the result is two separate objects that are "mirror images" of one another on both sides of the plane of reflection.

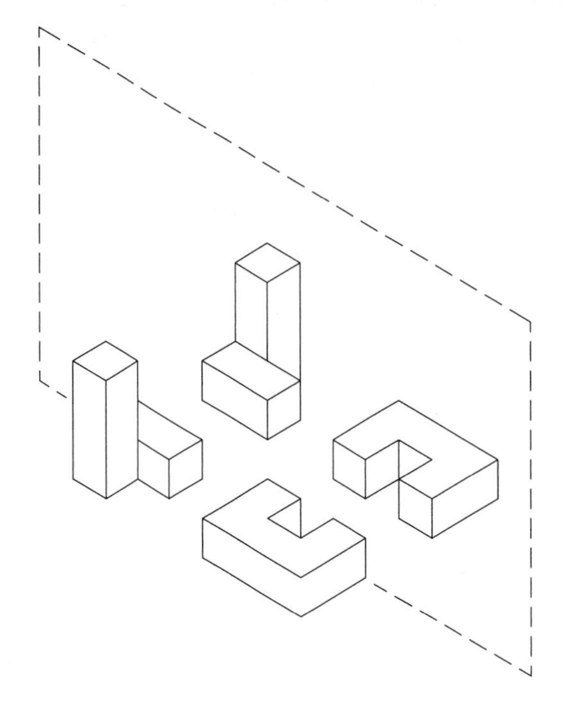

3-D Objects and their Reflections Across the Indicated Plane

Each point on the reflected object corresponds to a point on the original object.

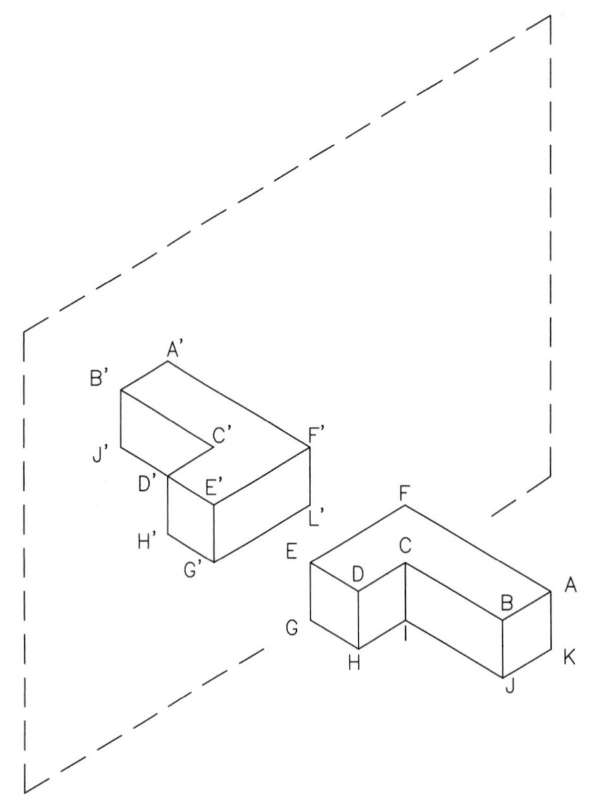


Object Points and Corresponding Points on Reflected Image

When drawing a reflection, imagine that the space on the opposite side of the plane of reflection is defined by the points from isometric dot paper. Points defining the object can be reflected through this plane, and corresponding edges and surfaces drawn. Points extend outward from the plane in both directions—points that are "closer" to you on the original object will be "farther" from you on the reflected image and vice versa.

An object is said to be symmetrical if a plane can cut it so that the part of the object on one side of the plane is a mirror image of the part on the other side of the plane.

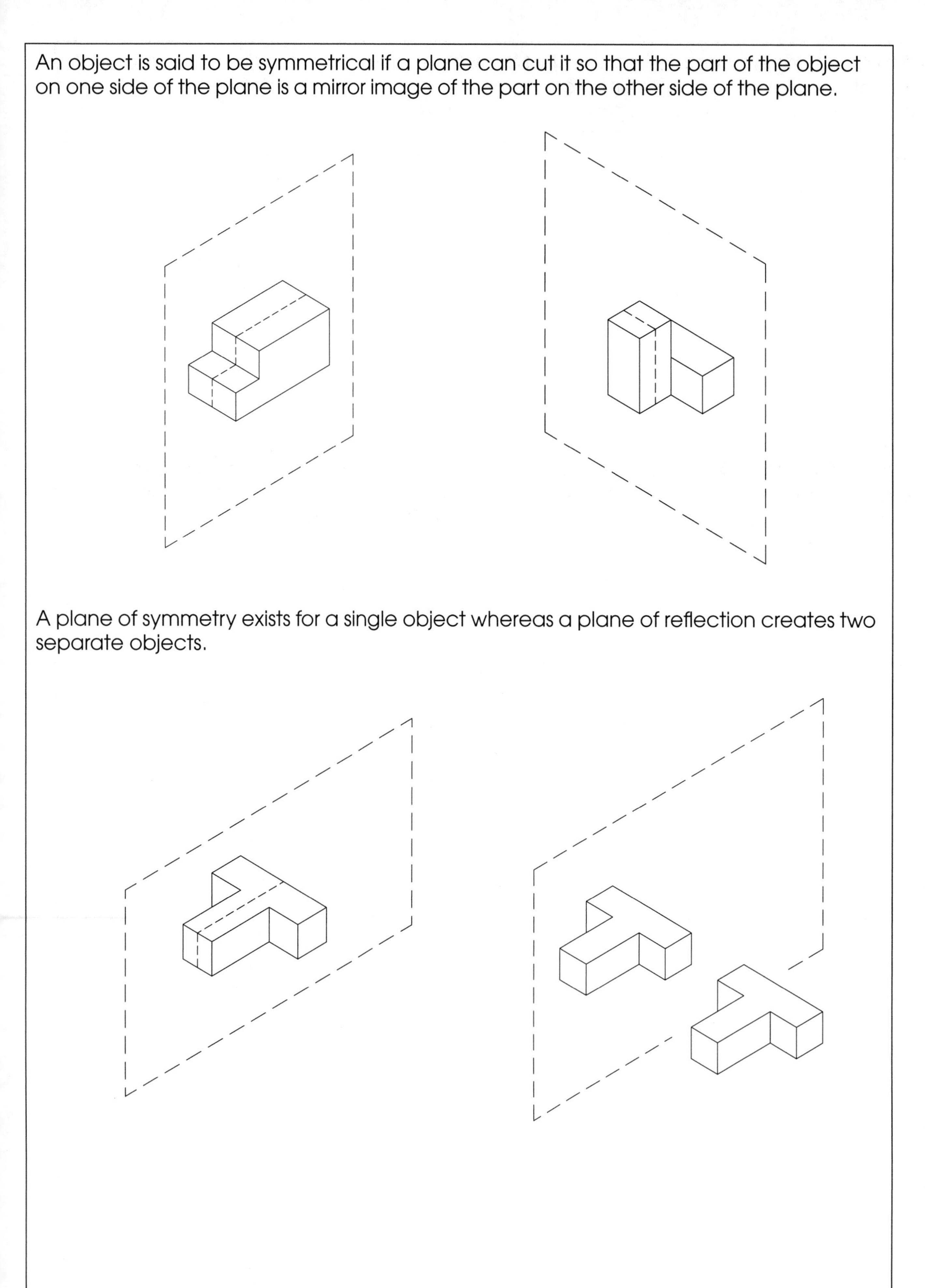

A plane of symmetry exists for a single object whereas a plane of reflection creates two separate objects.

An object can have multiple planes of symmetry.

When an object is symmetric, a rotation of 180 degrees can achieve the same result as a reflection. In this case, the plane of symmetry and the plane of reflection must be perpendicular to one another. The axis of rotation is the line formed by the intersection of these two planes.

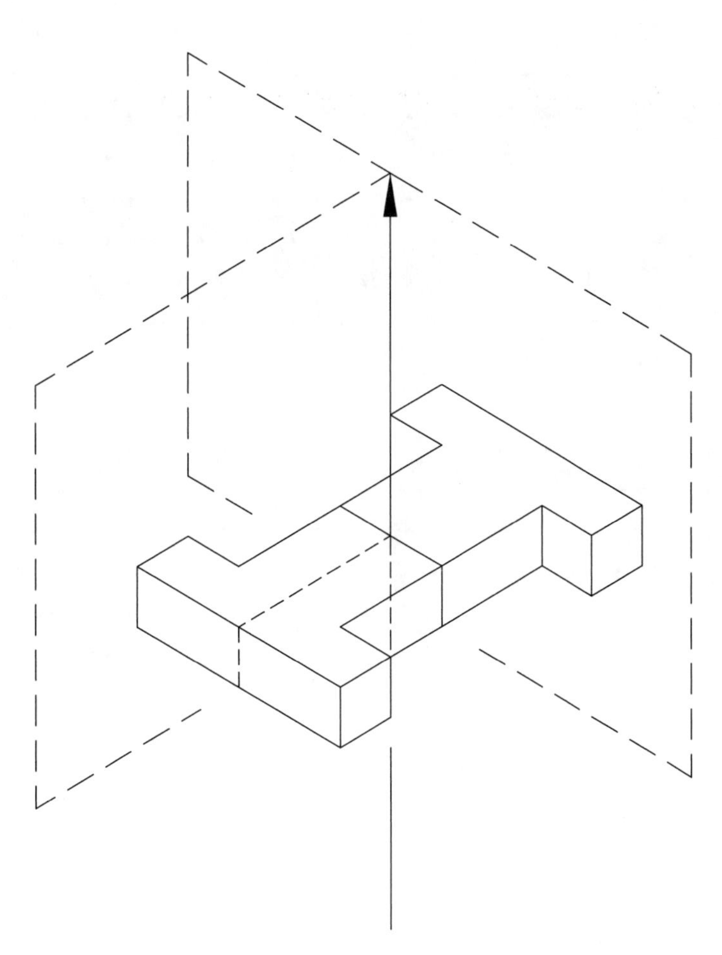

For the objects below, sketch the reflection across the indicated plan in the space provided. Sketch also the rotational axis about which the object could be rotated by 180° to achieve the same result, if feasible.

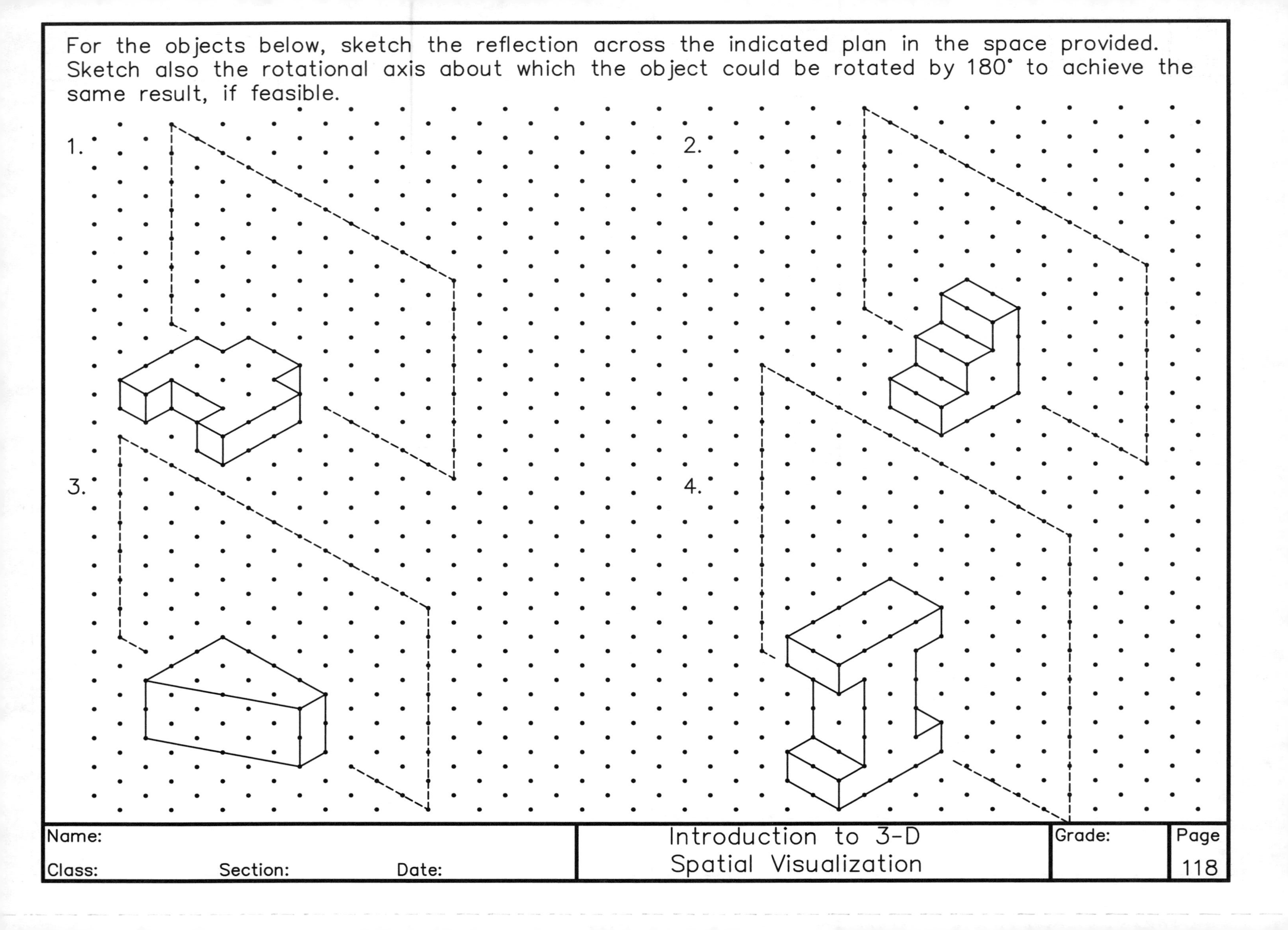

For the objects below, sketch the reflection across the indicated plan in the space provided. Sketch also the rotational axis about which the object could be rotated by 180° to achieve the same result, if feasible.

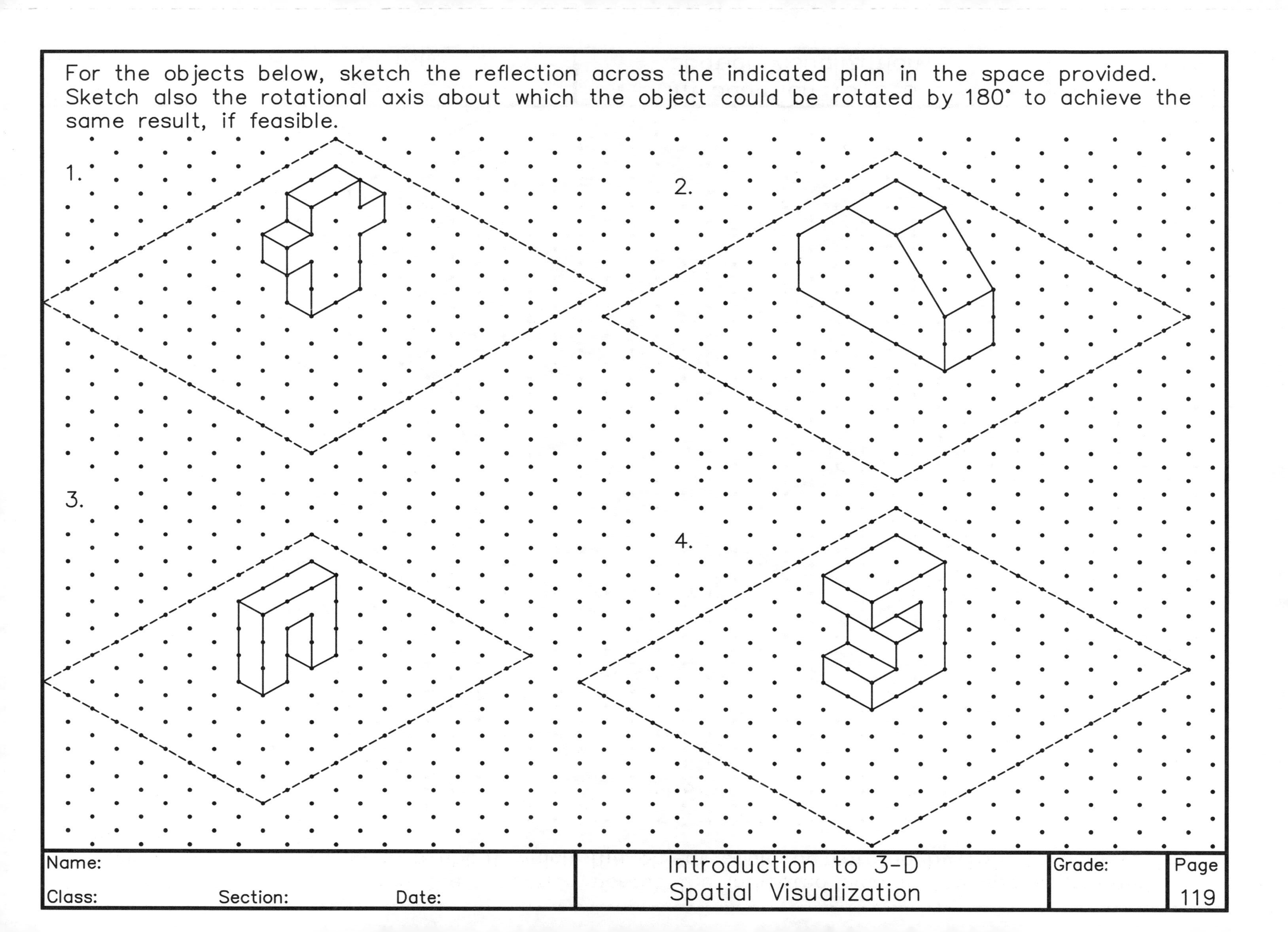


| Name: | Introduction to 3-D | Grade: | Page |
|---|---|---|---|
| Class: Section: Date: | Spatial Visualization | | 119 |

For the objects below, sketch the reflection across the indicated plan in the space provided. Sketch also the rotational axis about which the object could be rotated by 180° to achieve the same result, if feasible.

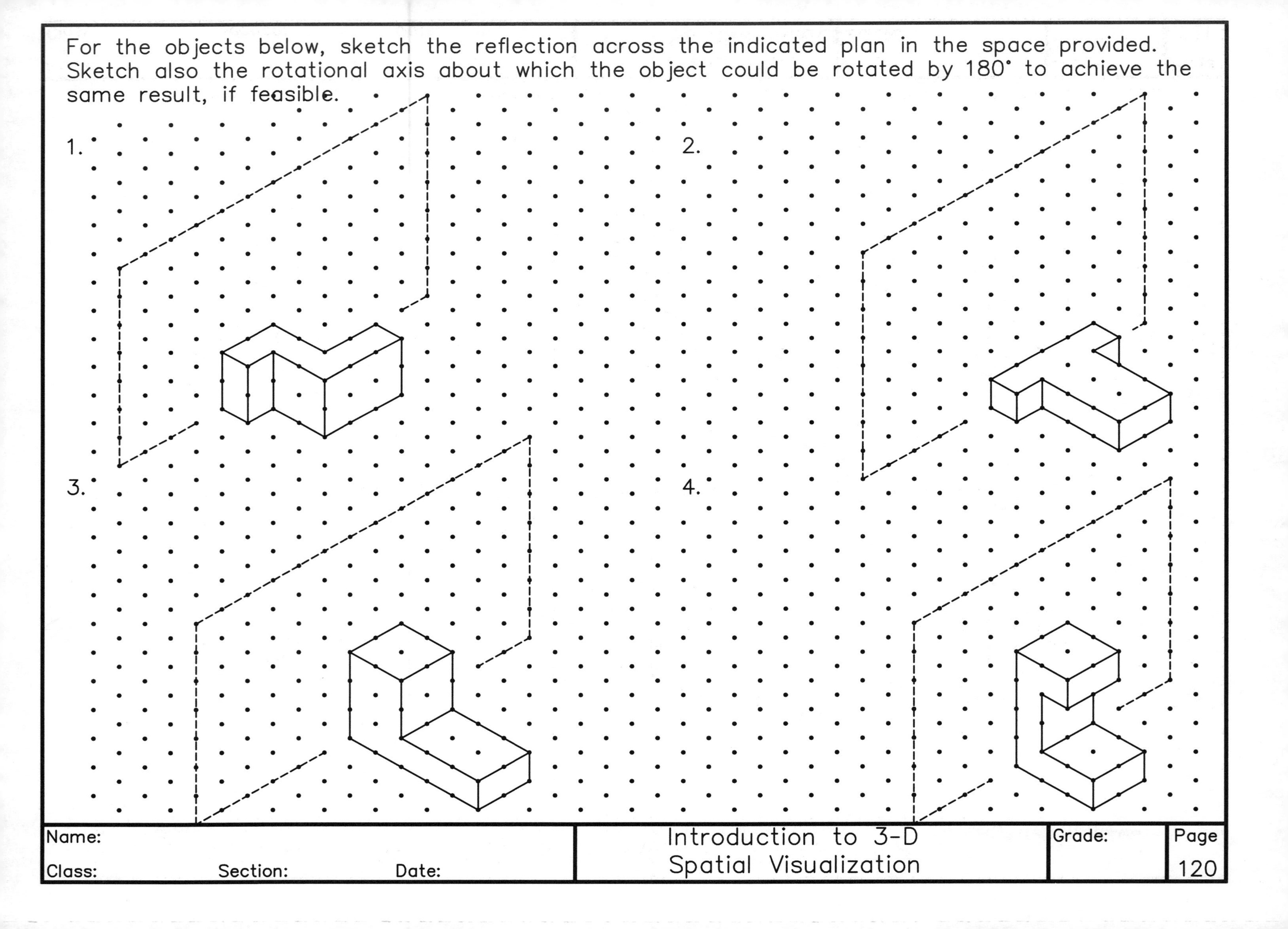


Name:
Class: Section: Date:

How many planes of symmetry do the objects shown below have?
Indicate your answer in the space provided. (Don't forget planes on the diagonal!)

1.

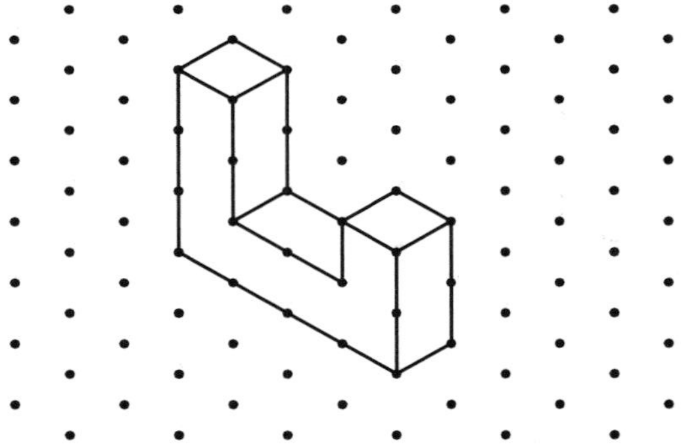

Planes of Symmetry=__________

2.

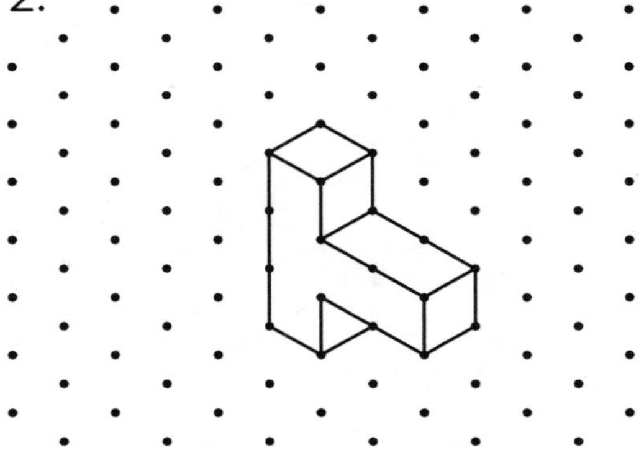

Planes of Symmetry=__________

3.

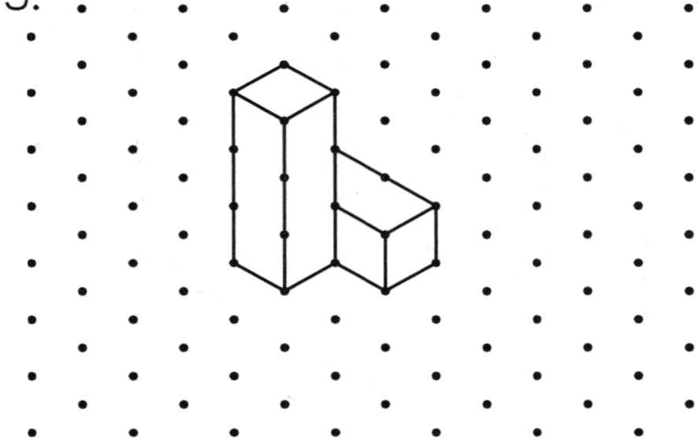

Planes of Symmetry=__________

4.

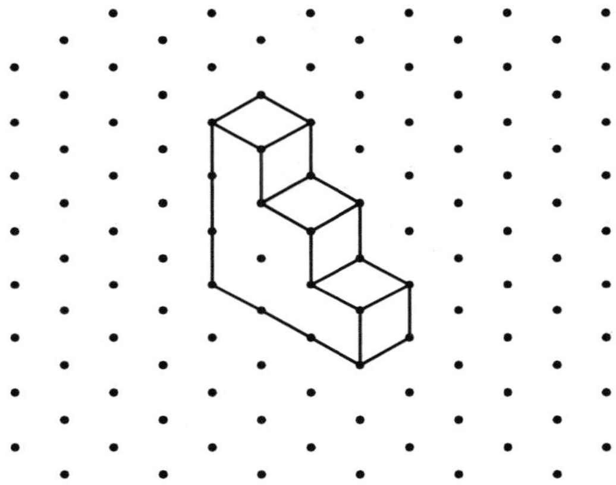

Planes of Symmetry=__________

5.

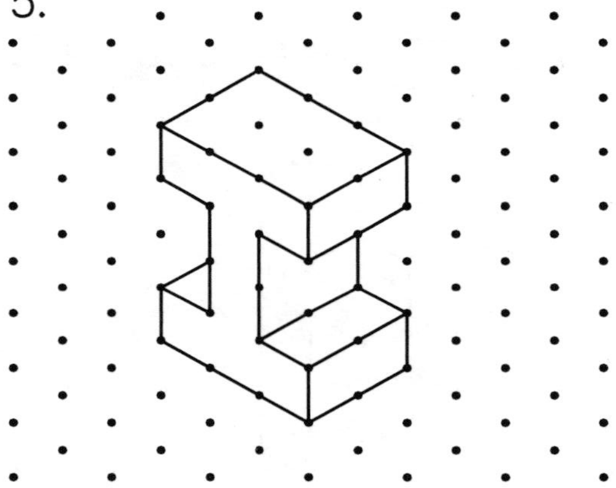

Planes of Symmetry=__________

6.

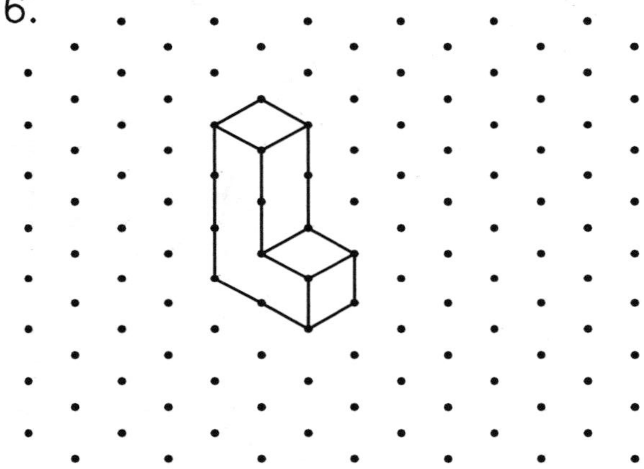

Planes of Symmetry=__________

How many planes of symmetry do the objects shown below have?
Indicate your answer in the space provided. (Don't forget planes on the diagonal!)

1\.

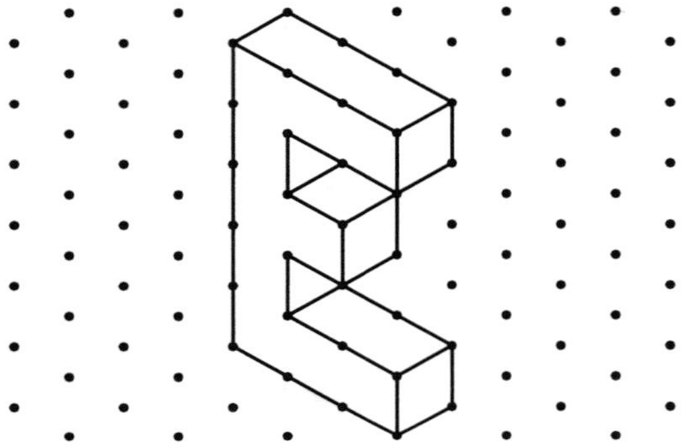

Planes of Symmetry=__________

2\.

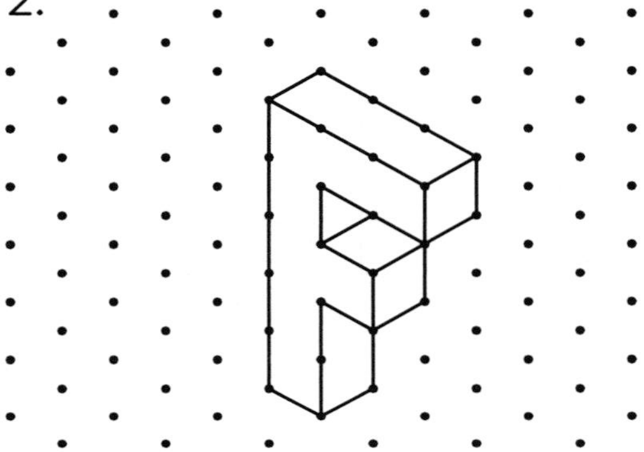

Planes of Symmetry=__________

3\.

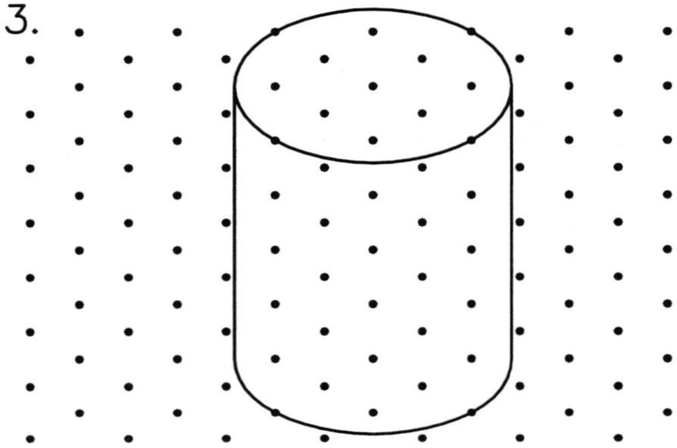

Planes of Symmetry=__________

4\.

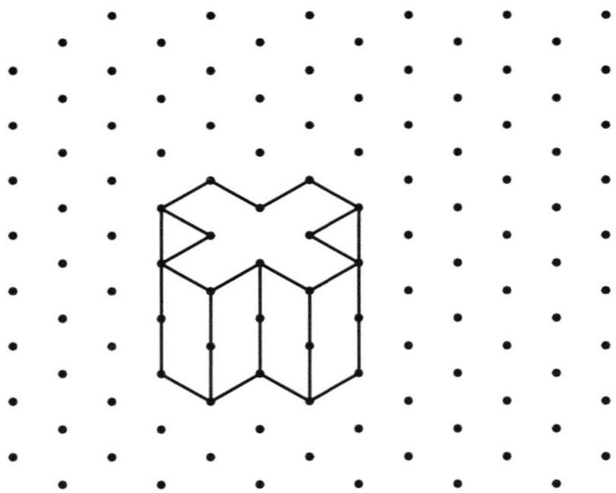

Planes of Symmetry=__________

5\.

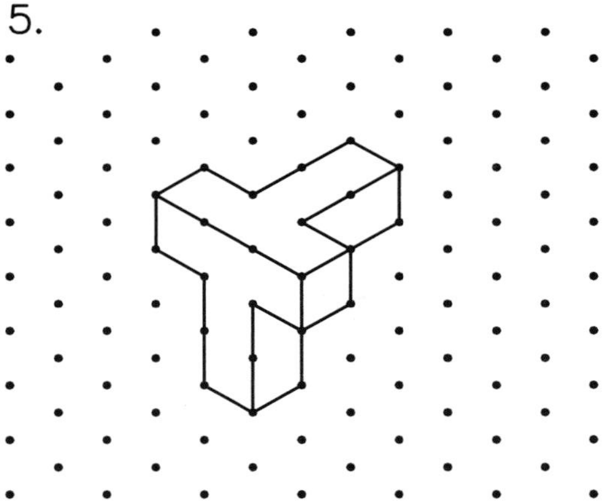

Planes of Symmetry=__________

6\.

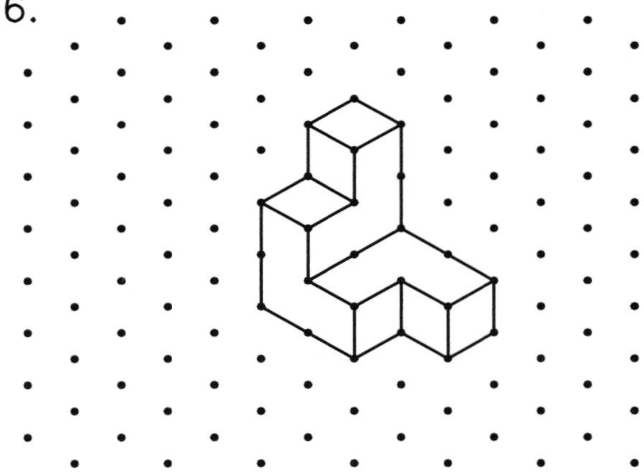

Planes of Symmetry=__________

How many planes of symmetry do the objects shown below have?
Indicate your answer in the space provided. (Don't forget planes on the diagonal!)

1.

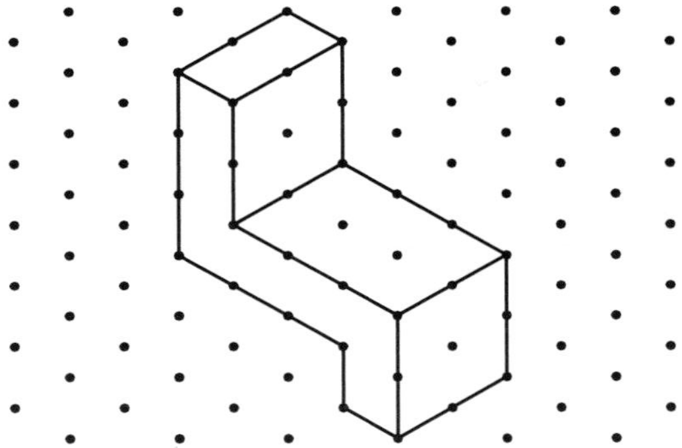

Planes of Symmetry=__________

2.

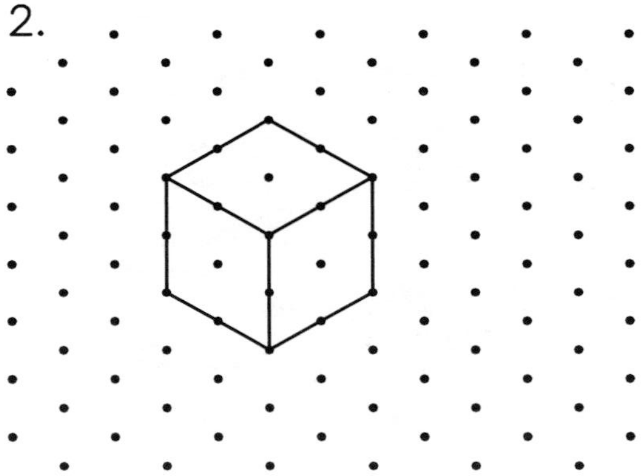

Planes of Symmetry=__________

3.

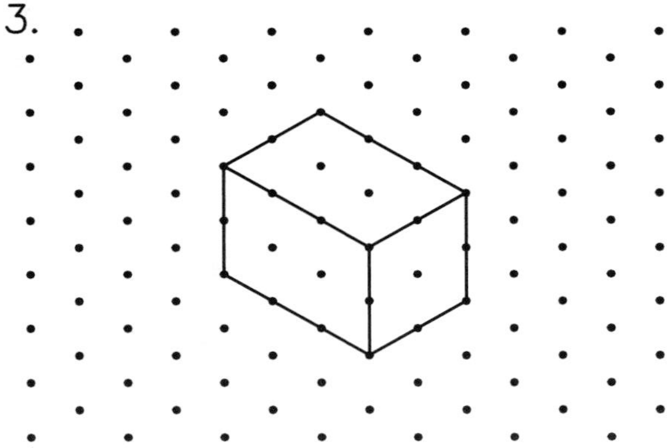

Planes of Symmetry=__________

4.

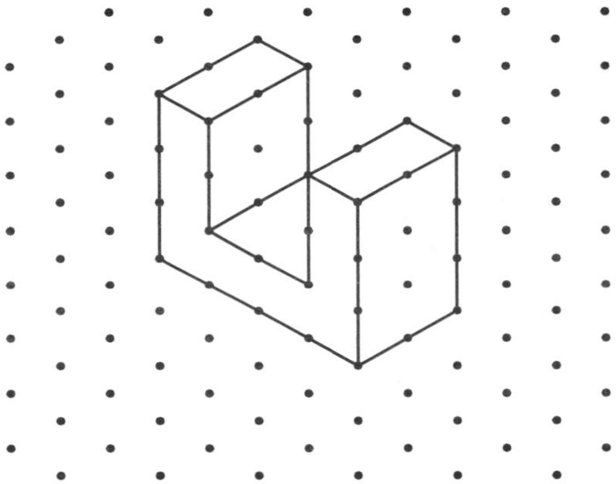

Planes of Symmetry=__________

5.

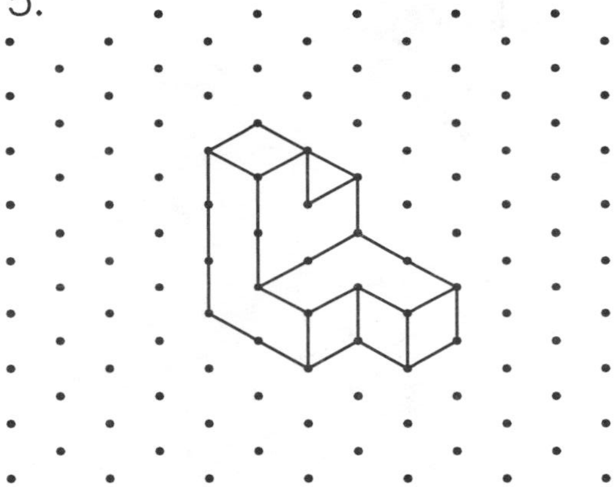

Planes of Symmetry=__________

6.

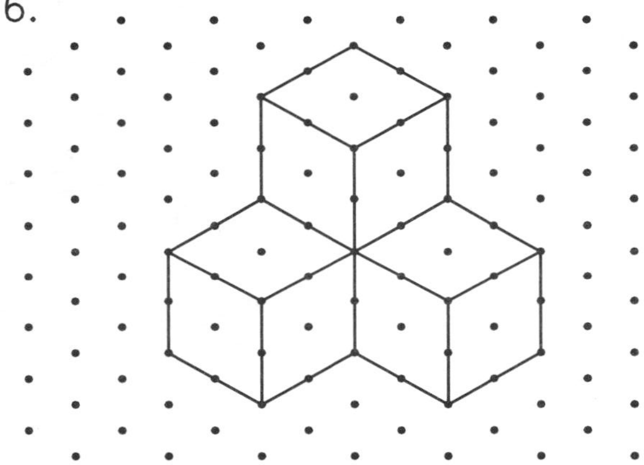

Planes of Symmetry=__________

| Name: Class: Section: Date: | Introduction to 3-D Spatial Visualization | Grade: | Page 123 |
|---|---|---|---|

Sketch the reflected image of the objects below across the indicated plane.
The sketch of the reflection is started for you.

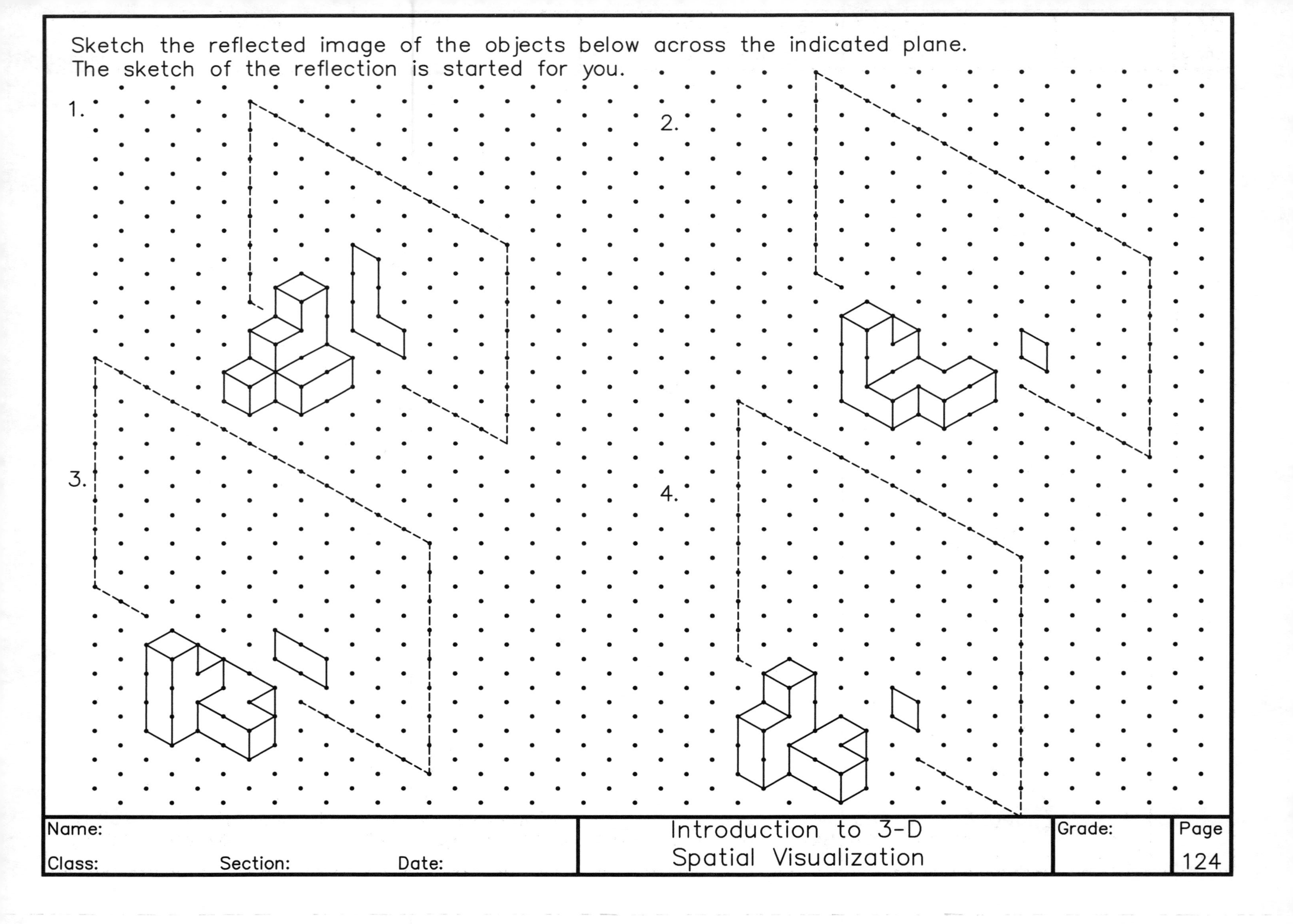

Sketch the reflected image of the objects below across the indicated plane.
The sketch of the reflection is started for you.

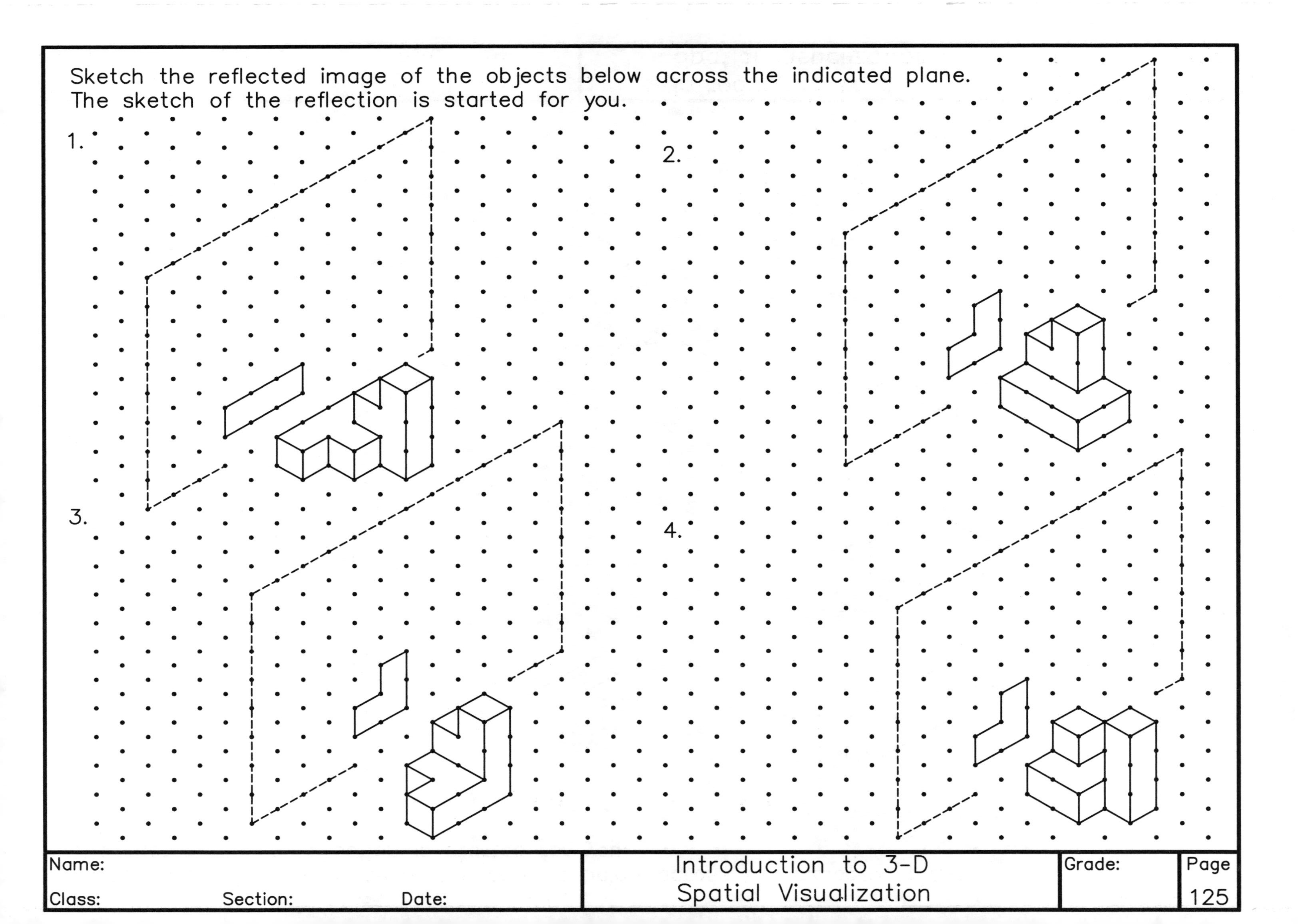


| Name: | Introduction to 3-D | Grade: | Page |
|---|---|---|---|
| Class: Section: Date: | Spatial Visualization | | 125 |

Sketch the reflected image of the objects below across the indicated plane.
The sketch of the reflection is started for you.

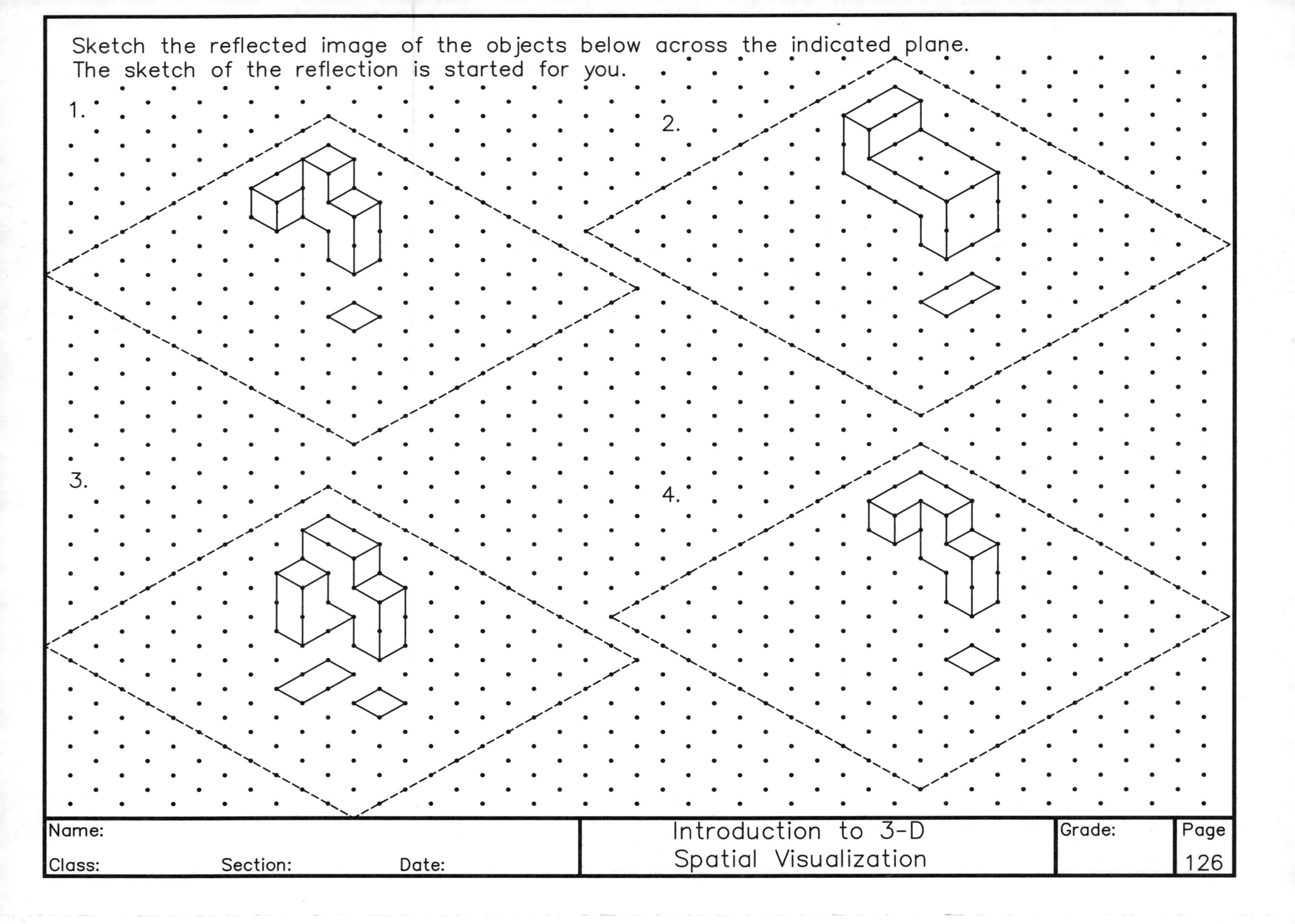


| Name: | Introduction to 3-D | Grade: | Page |
|---|---|---|---|
| Class: Section: Date: | Spatial Visualization | | 126 |

Sketch the reflected image of the objects below across the indicated plane.
The sketch of the reflection is started for you.

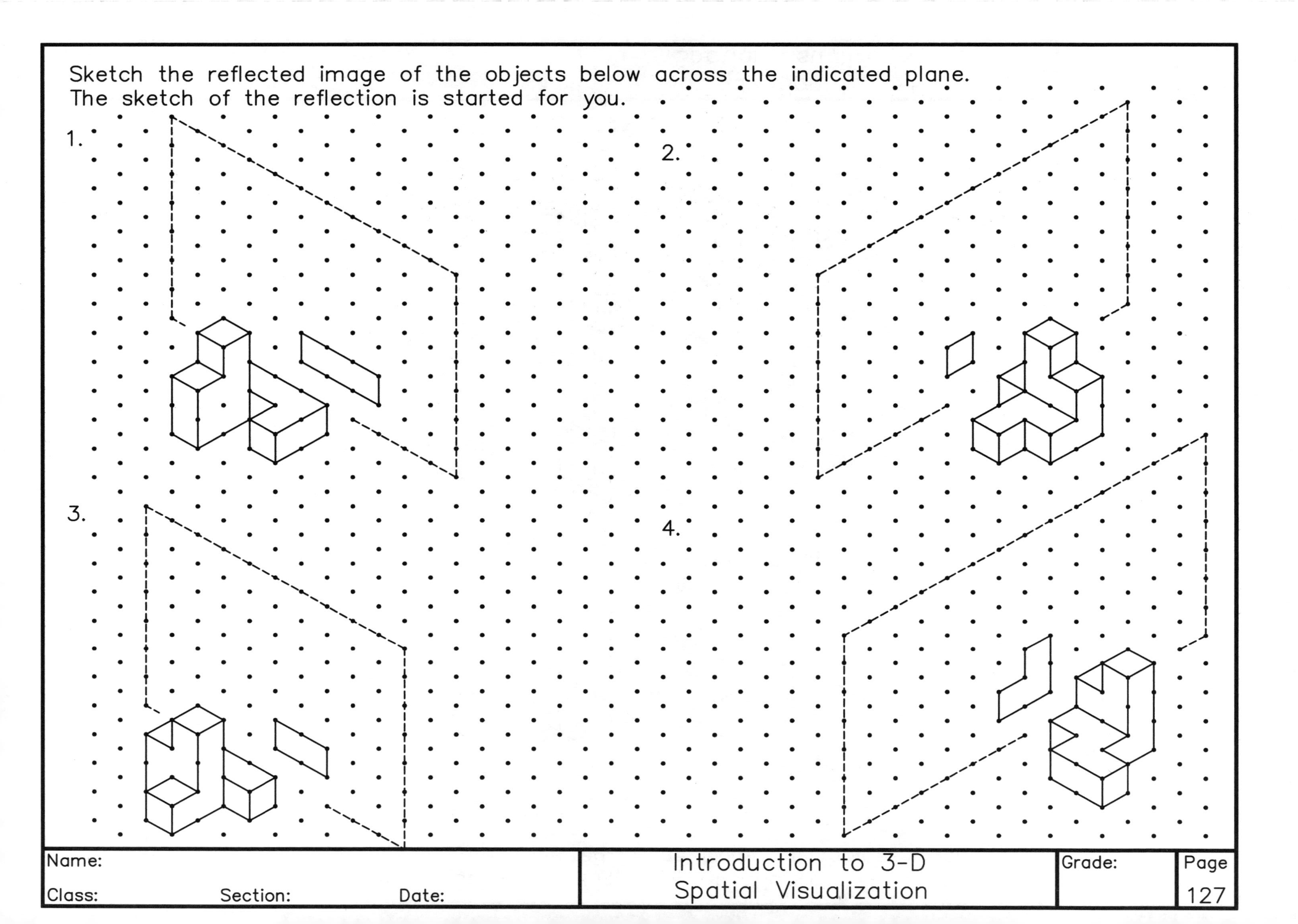

An object and a plane of reflection are shown on the left below.
Select the reflected image from the choices given.

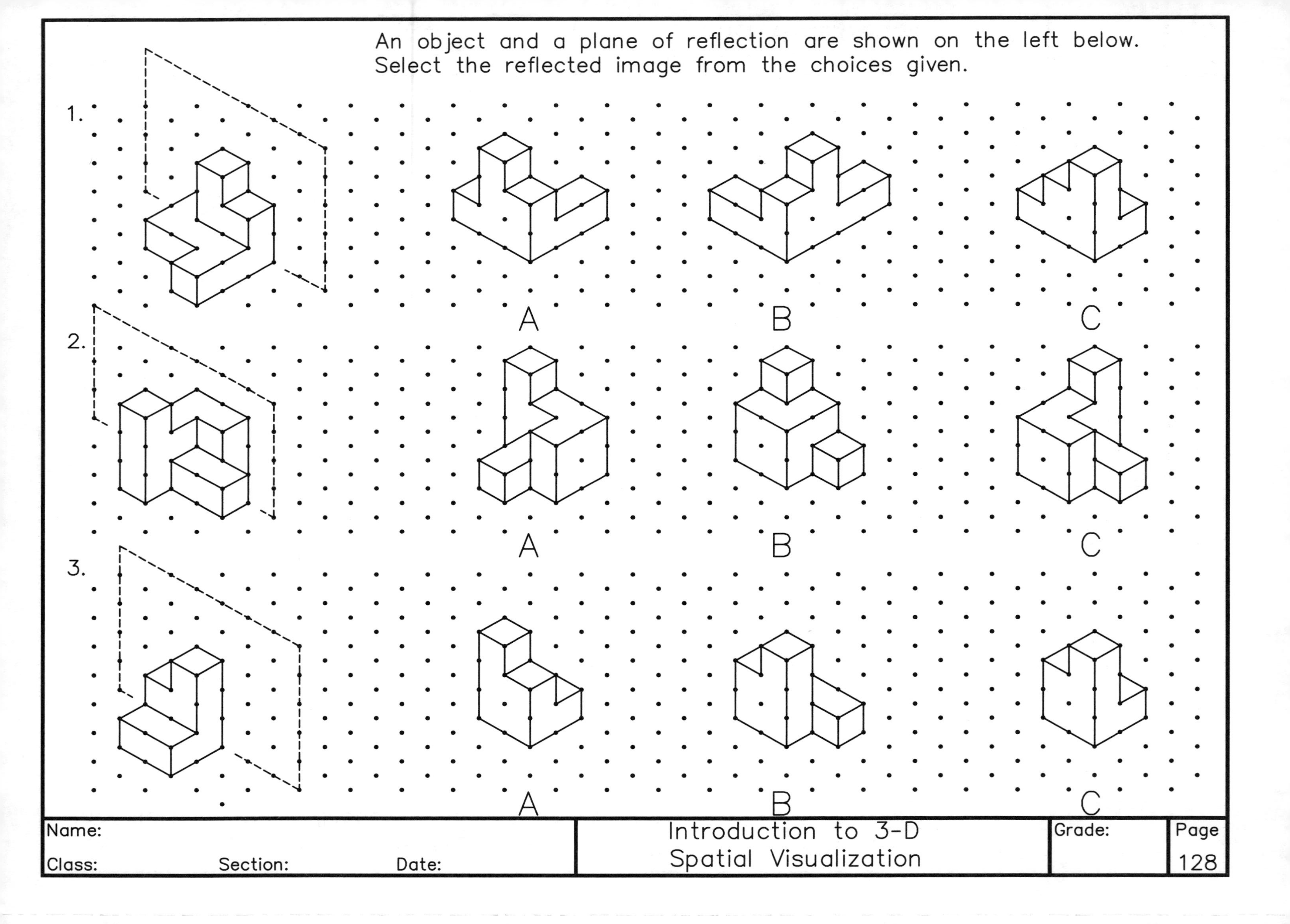

An object and a plane of reflection are shown on the left below.
Select the reflected image from the choices given.

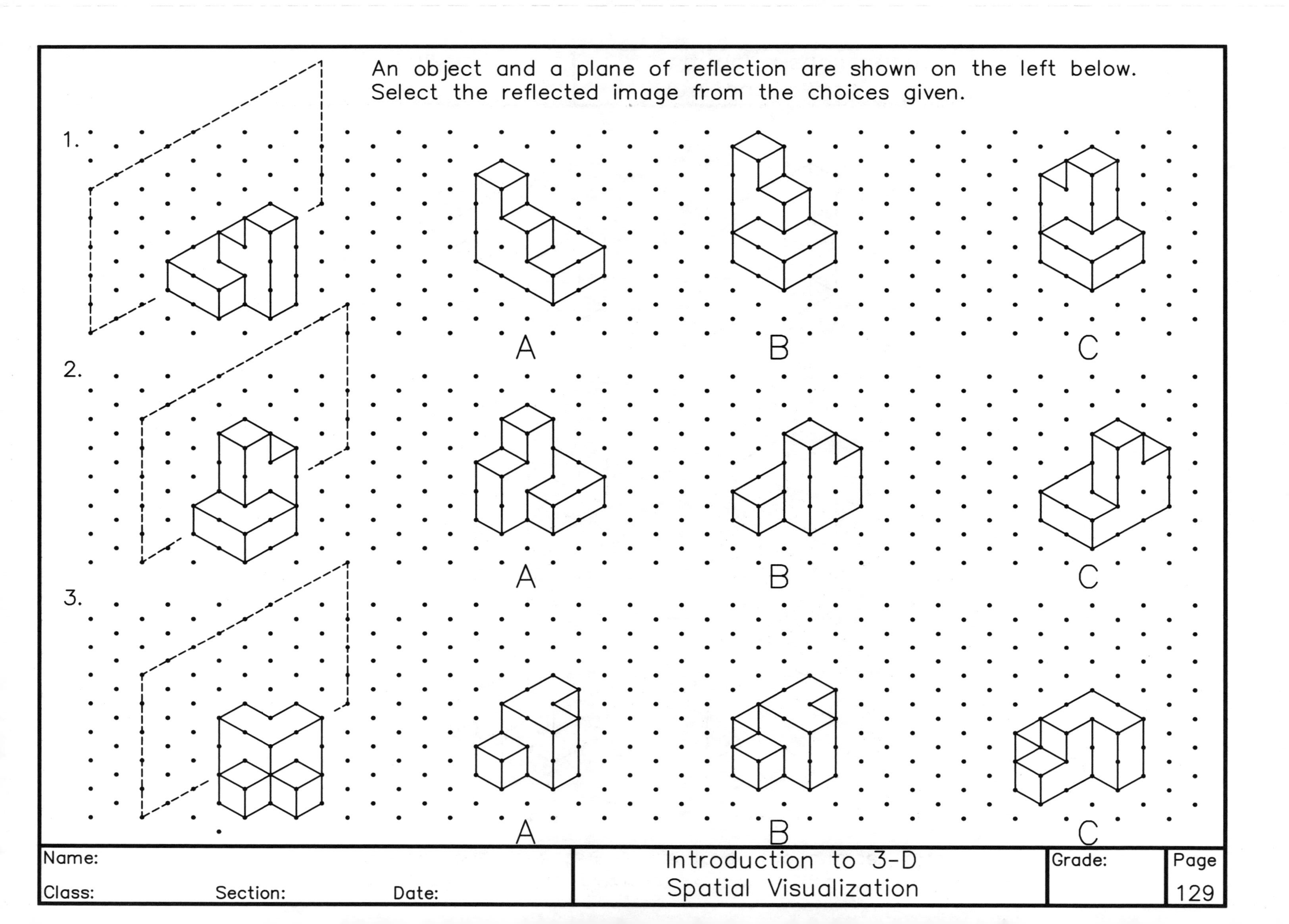

An object and a plane of reflection are shown on the left below. Select the reflected image from the choices given.

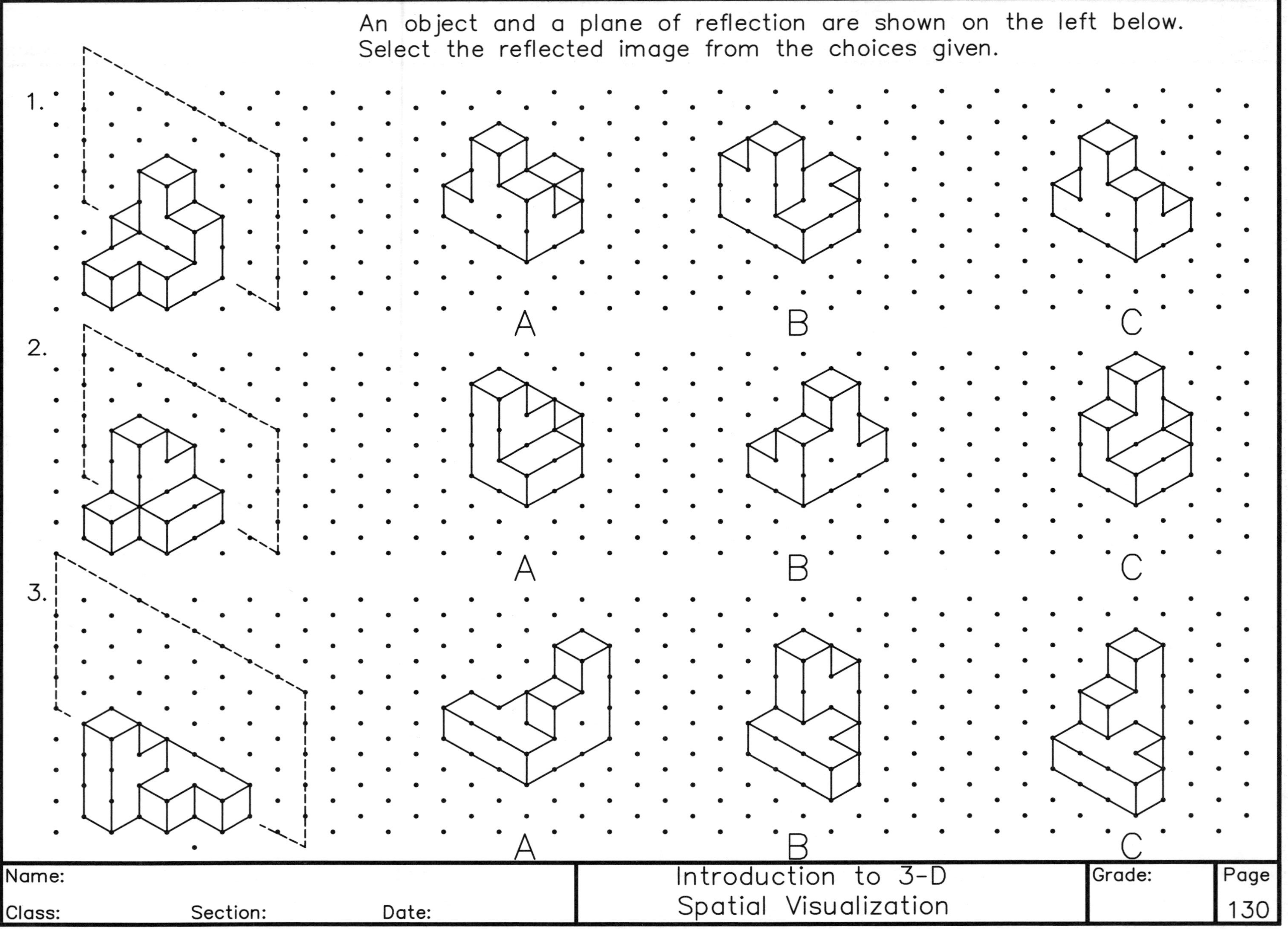

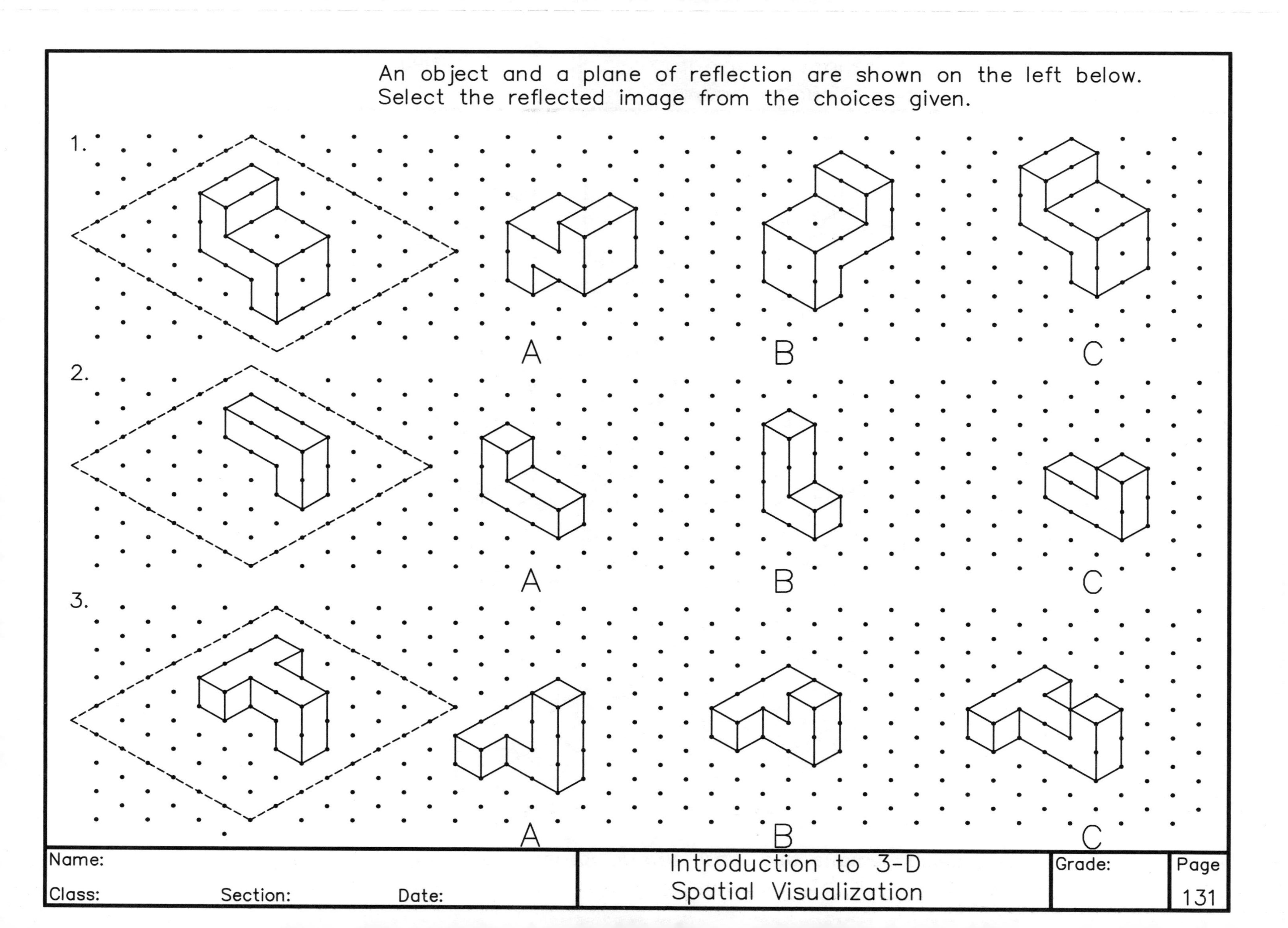
An object and a plane of reflection are shown on the left below.
Select the reflected image from the choices given.
1.
A
B
C
2.
A
B
C
3.
A
B
C
Name:
Class:
Section:
Date:
Introduction to 3-D
Spatial Visualization
Grade:
Page
131

An object and a plane of reflection are shown on the left below. Select the reflected image from the choices given.

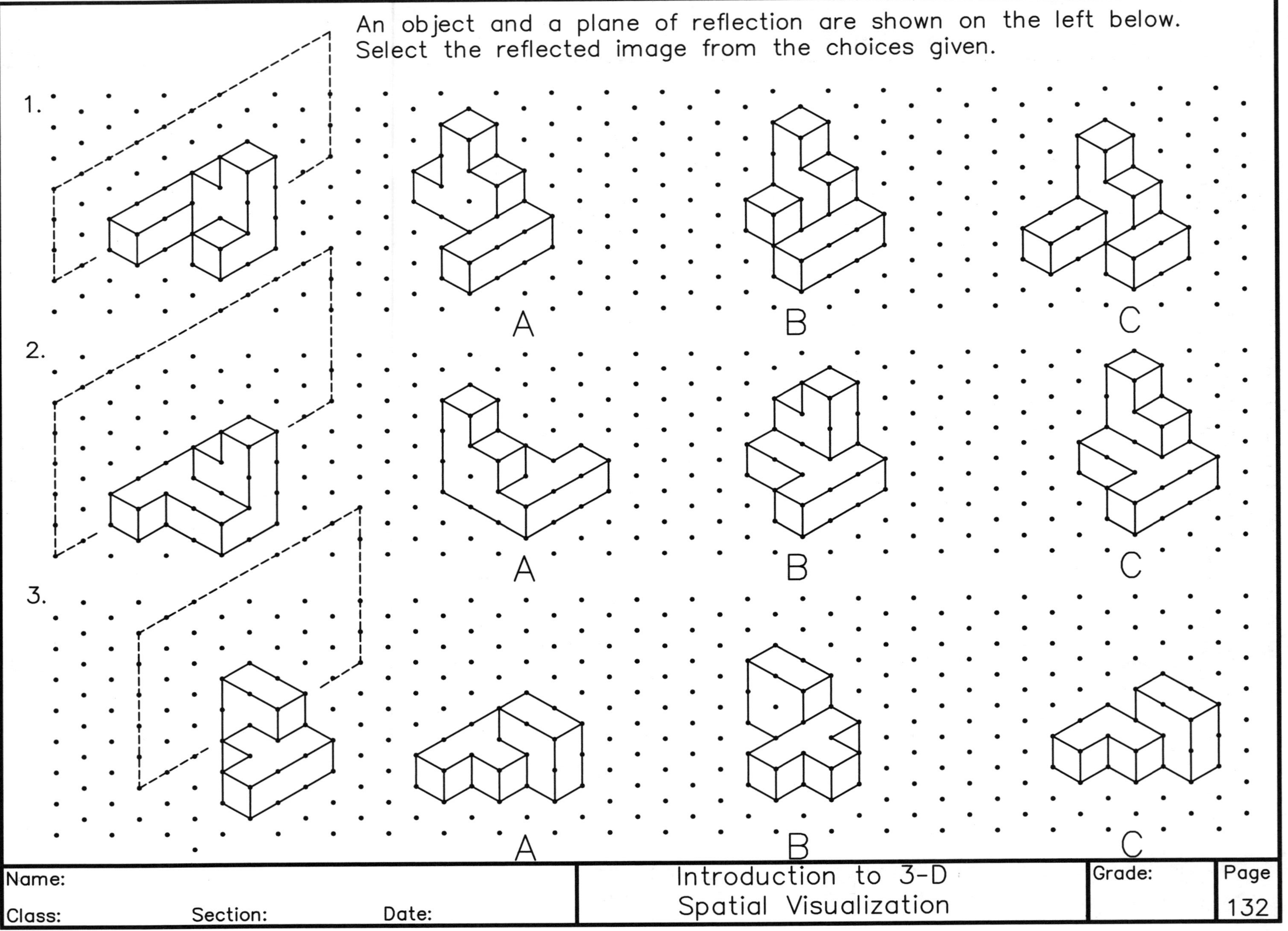


| Name: | Introduction to 3-D Spatial Visualization | Grade: | Page 132 |
| --- | --- | --- | --- |
| Class: Section: Date: | | | |

## Cutting Planes and Cross Sections

A cross section is the intersection between a cutting plane and a solid object. The result is a two-dimensional shape whose boundaries are defined by the edges and the surfaces of the original object.

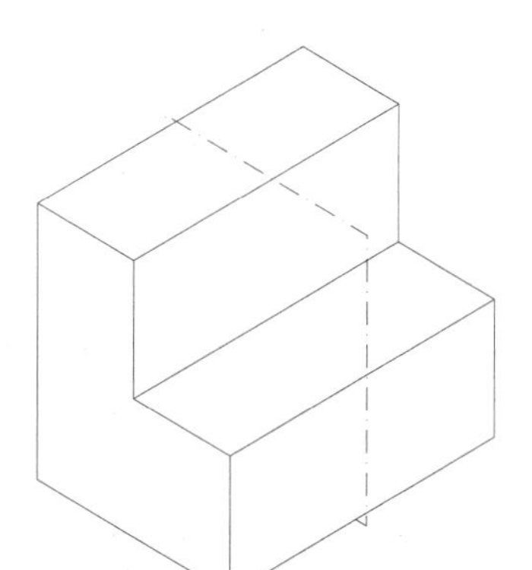

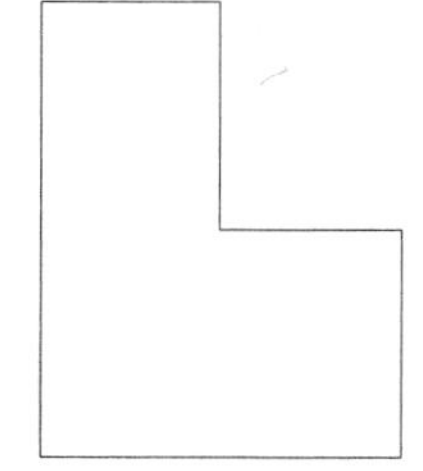

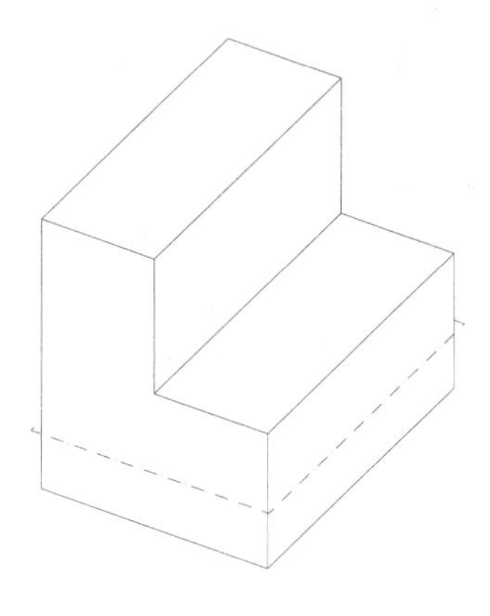

Object and Cutting Plane

Resulting Cross Section

When a plane cuts an object, the shape of the resulting cross section depends on the orientation of the cutting plane and the object with respect to one another.

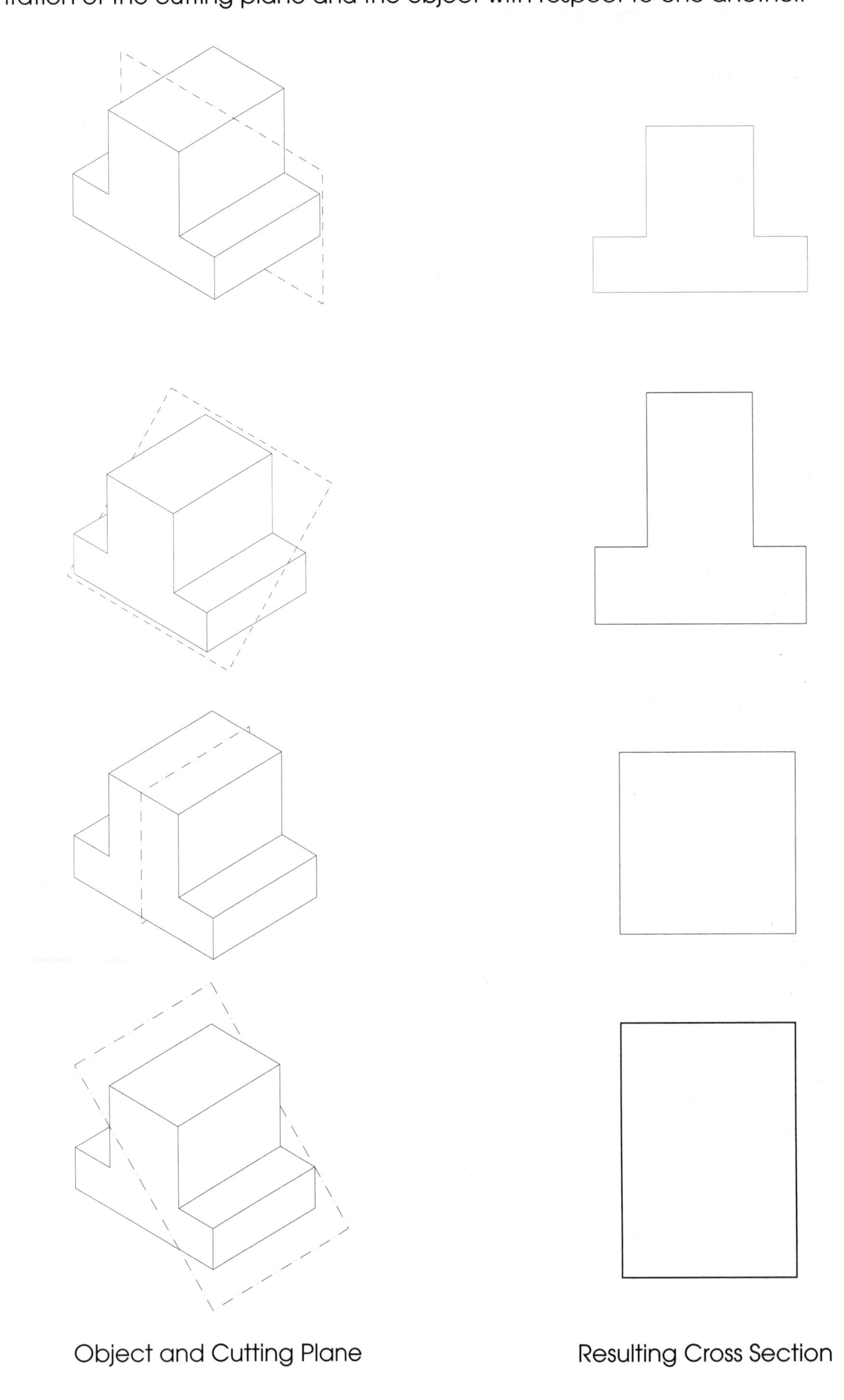

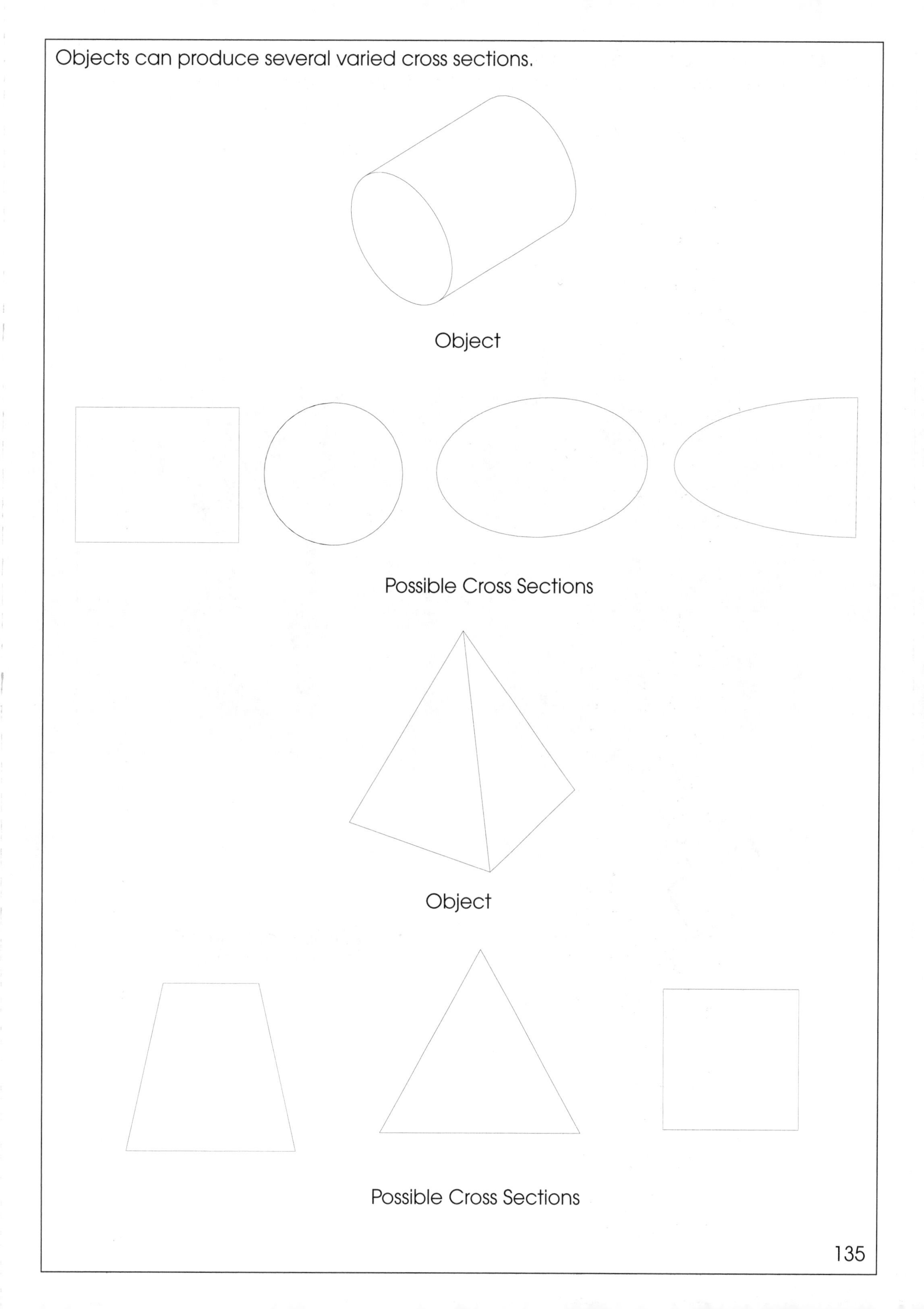
Objects can produce several varied cross sections.
Object
Possible Cross Sections
Object
Possible Cross Sections

To visualize the cross section that a cutting plane produces with an object, imagine the cutting plane extending through the object. As the plane "cuts" through a surface, it will intersect with it and a "boundary" of the cross section will be created. The boundary edges must be on the surfaces of the object itself and parallel to the edges of the cutting plane. After you have constructed the boundaries defining the cross section, mentally rotate the result so that it lies in the viewing plane.

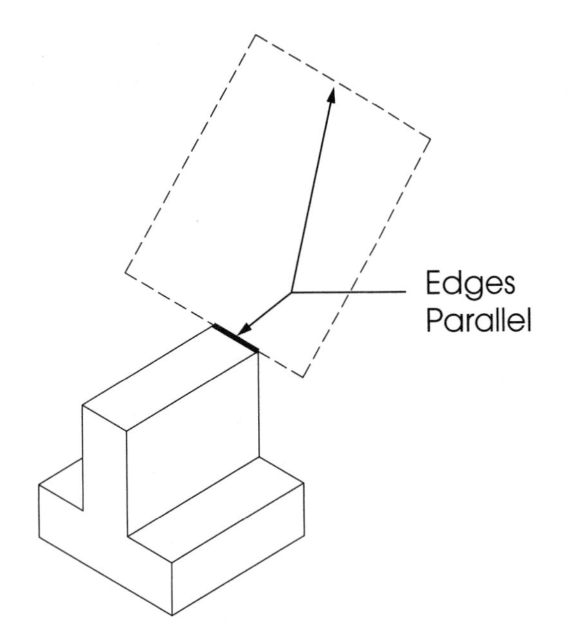


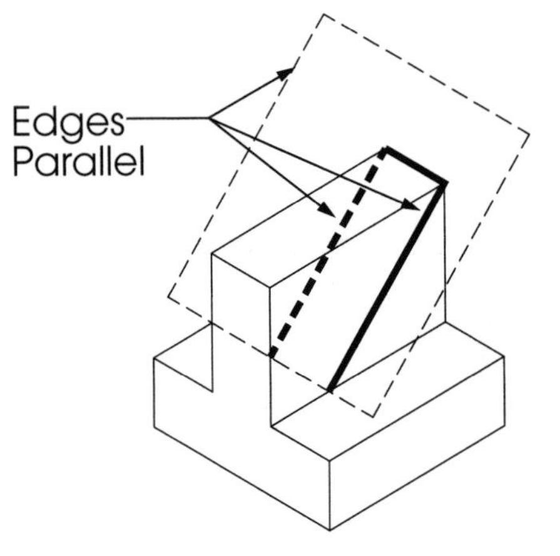


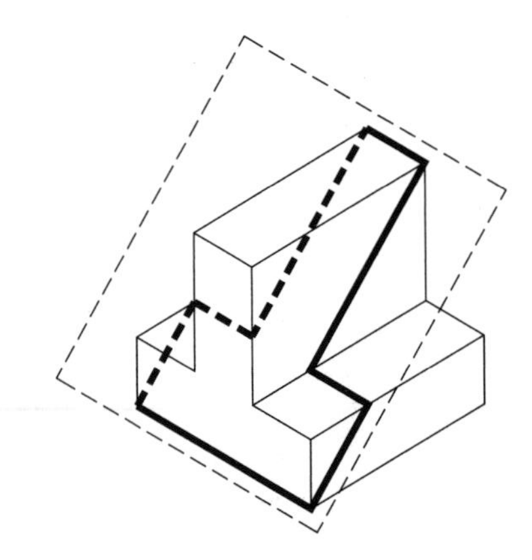

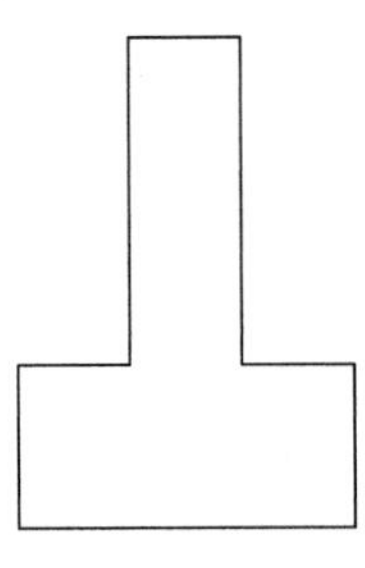

Cross Section Rotated to Viewing Plane

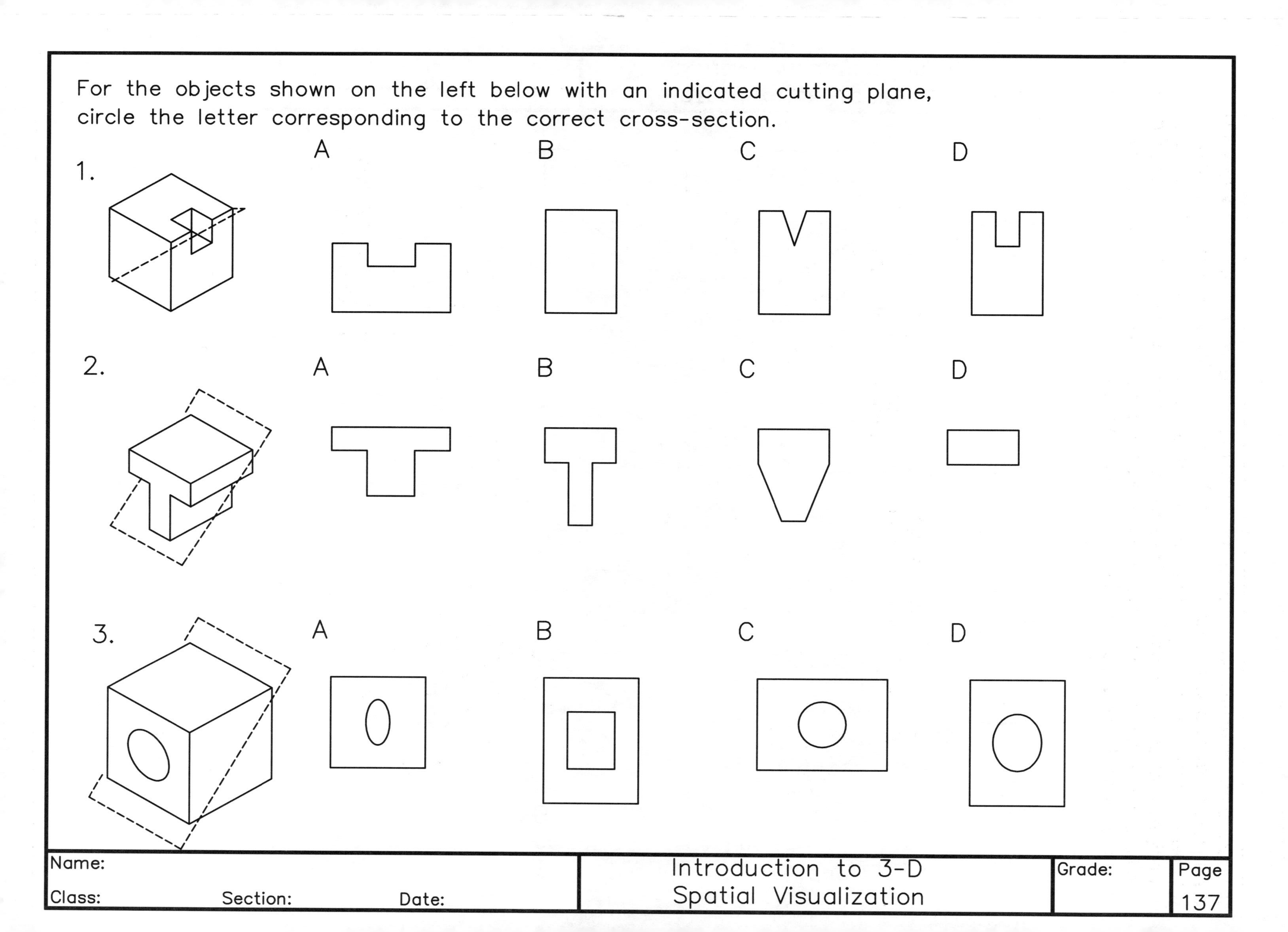
For the objects shown on the left below with an indicated cutting plane,
circle the letter corresponding to the correct cross-section.
1.
A
B
C
D
2.
A
B
C
D
3.
A
B
C
D
Name:
Class:
Section:
Date:
Introduction to 3-D
Spatial Visualization
Grade:
Page
137

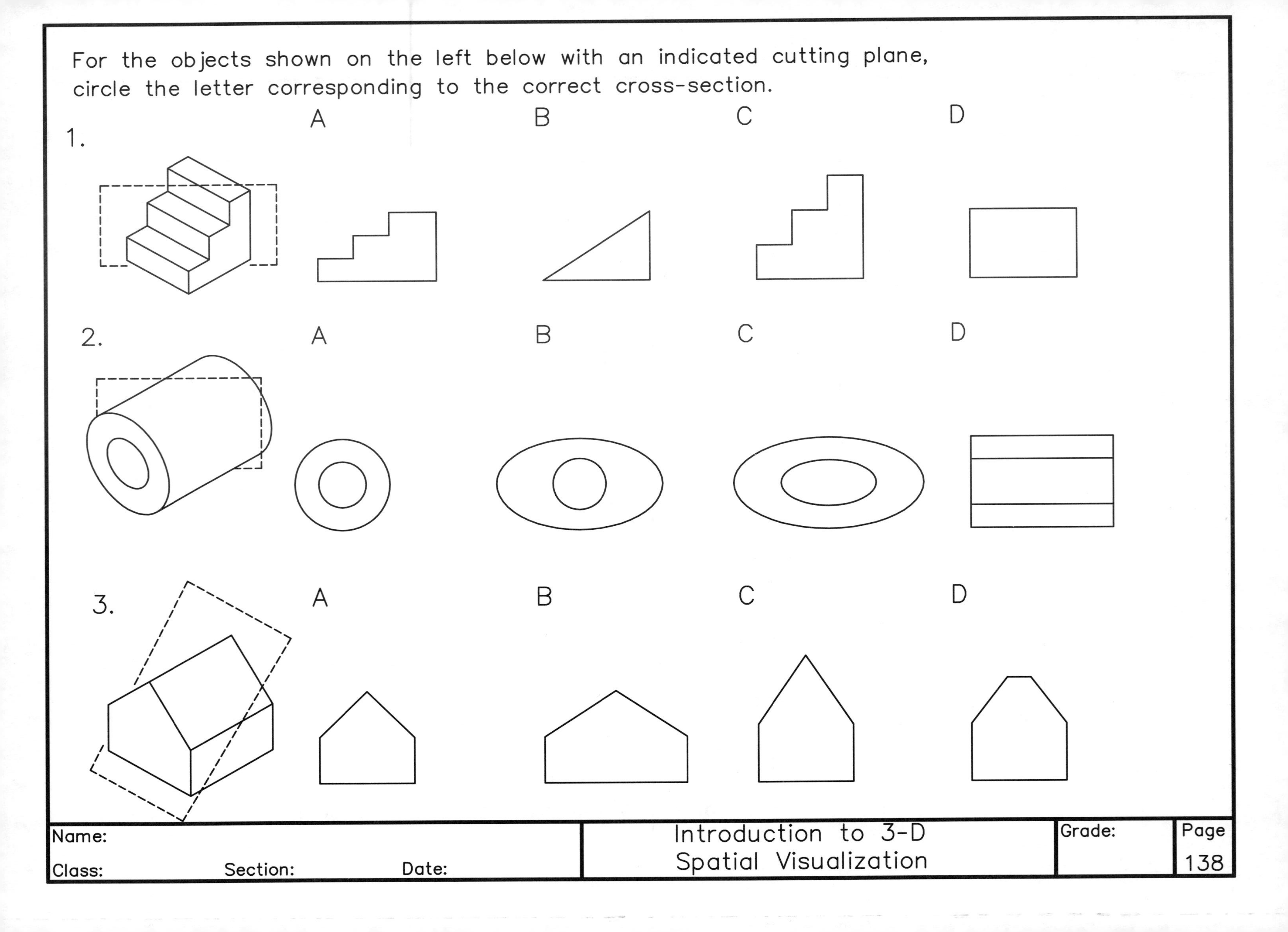

For the objects shown on the left below with an indicated cutting plane,
circle the letter corresponding to the correct cross-section.
1.
A
B
C
D
2.
A
B
C
D
3.
A
B
C
D
Name:
Class:
Section:
Date:
Introduction to 3-D
Spatial Visualization
Grade:
Page
138

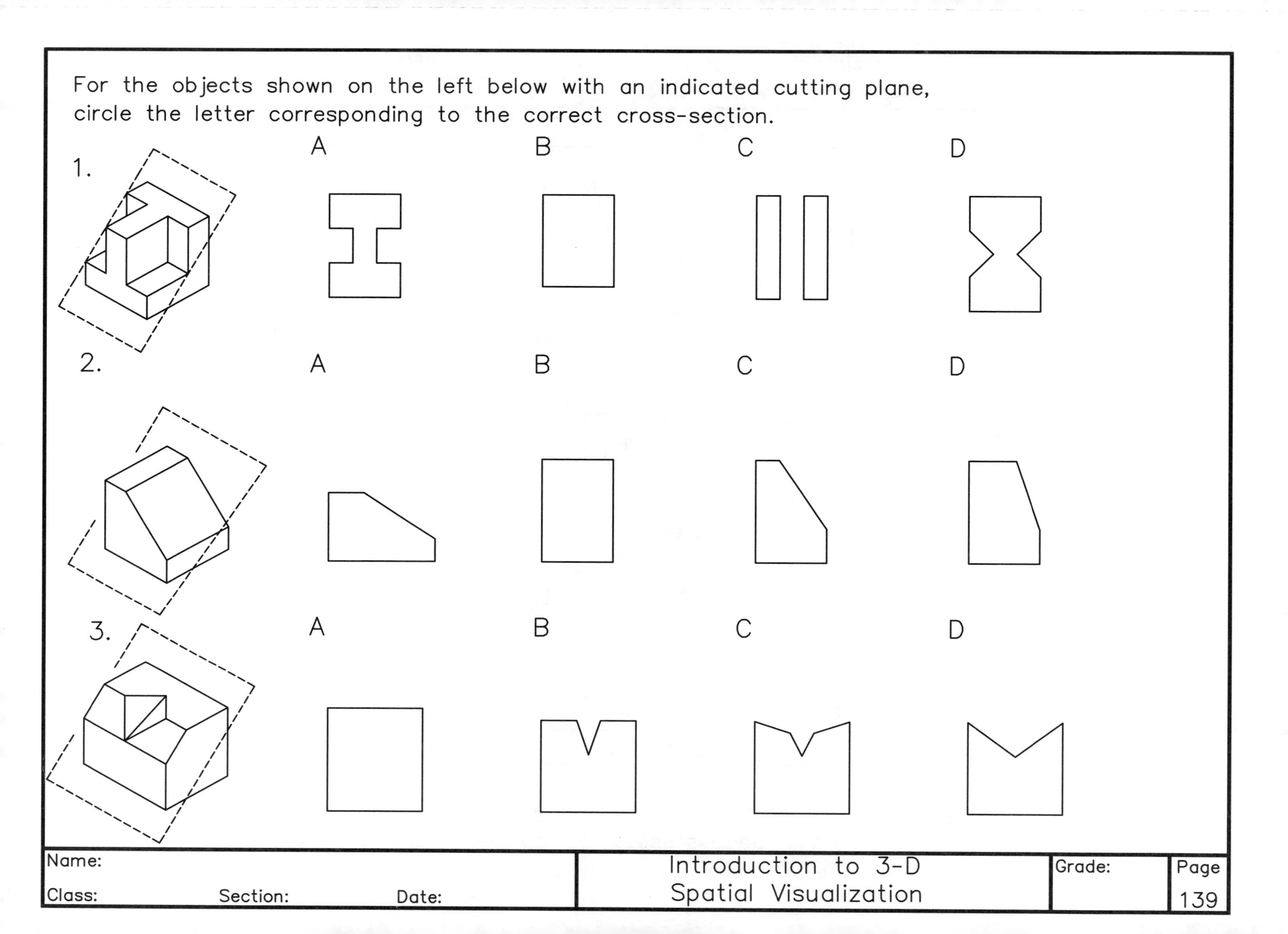
For the objects shown on the left below with an indicated cutting plane,
circle the letter corresponding to the correct cross-section.
1.
A
B
C
D
2.
A
B
C
D
3.
A
B
C
D
Name:
Class:
Section:
Date:
Introduction to 3-D
Spatial Visualization
Grade:
Page
139

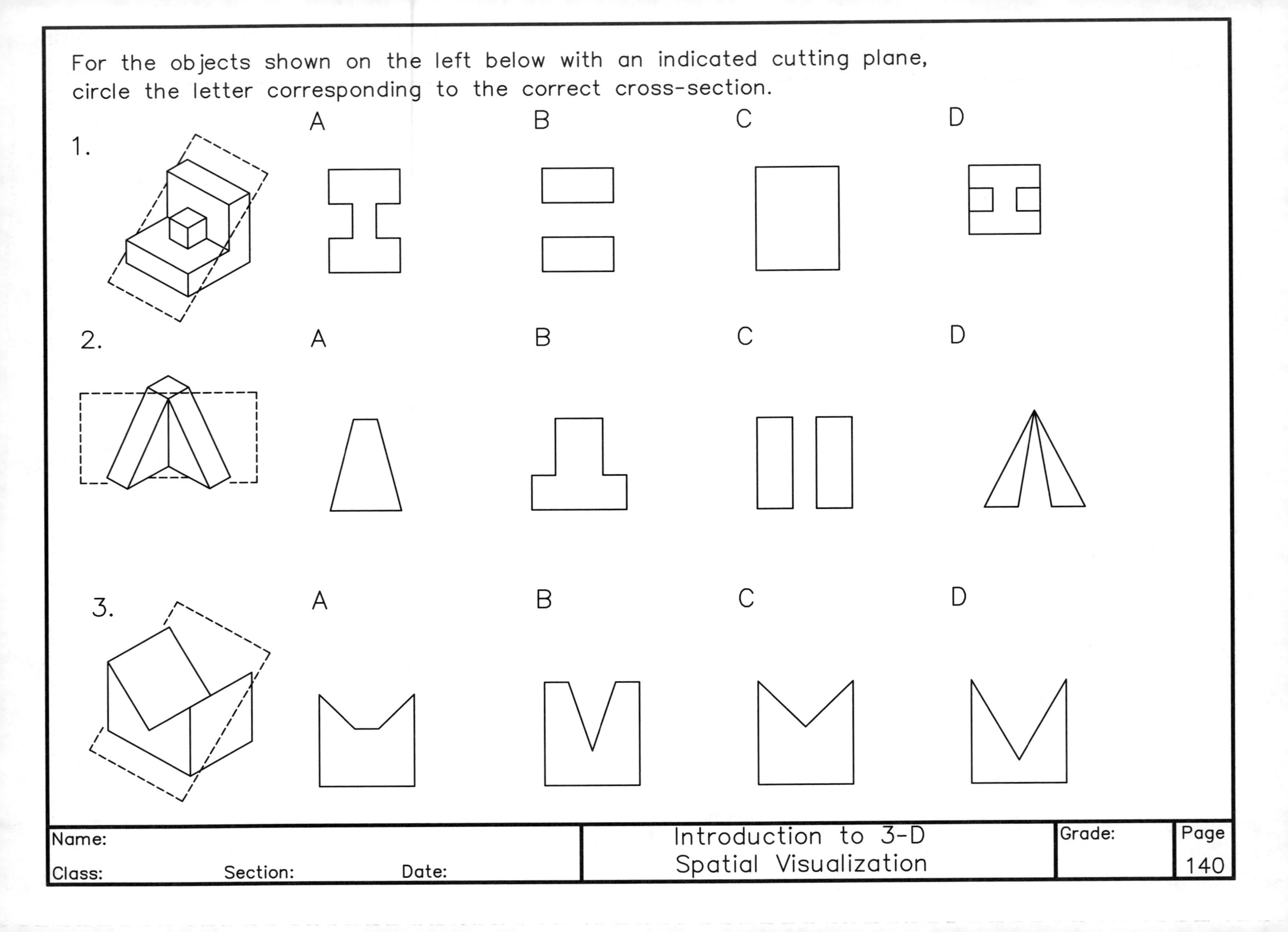

For the objects shown on the left below with an indicated cutting plane, circle the letter corresponding to the correct cross-section.
1.
A
B
C
D
2.
A
B
C
D
3.
A
B
C
D
Name:
Class:
Section:
Date:
Introduction to 3-D Spatial Visualization
Grade:
Page
140

Match the cross-sections with the appropriate object and cutting plane for the problems shown below.

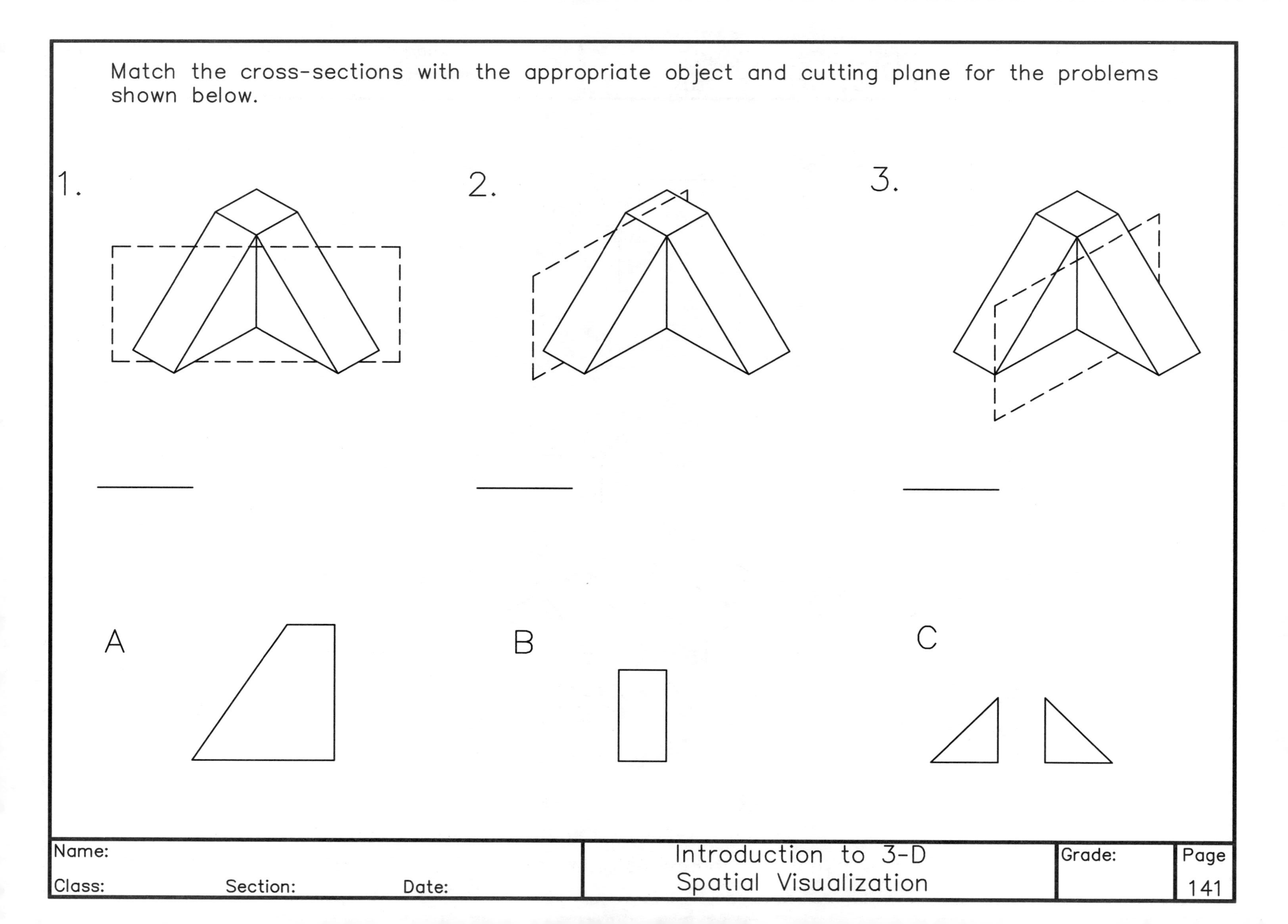


| Name: | | | Introduction to 3-D Spatial Visualization | Grade: | Page |
|---|---|---|---|---|---|
| Class: | Section: | Date: | | | 141 |

Match the cross-sections with the appropriate object and cutting plane for the problems shown below.

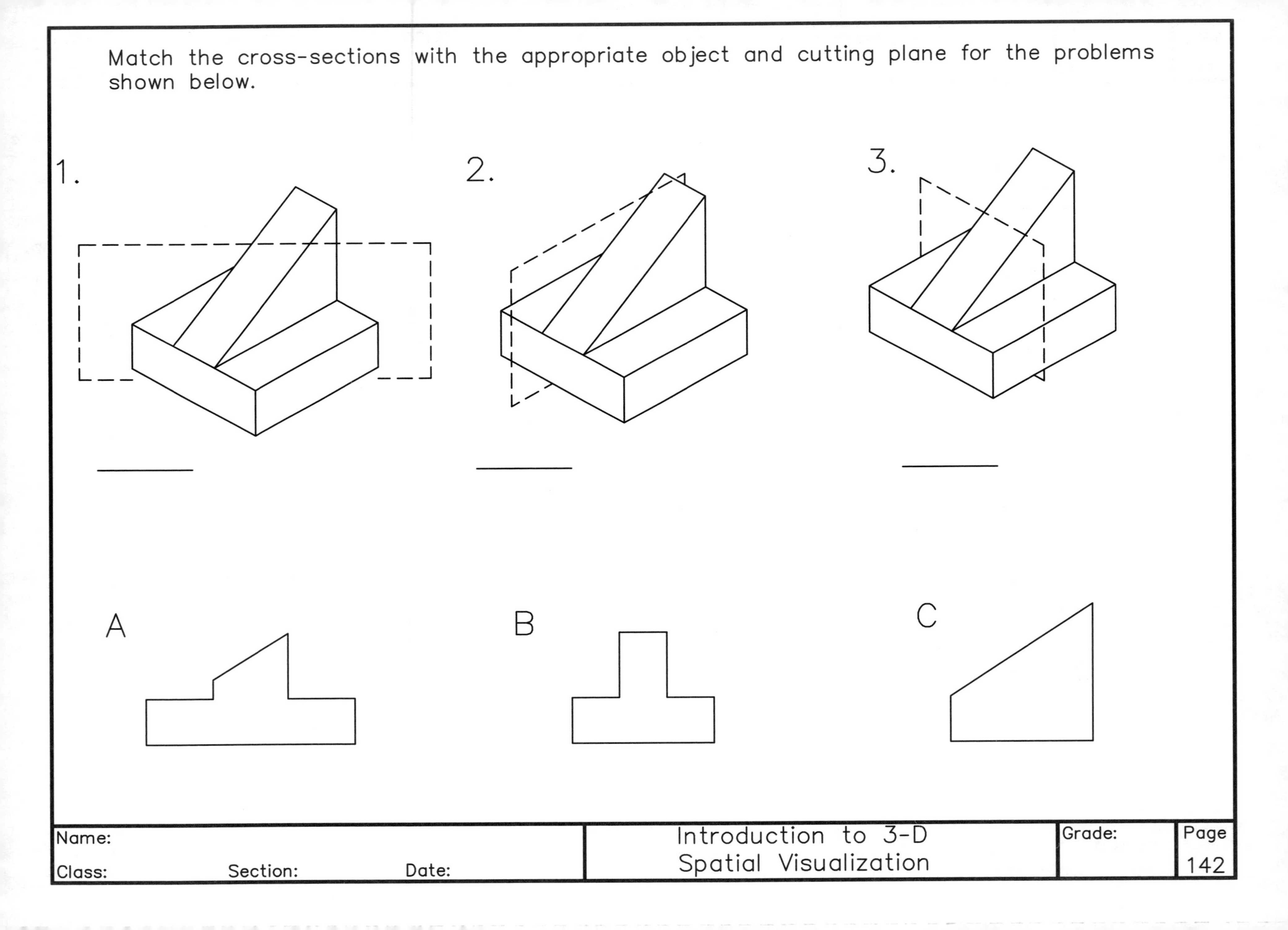

Match the cross-sections with the appropriate object and cutting plane for the problems shown below.

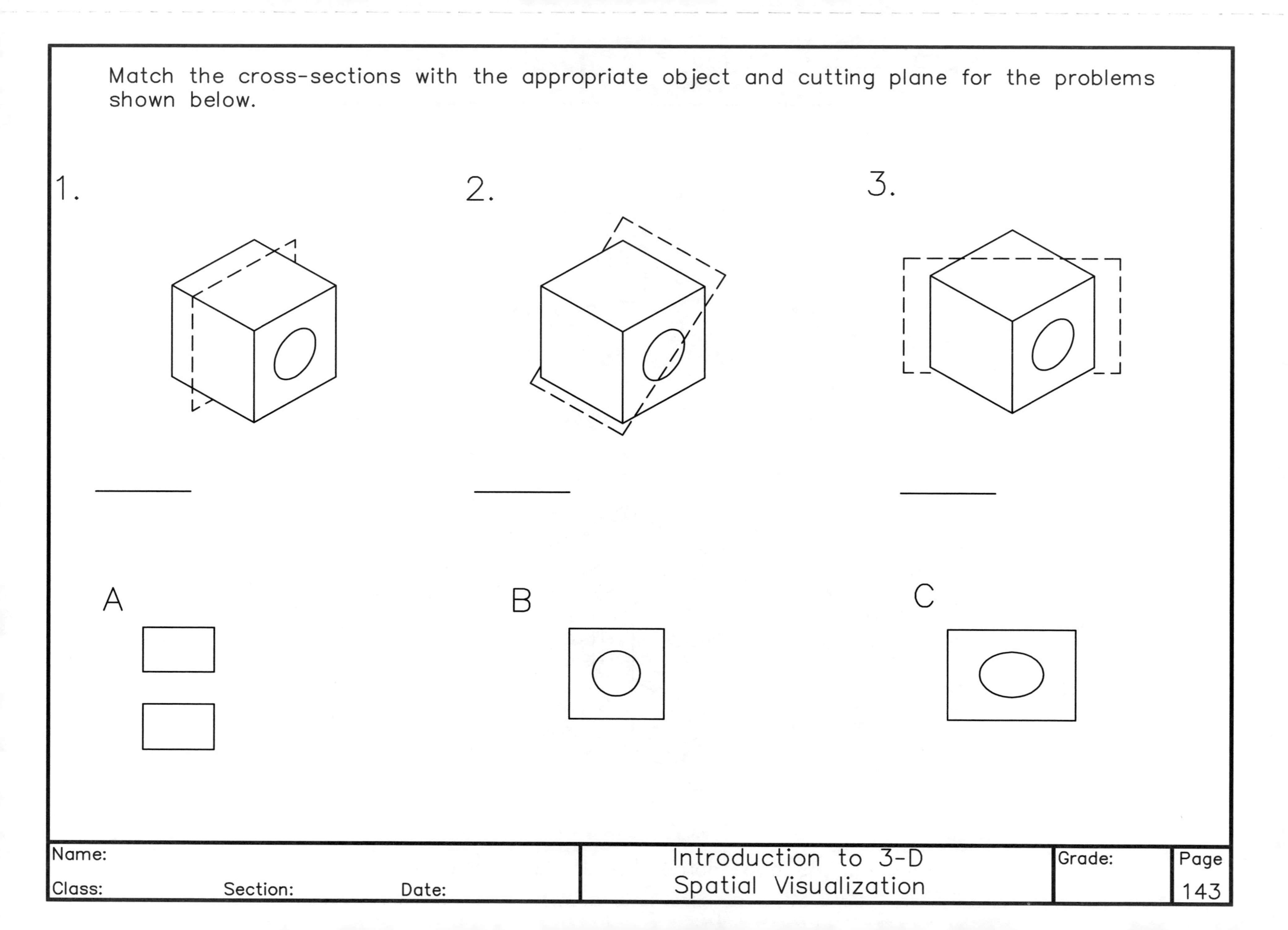


| Name: | | | Introduction to 3-D Spatial Visualization | Grade: | Page |
|---|---|---|---|---|---|
| Class: | Section: | Date: | | | 143 |

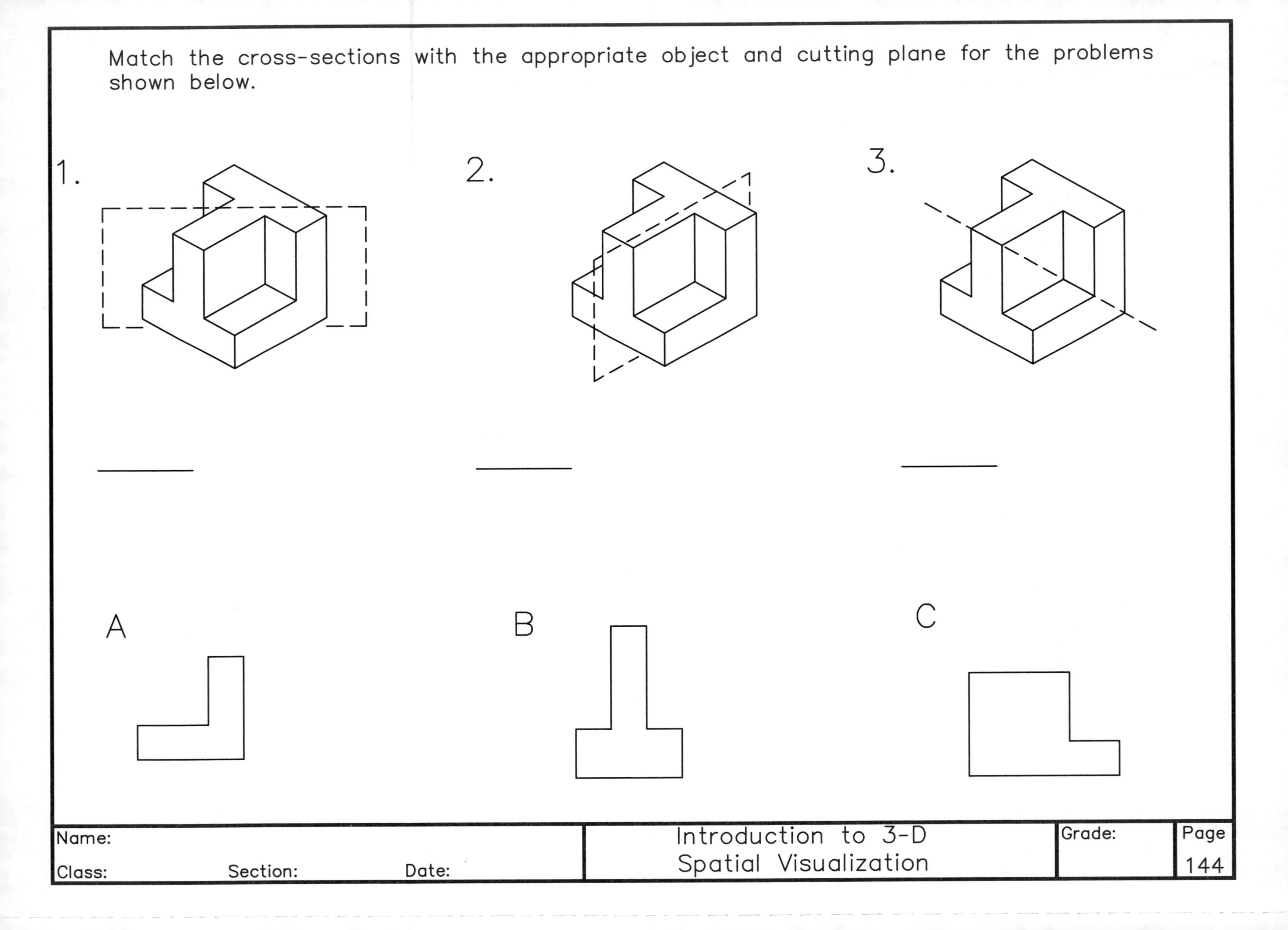
Match the cross-sections with the appropriate object and cutting plane for the problems shown below.
1.
2.
3.
A
B
C
Name:
Class:
Section:
Date:
Introduction to 3-D Spatial Visualization
Grade:
Page
144

Match the cross-sections with the appropriate object and cutting plane for the problems shown below.

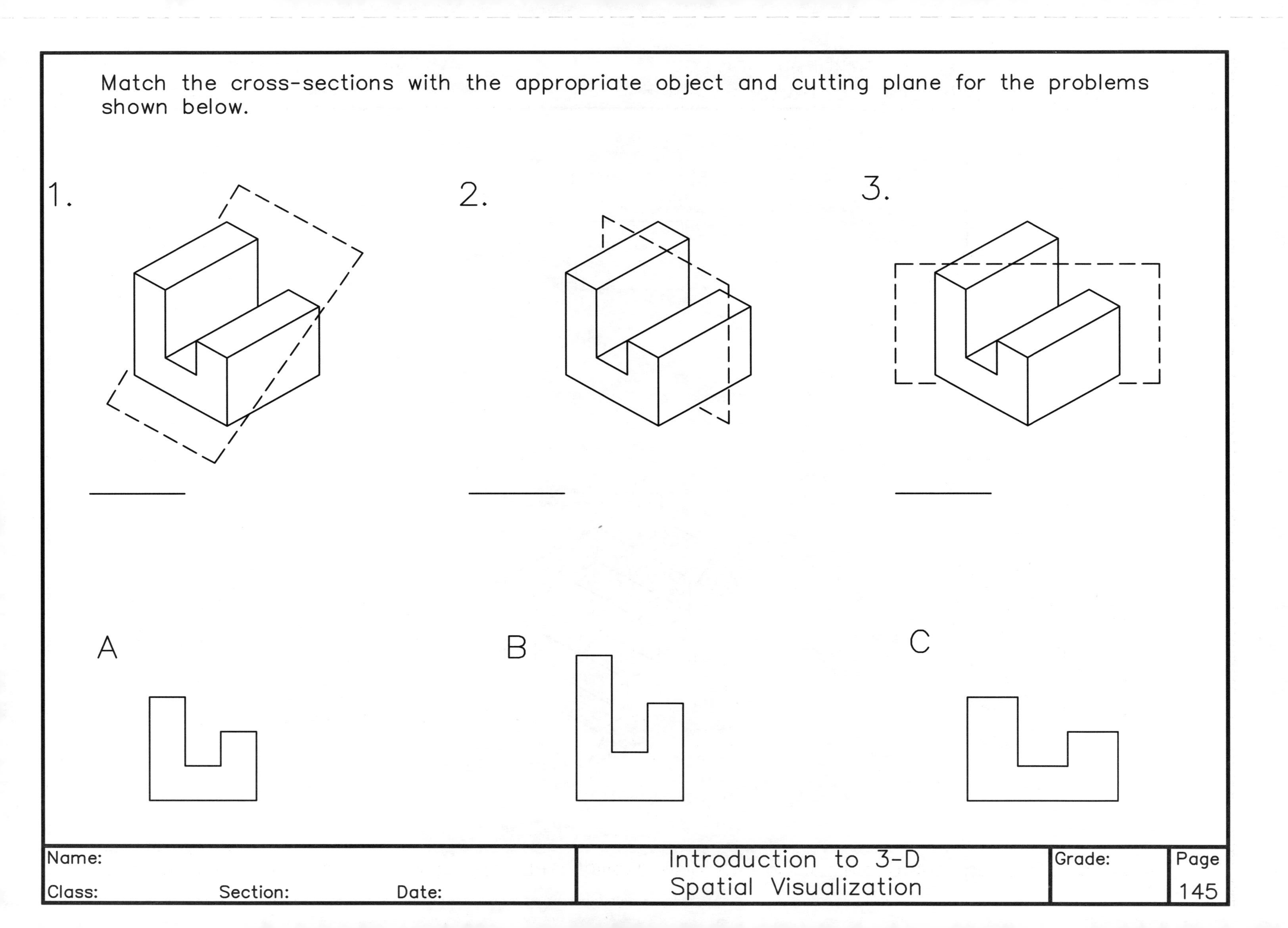

Match the cross-sections with the appropriate object and cutting plane for the problems shown below.

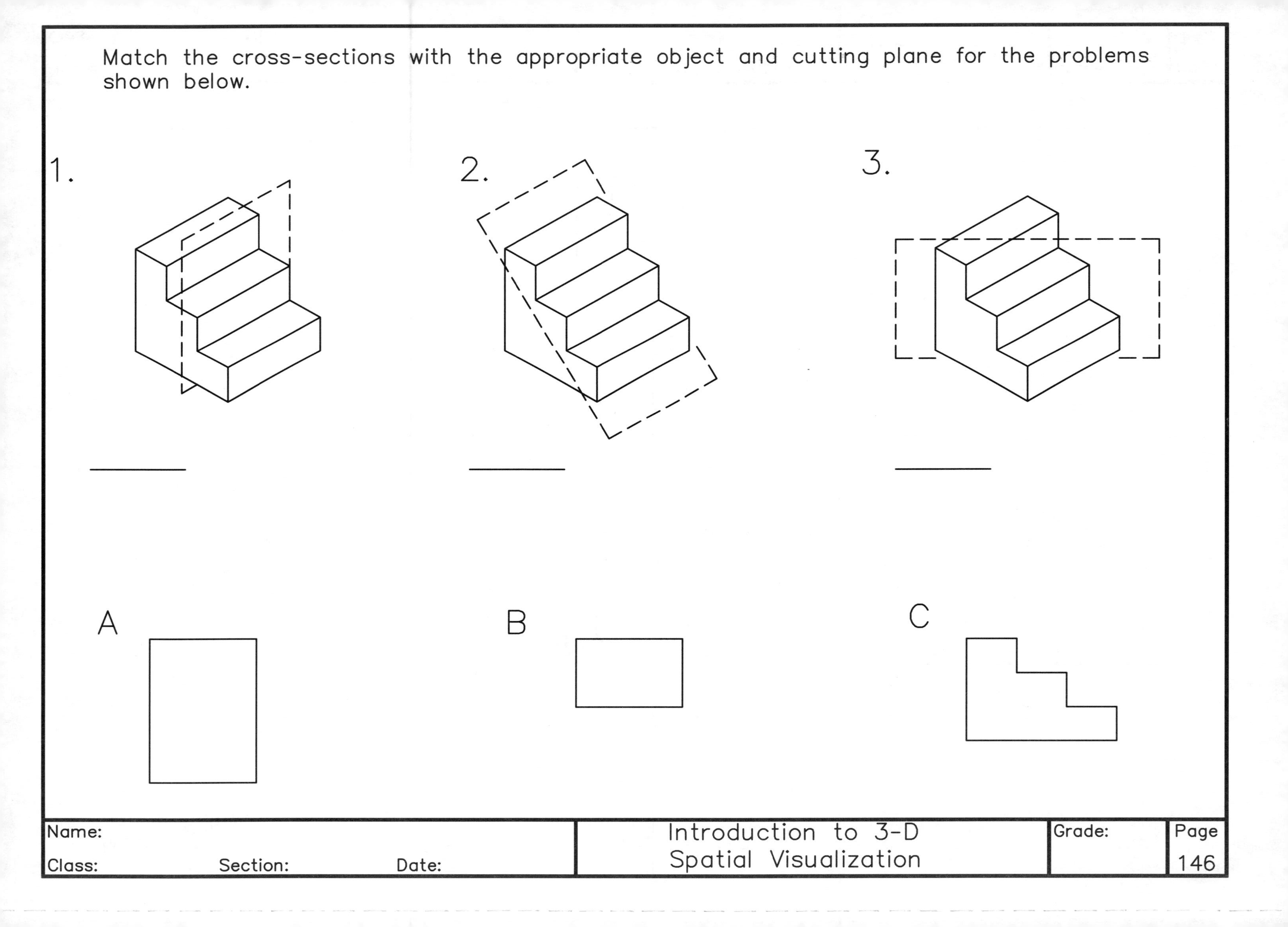


| Name: | Introduction to 3-D | Grade: | Page |
| --- | --- | --- | --- |
| Class: Section: Date: | Spatial Visualization | | 146 |

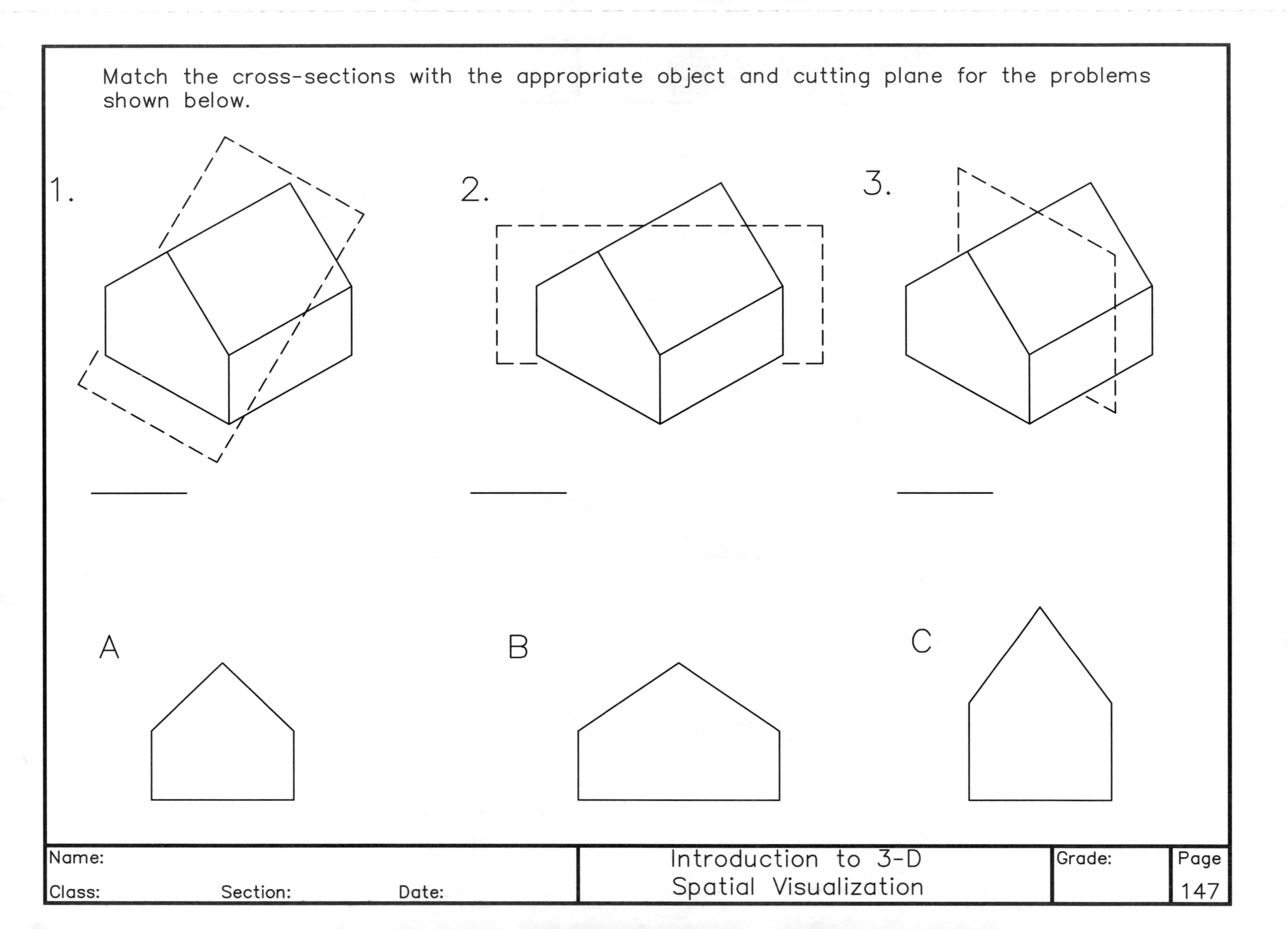
Match the cross-sections with the appropriate object and cutting plane for the problems shown below.
1.
2.
3.
A
B
C
Name:
Class:
Section:
Date:
Introduction to 3-D
Spatial Visualization
Grade:
Page
147

For the objects shown on the left below, circle the letters corresponding to all possible cross-sections that could be obtained by slicing them with an appropriate cutting plane. There could be more than one correct answer.

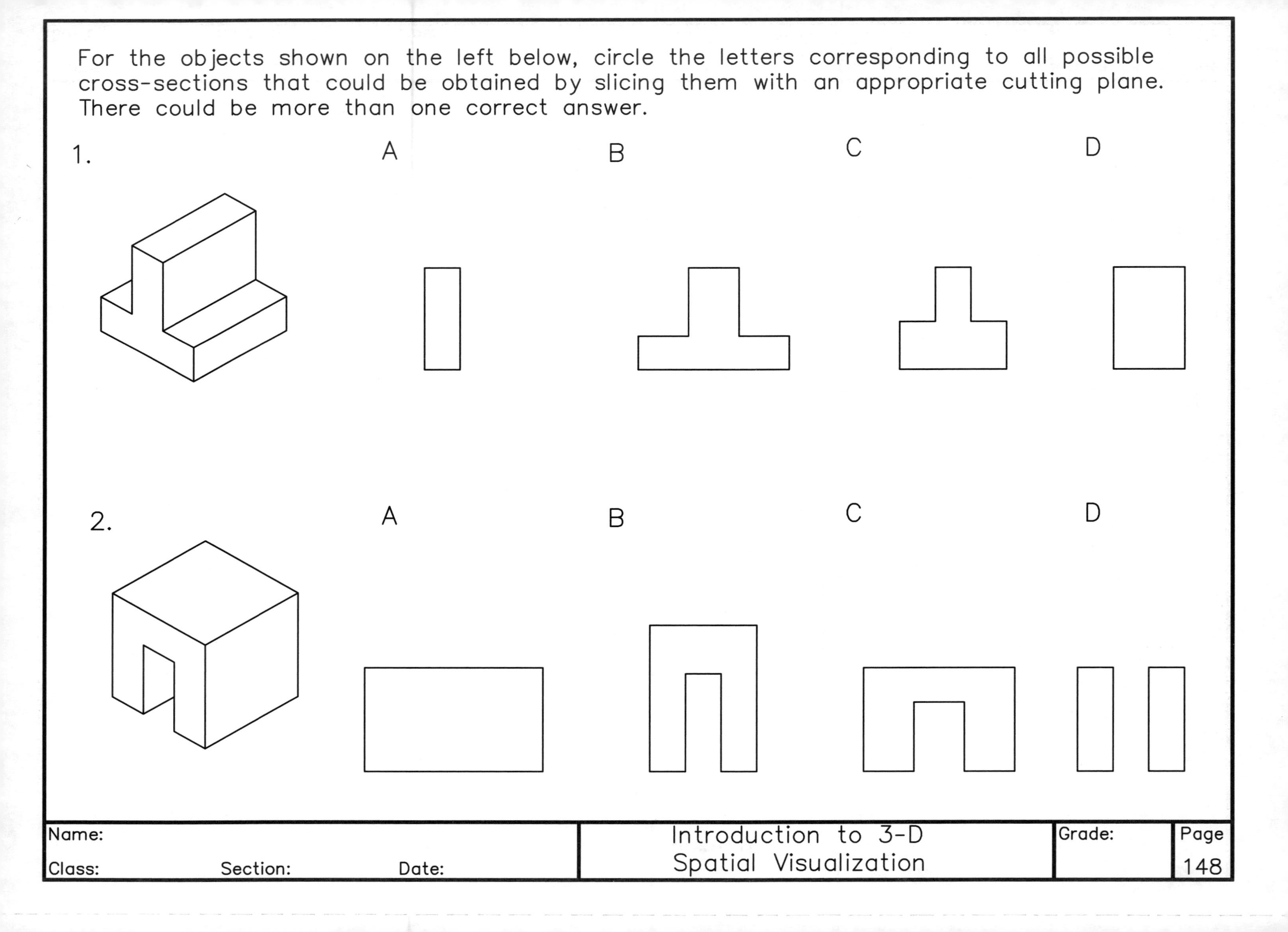

For the objects shown on the left below, circle the letters corresponding to all possible cross-sections that could be obtained by slicing them with an appropriate cutting plane. There could be more than one correct answer.

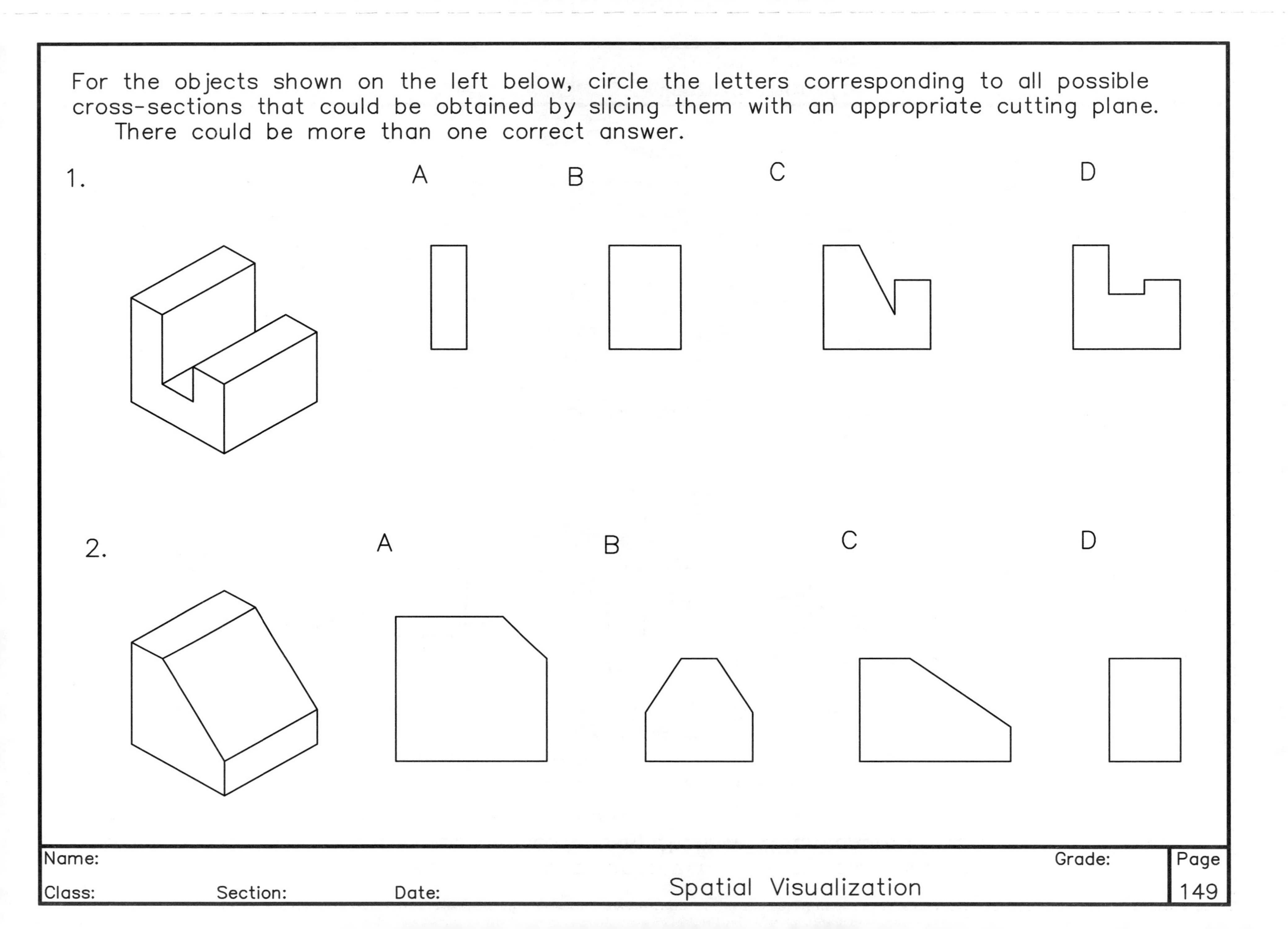

For the objects shown on the left below, circle the letters corresponding to all possible cross-sections that could be obtained by slicing them with an appropriate cutting plane. There could be more than one correct answer.

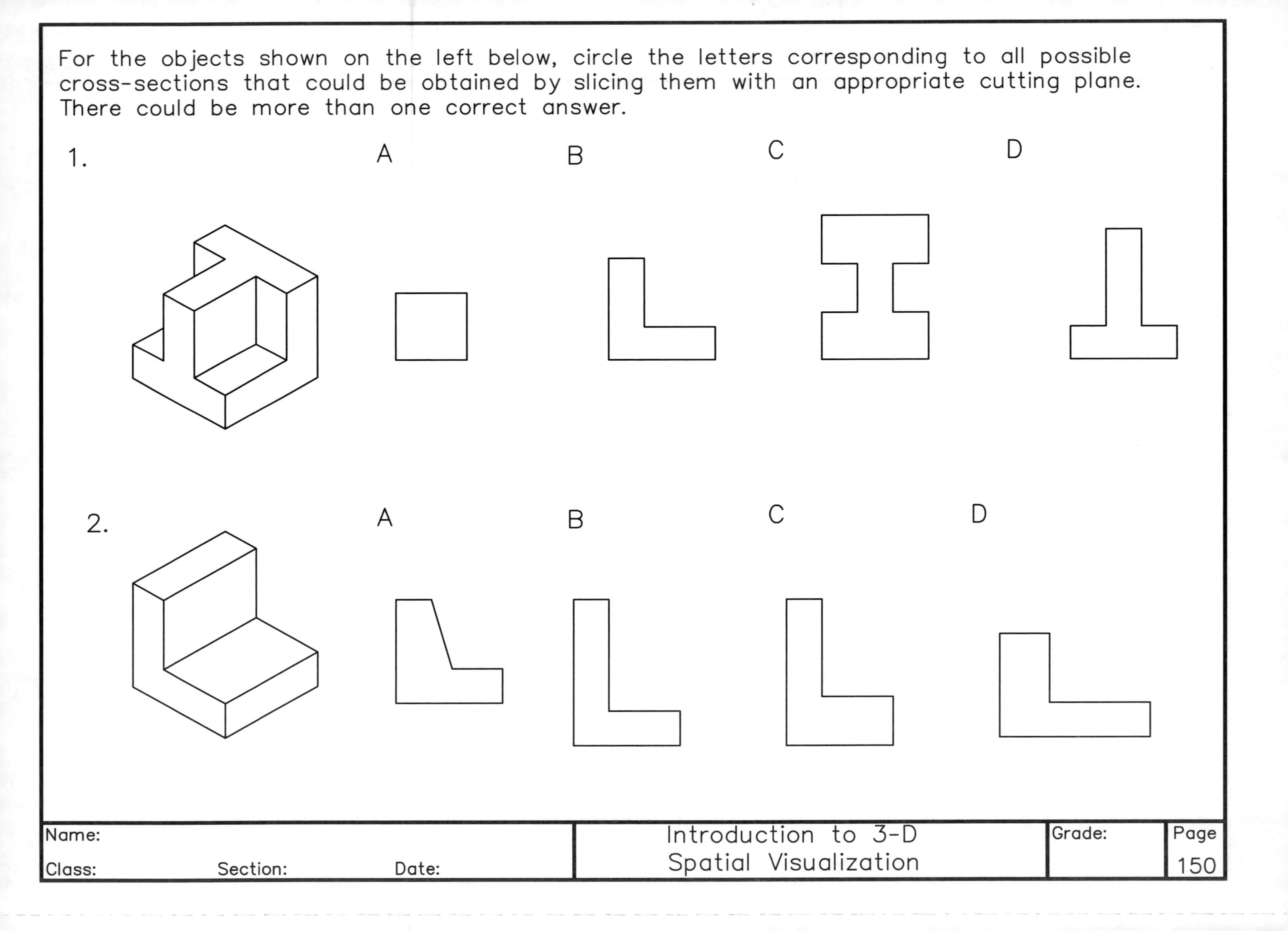


| Name: | Introduction to 3-D | Grade: | Page |
|---|---|---|---|
| Class: Section: Date: | Spatial Visualization | | 150 |

For the objects shown on the left below, circle the letters corresponding to all possible cross-sections that could be obtained by slicing them with an appropriate cutting plane. There could be more than one correct answer.

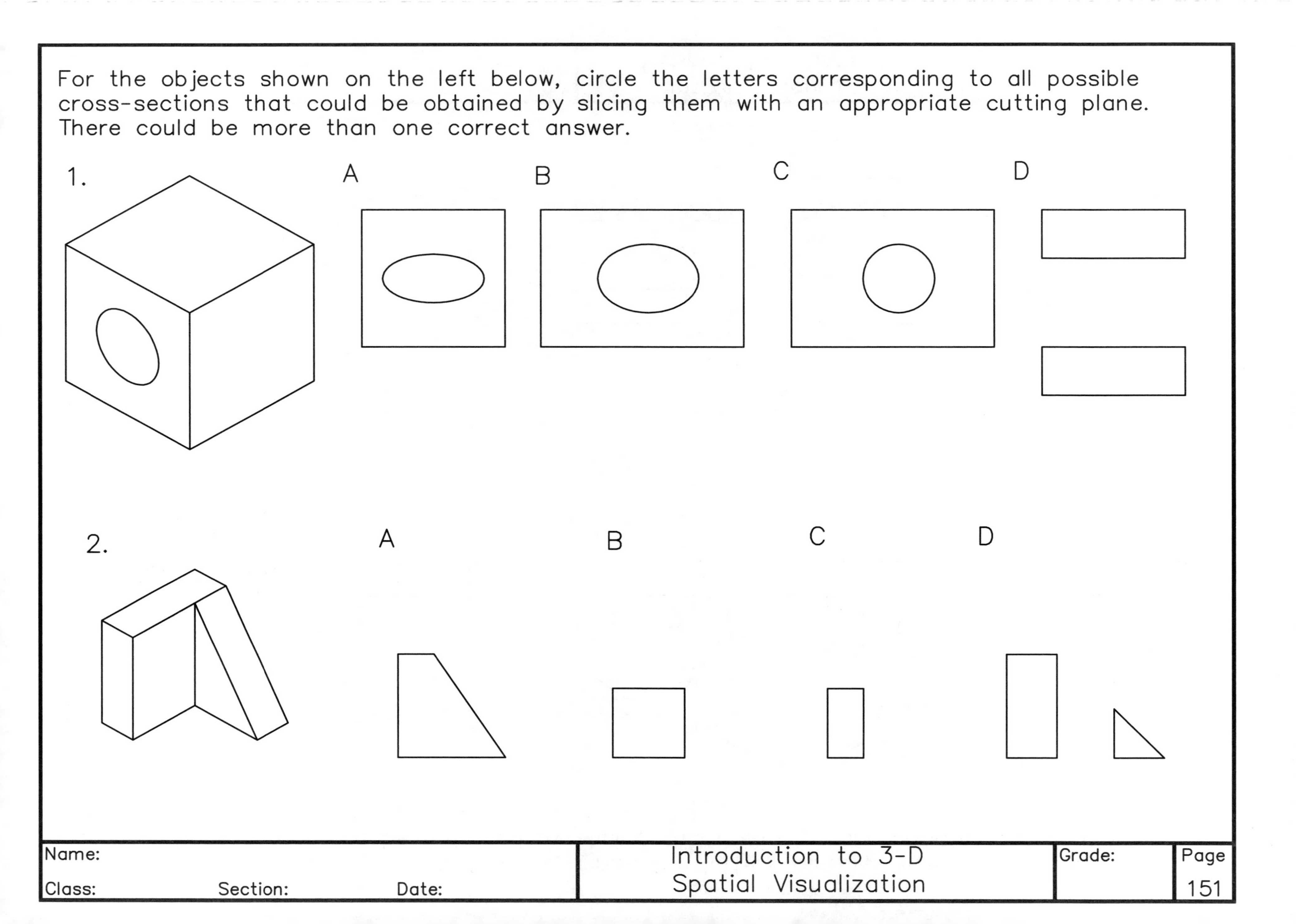

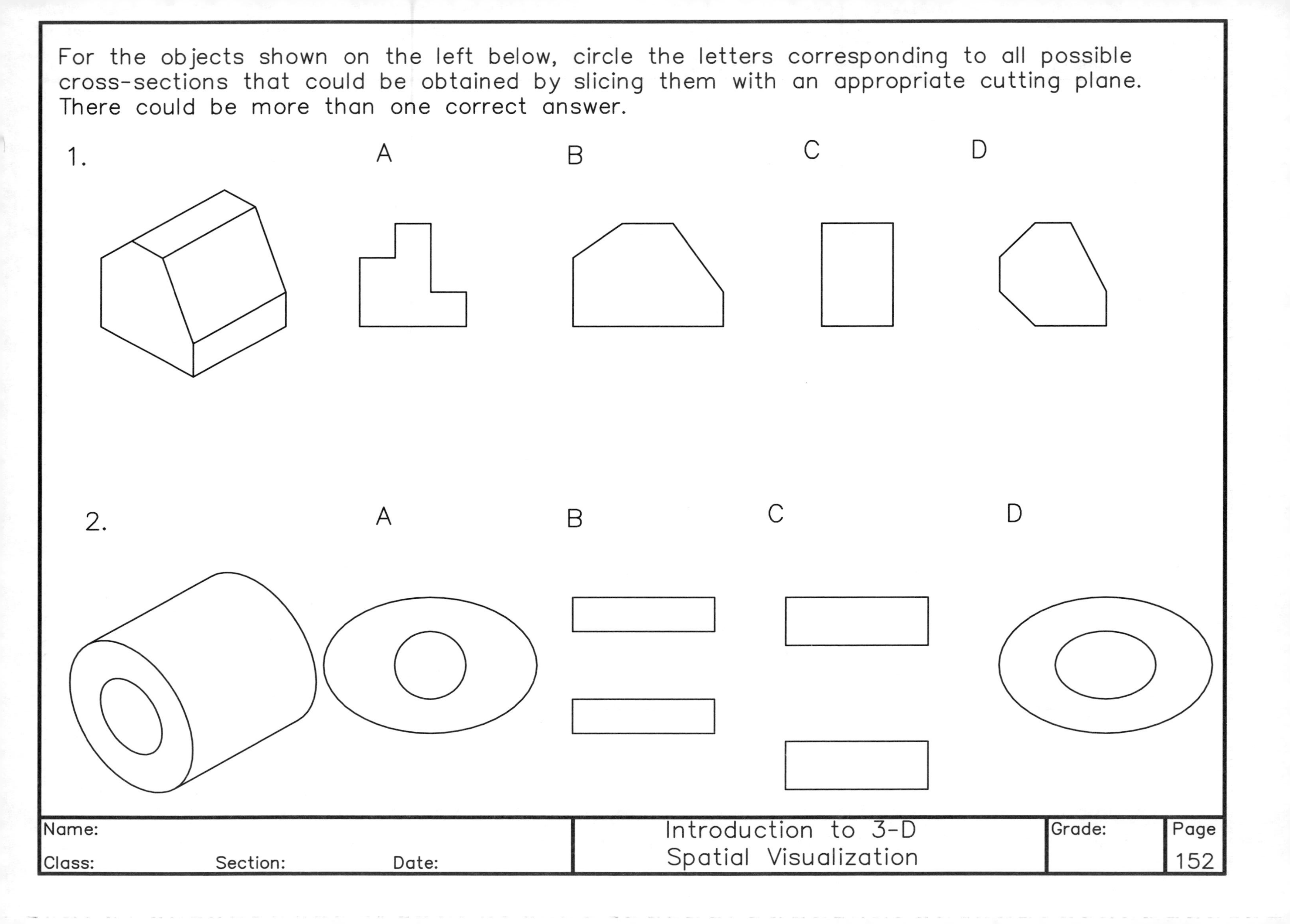

For the objects shown on the left below, circle the letters corresponding to all possible cross-sections that could be obtained by slicing them with an appropriate cutting plane. There could be more than one correct answer.

For the objects shown on the left below, circle the letters corresponding to all possible cross-sections that could be obtained by slicing them with an appropriate cutting plane. There could be more than one correct answer.

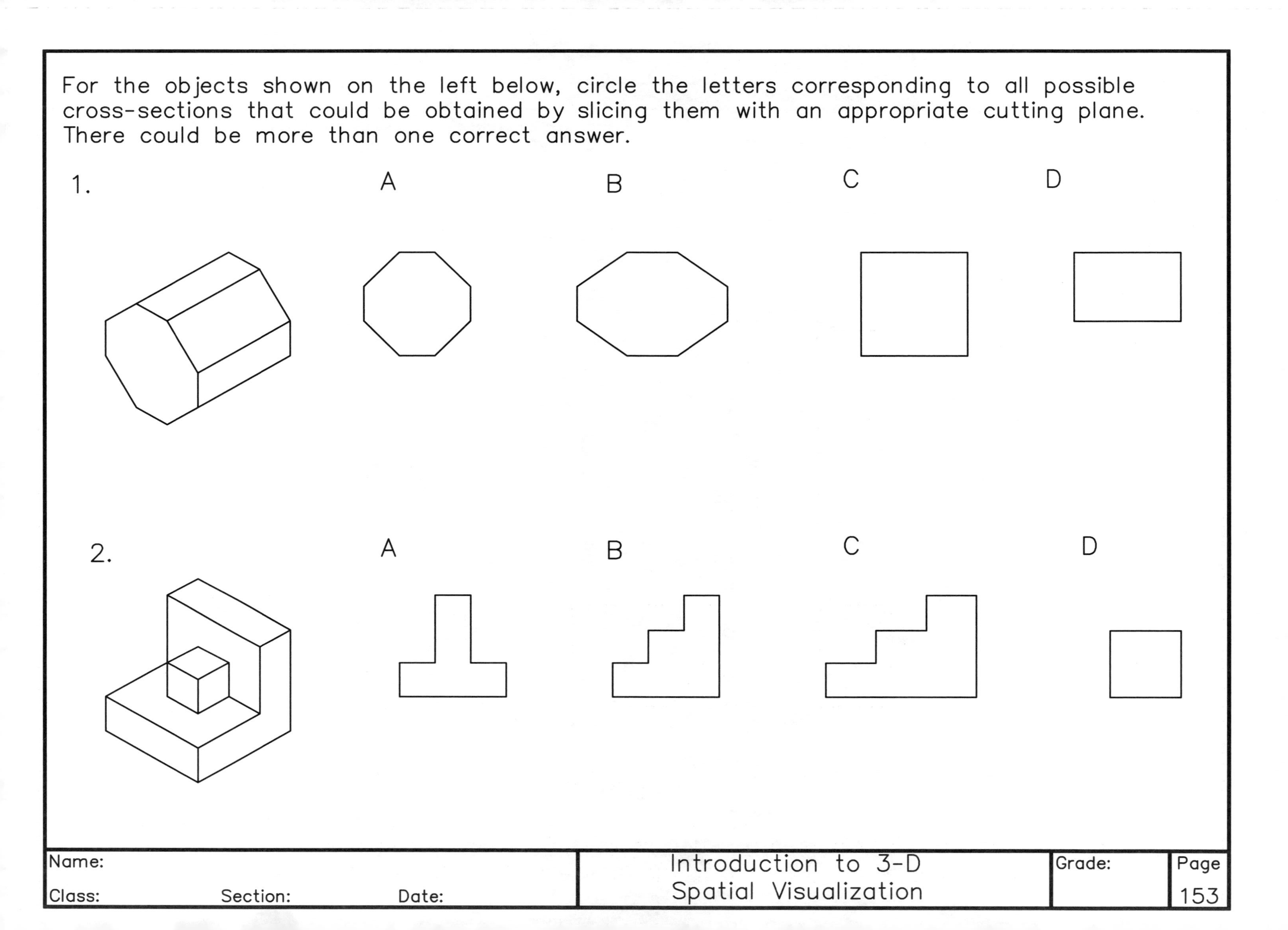

## Surfaces and Solids of Revolution

A surface of revolution is like a hollow shell created by revolving a set of 2-D curves about a coordinate axis or about another line in 3-D space.

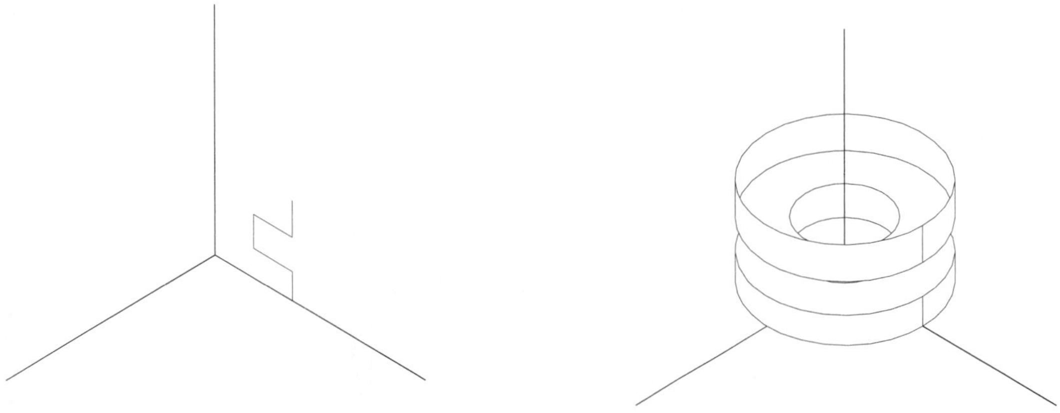

2-D Shape and Surface of Revolution

A solid of revolution is a 3-D object of a finite volume. It is generated by revolving a closed 2-D shape about a coordinate axis of another line in 3-D space.

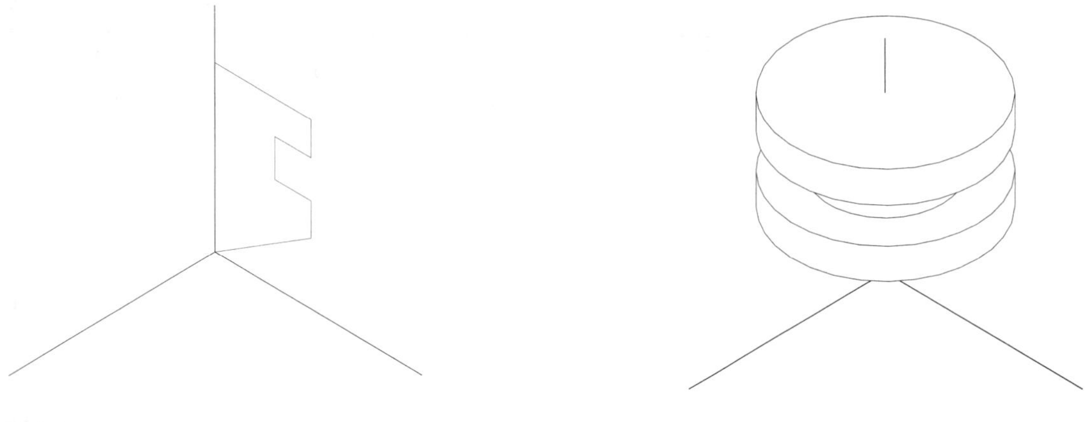

Closed 2-D Shape and Solid of Revolution

The resulting solid or surface of revolution will vary depending on the chosen axis of revolution.

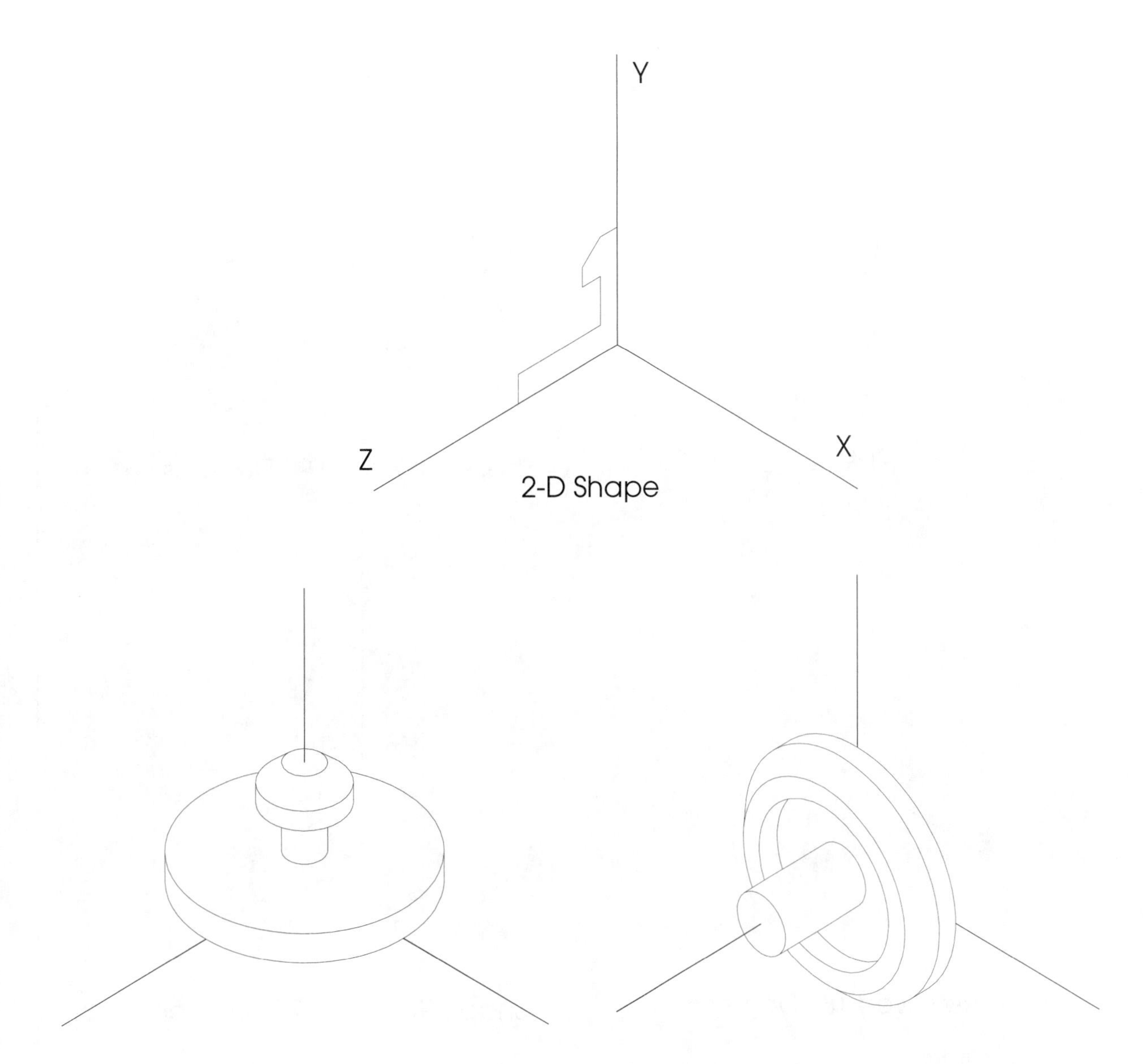


2-D Shape

Shape Revolved About Y

Shape Revolved About Z

When 2-D shapes are revolved about an axis, you can choose to revolve them less than 360 degrees with different results.

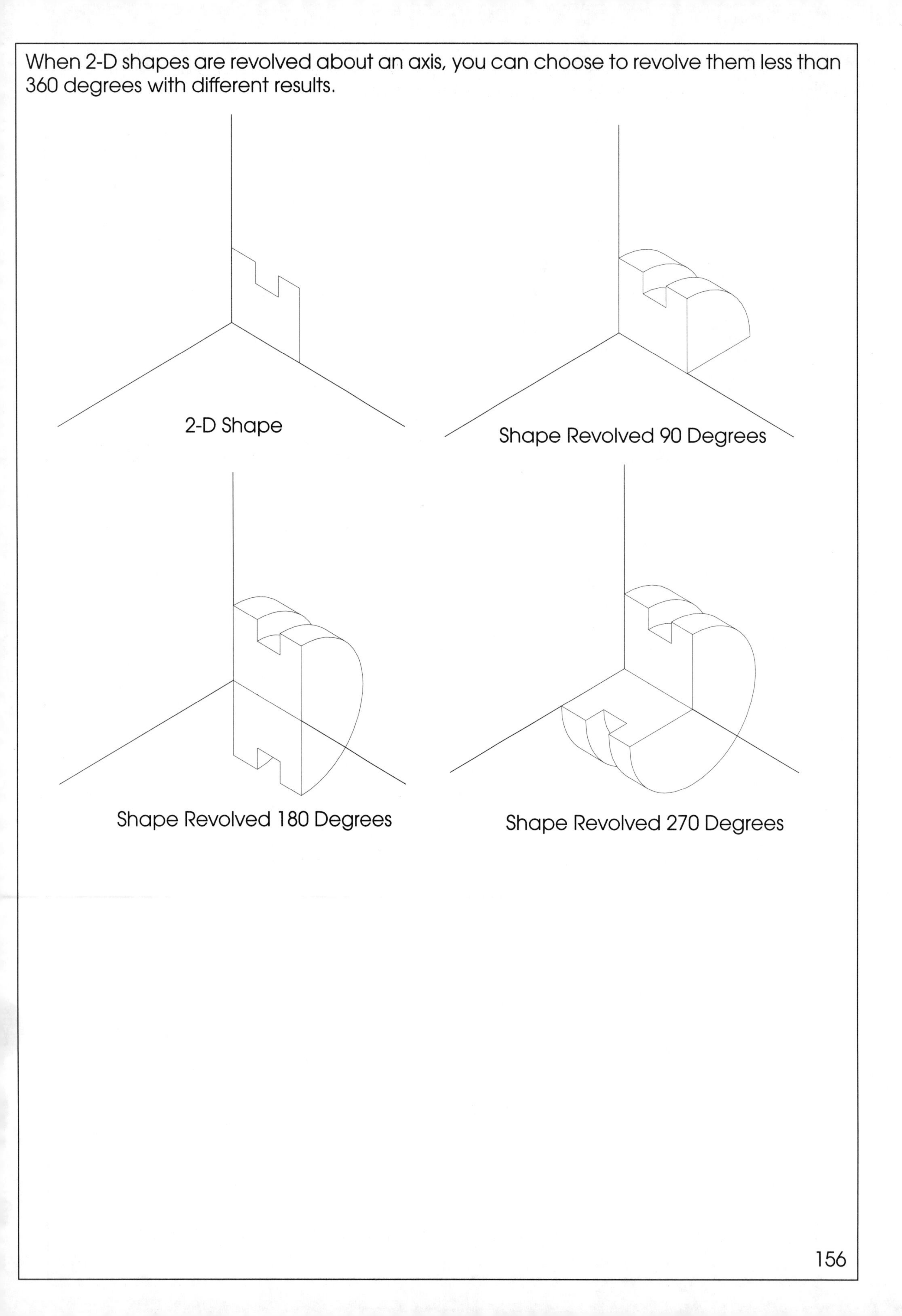

Different objects will be formed depending on a combination of the choice for the axis of revolution and the choice of the angle of revolution.

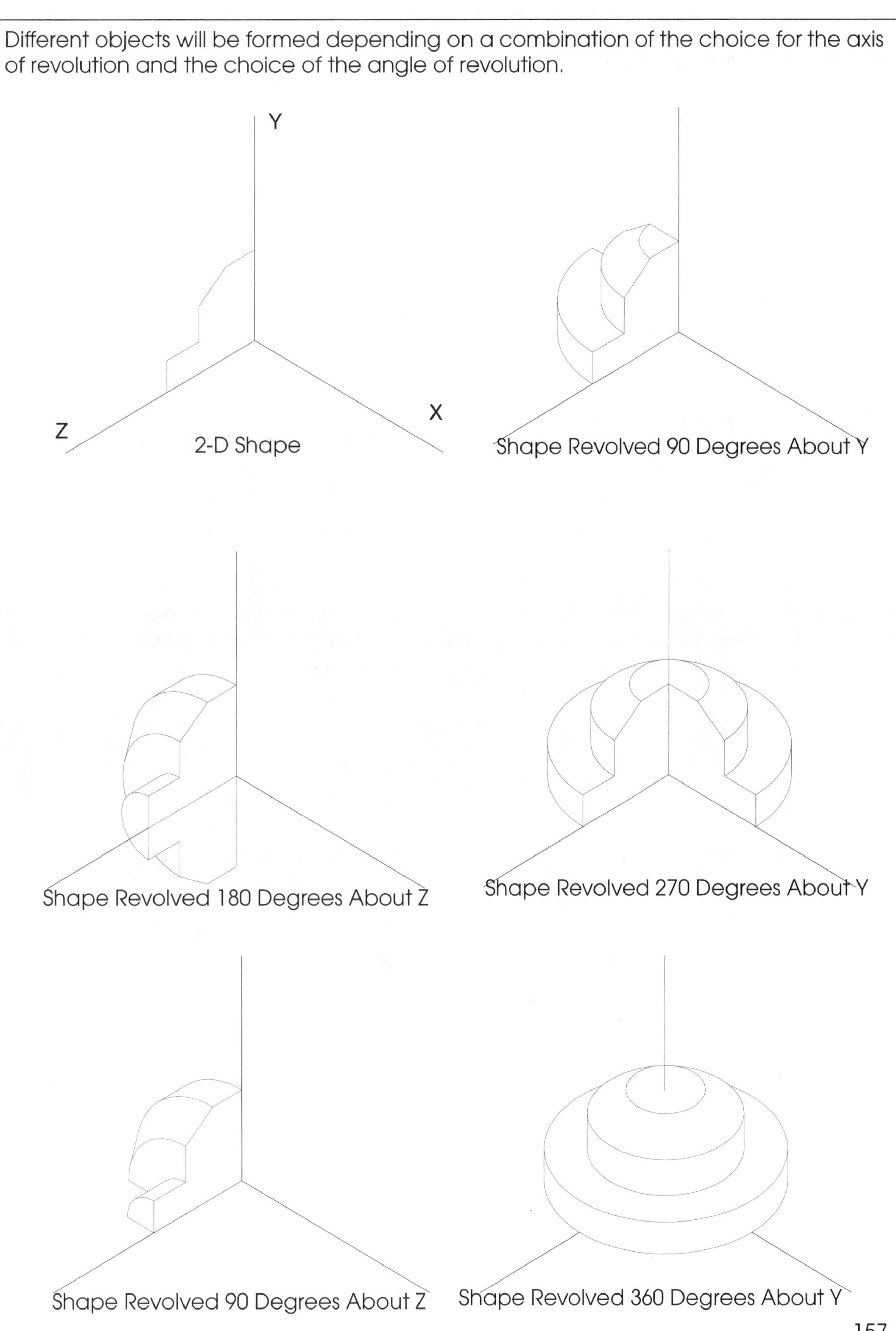

If the axis of revolution is not located on the 2-D shape itself but some distance, x, away from it, then a solid of revolution will be created with a cylindrical hole of diameter 2x will be created from it.

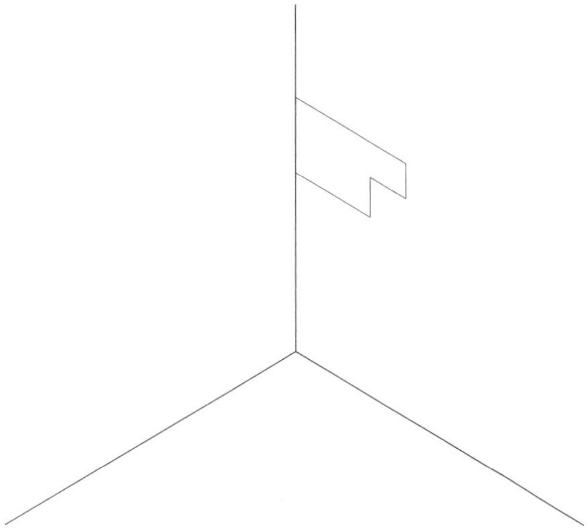

2-D Shape Located Away From Axis of Revolution

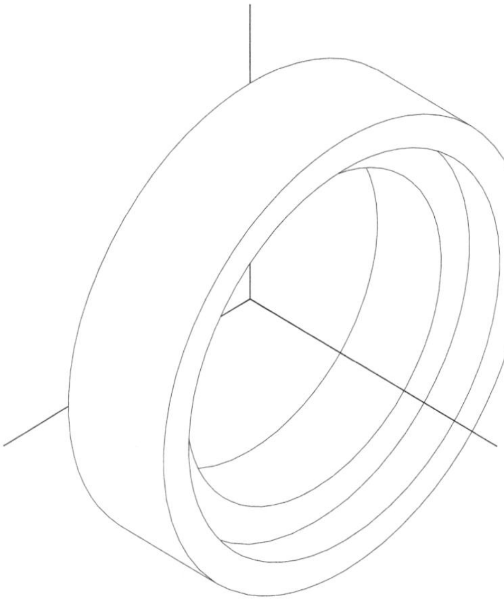

"Hollow" 3-D Part

To visualize what a shape will look like after it has been revolved around an axis, first think about mirroring the shape on the opposite side of the axis of revolution. Then form an object with a basic cylindrical outline through the two shapes.

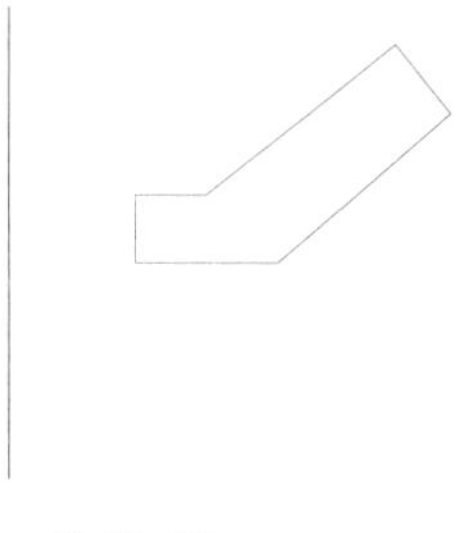

2-D Shape

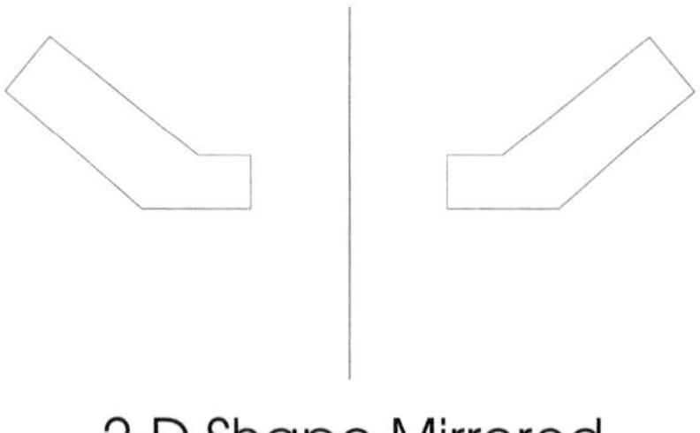

2-D Shape Mirrored About Axis of Revolution

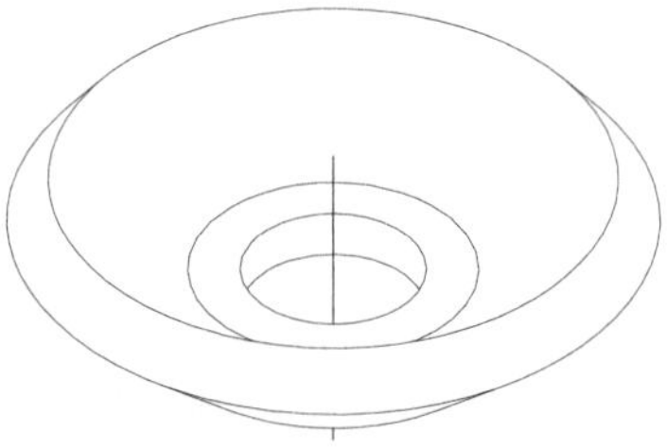

Solid of Revolution

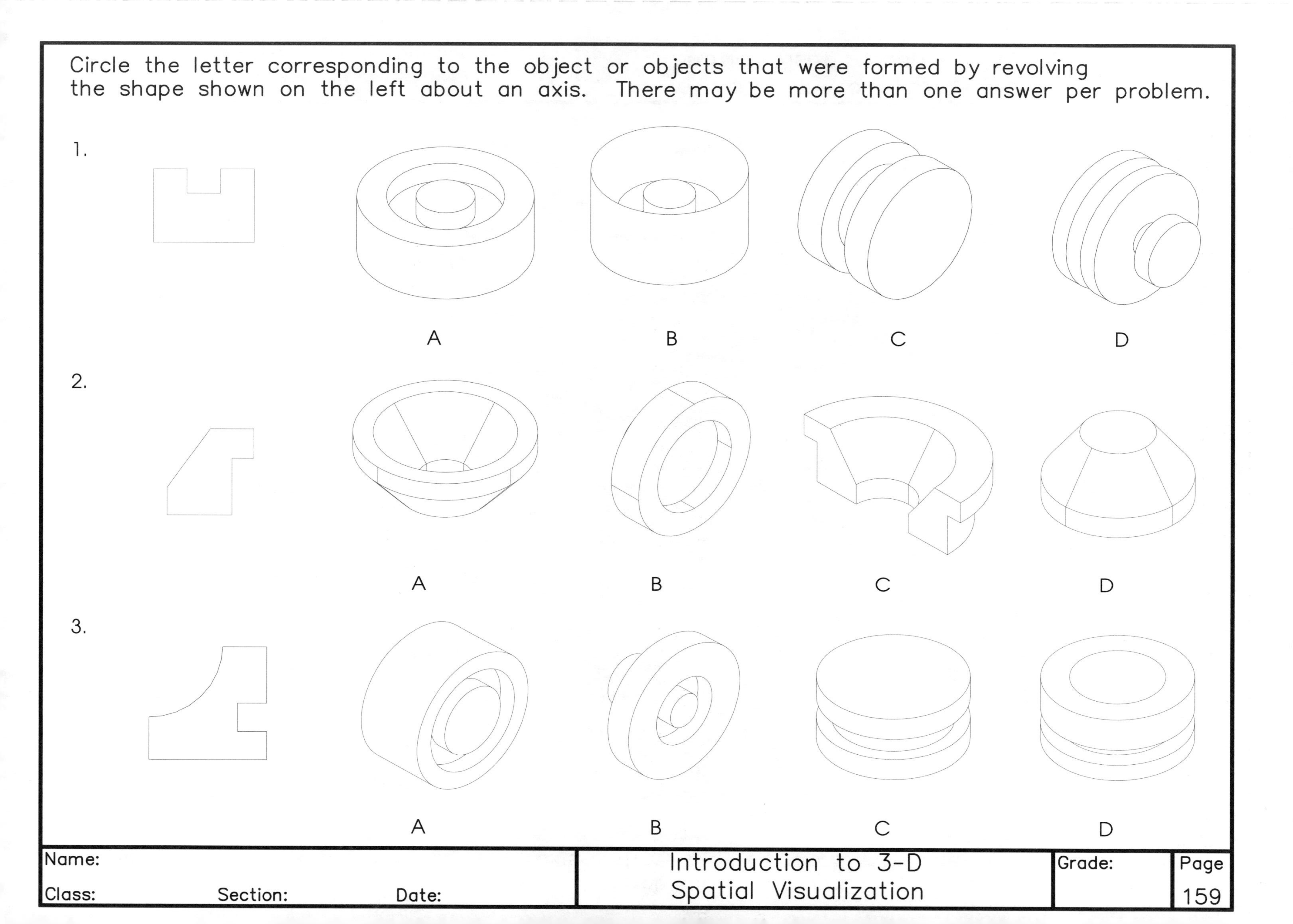
Circle the letter corresponding to the object or objects that were formed by revolving the shape shown on the left about an axis. There may be more than one answer per problem.
1.
A
B
C
D
2.
A
B
C
D
3.
A
B
C
D
Name:
Class:
Section:
Date:
Introduction to 3-D Spatial Visualization
Grade:
Page
159

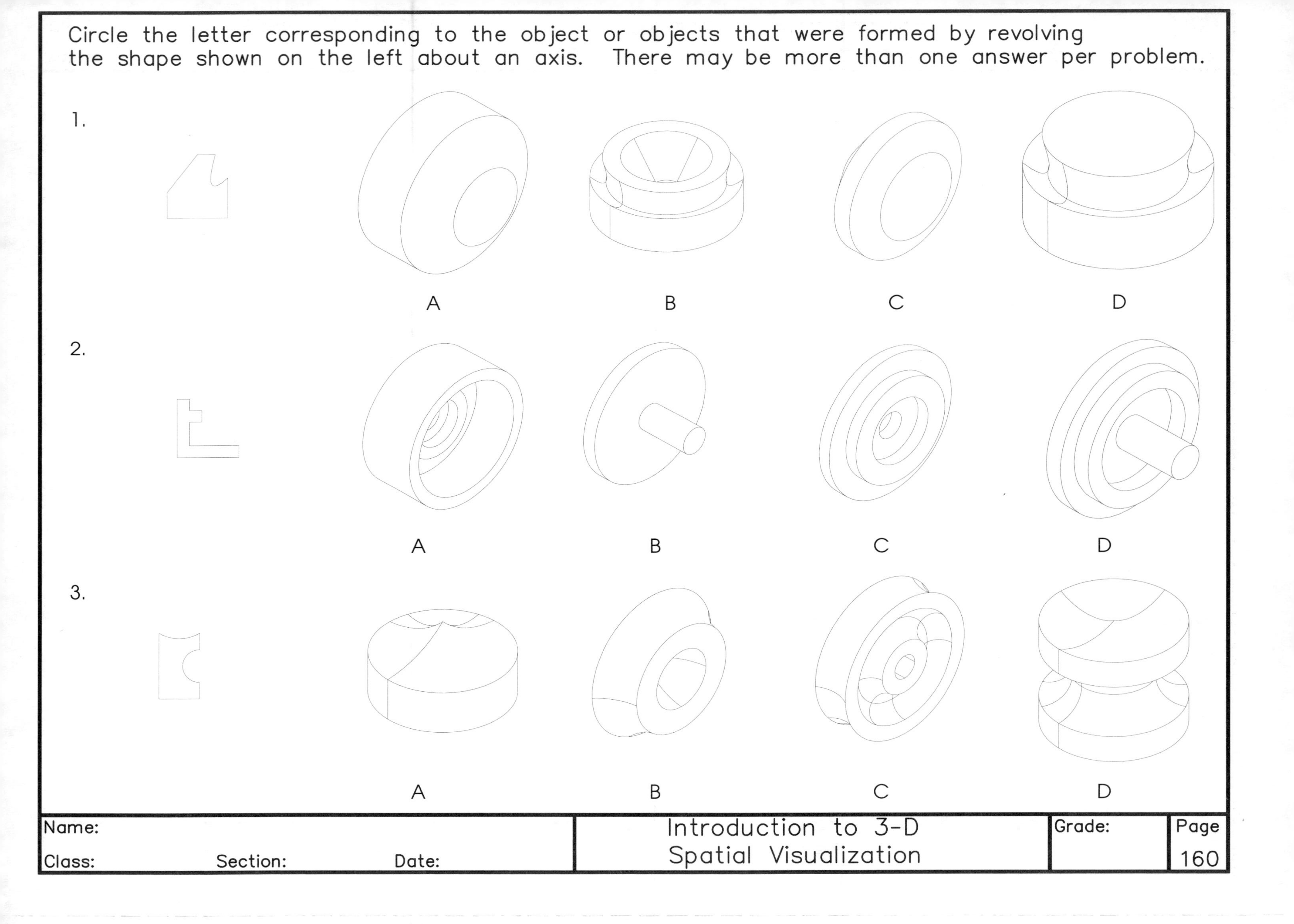

Circle the letter corresponding to the object or objects that were formed by revolving the shape shown on the left about an axis. There may be more than one answer per problem.



| Name:<br>Class: Section: Date: | Introduction to 3-D<br>Spatial Visualization | Grade: | Page<br>160 |
|---|---|---|---|

Circle the letter corresponding to the object or objects that were formed by revolving the shape shown on the left about an axis. There may be more than one answer per problem.

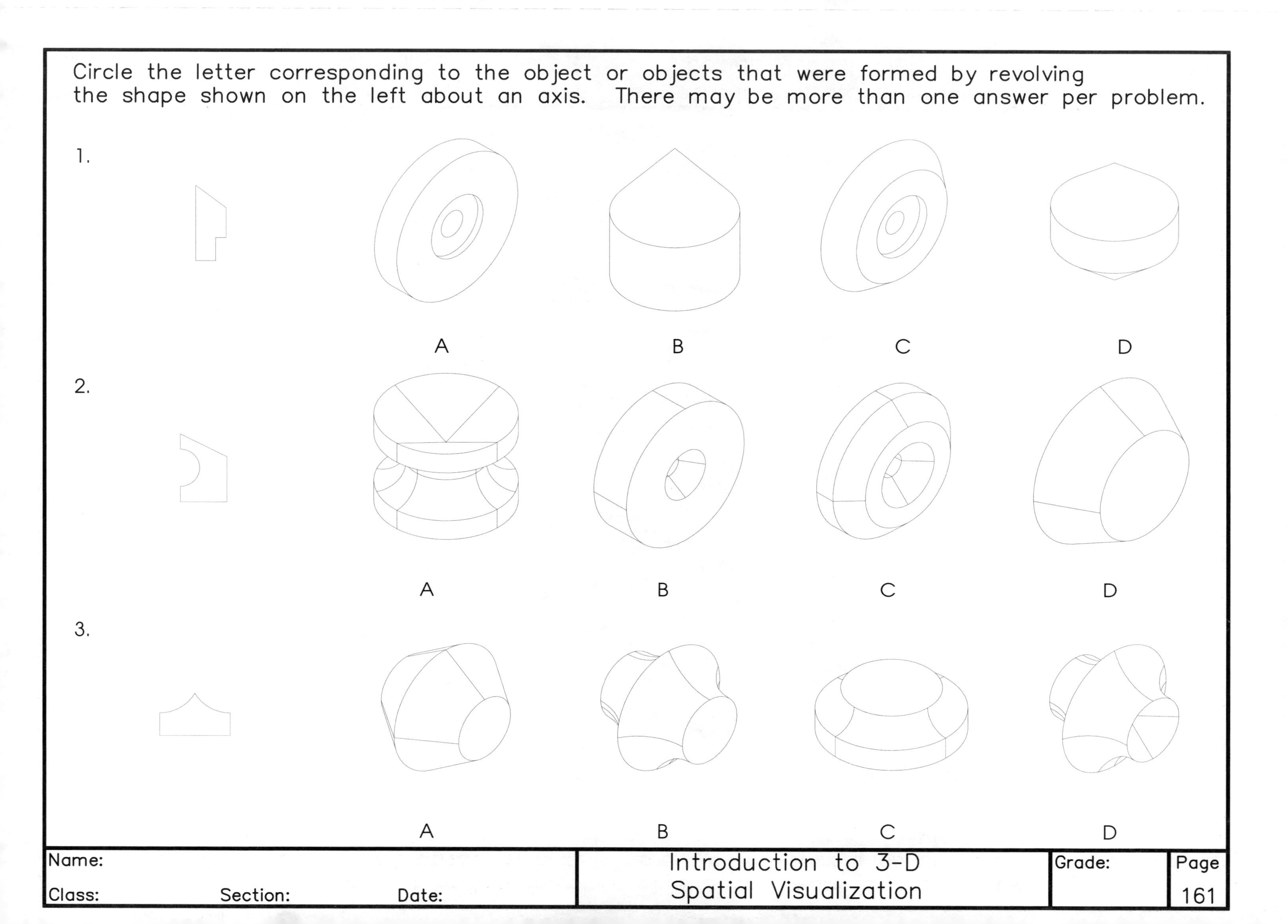

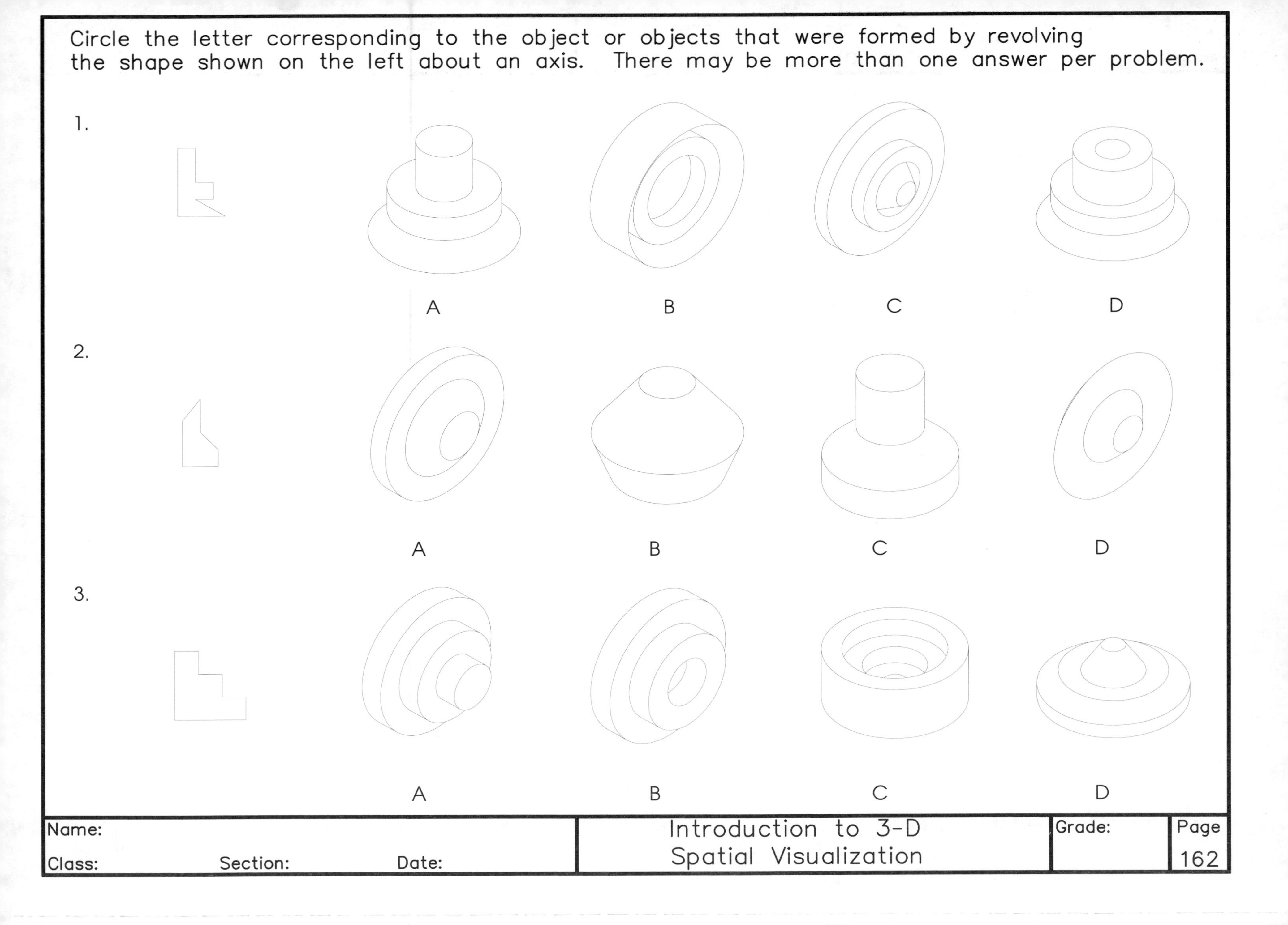
Circle the letter corresponding to the object or objects that were formed by revolving the shape shown on the left about an axis. There may be more than one answer per problem.
1.
A
B
C
D
2.
A
B
C
D
3.
A
B
C
D
Name:
Class:
Section:
Date:
Introduction to 3-D
Spatial Visualization
Grade:
Page
162

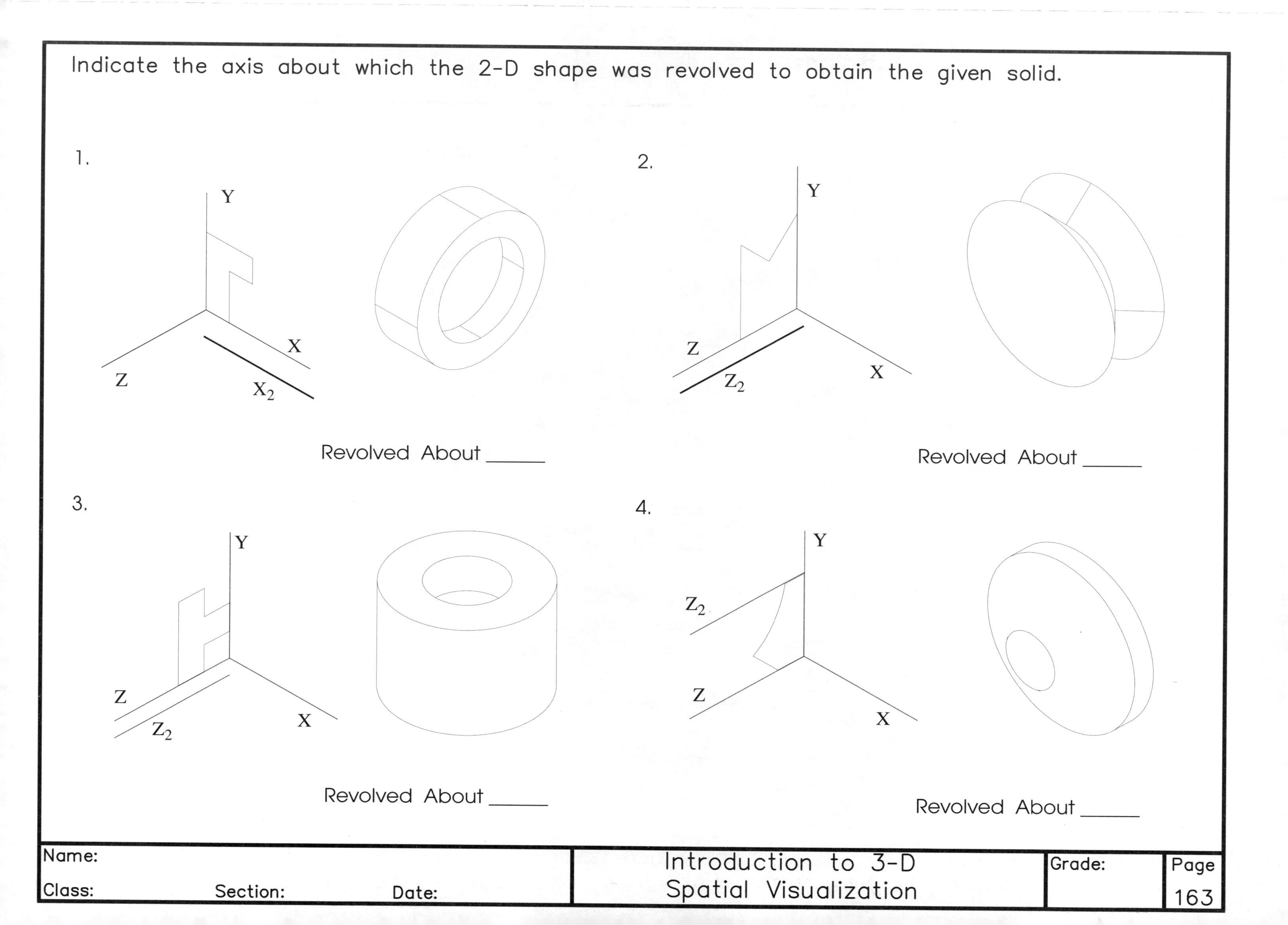
Indicate the axis about which the 2-D shape was revolved to obtain the given solid.
1.
Y
X
Z
X2
Revolved About ______
2.
Y
Z
Z2
X
Revolved About ______
3.
Y
Z
Z2
X
Revolved About ______
4.
Y
Z2
Z
X
Revolved About ______
Name:
Class:
Section:
Date:
Introduction to 3-D
Spatial Visualization
Grade:
Page
163

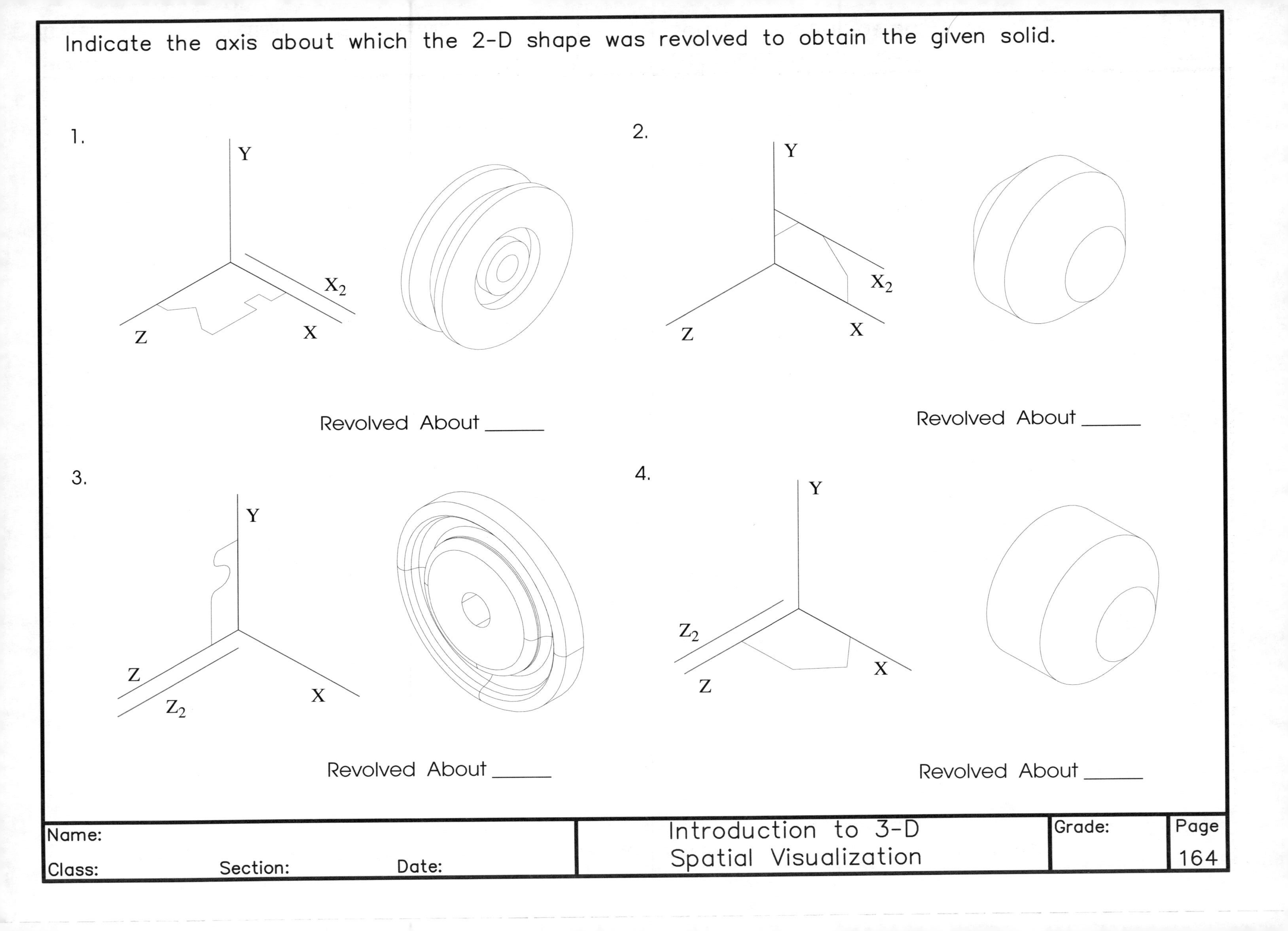
Indicate the axis about which the 2-D shape was revolved to obtain the given solid.
1.
Y
$X_2$
X
Z
Revolved About ______
2.
Y
$X_2$
X
Z
Revolved About ______
3.
Y
Z
$Z_2$
X
Revolved About ______
4.
Y
$Z_2$
Z
X
Revolved About ______
Name:
Class:
Section:
Date:
Introduction to 3-D
Spatial Visualization
Grade:
Page
164

Indicate the axis about which the 2-D shape was revolved to obtain the given solid.

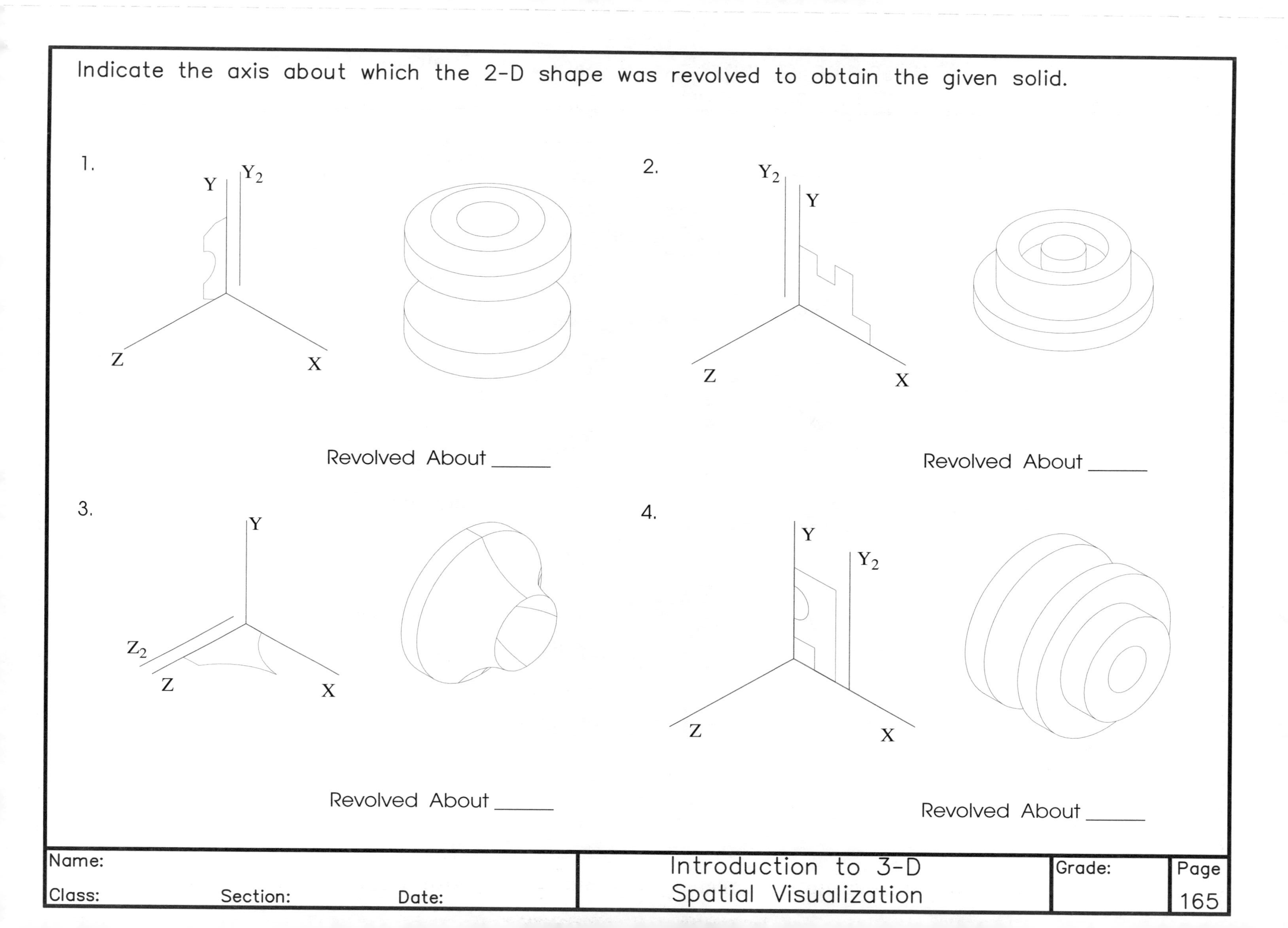


| Name: | Introduction to 3-D Spatial Visualization | Grade: | Page |
|---|---|---|---|
| Class: Section: Date: | | | 165 |

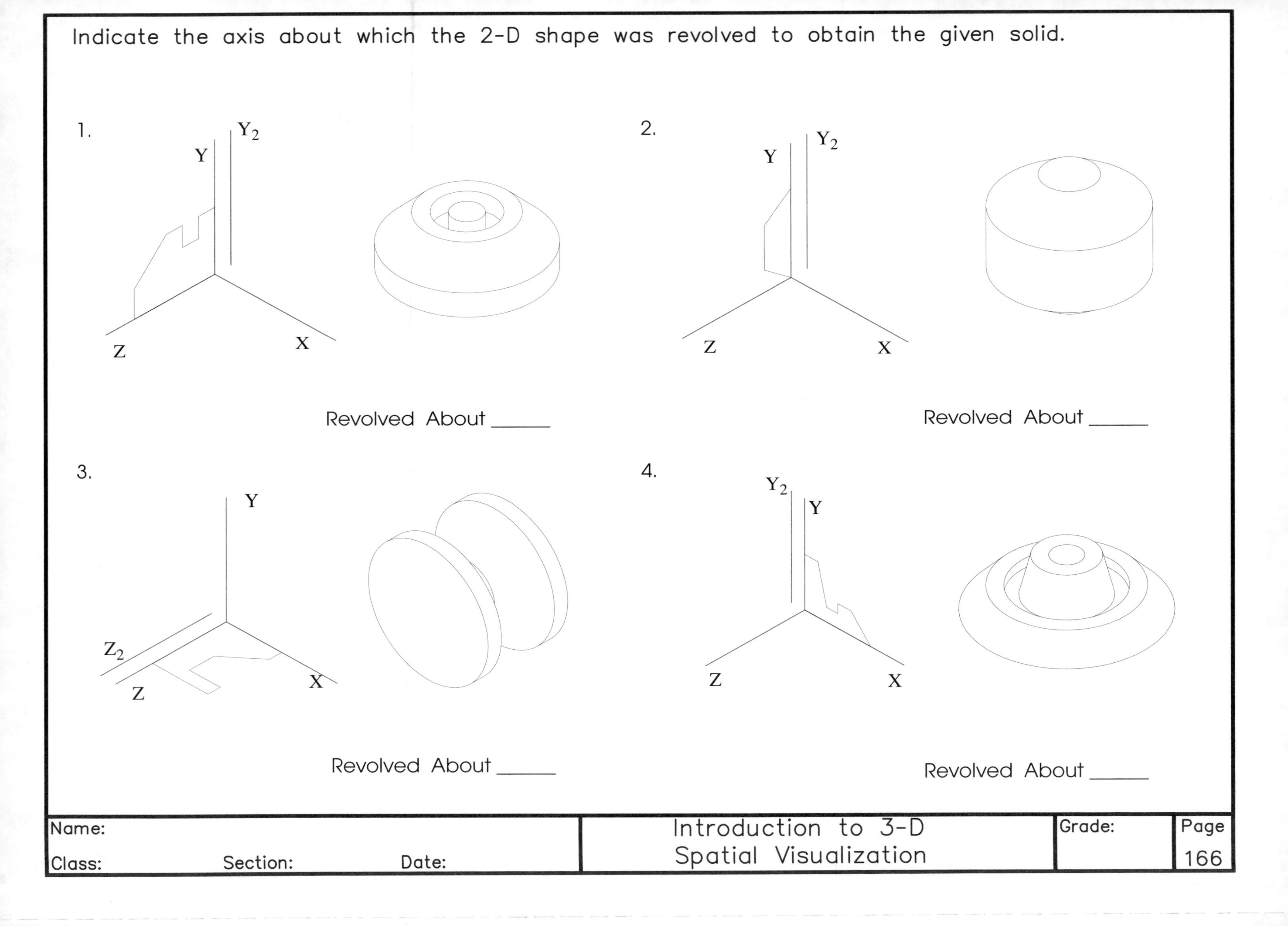
Indicate the axis about which the 2-D shape was revolved to obtain the given solid.
1.
Y
Y2
Z
X
Revolved About _____
2.
Y
Y2
Z
X
Revolved About _____
3.
Y
Z2
Z
X
Revolved About _____
4.
Y2
Y
Z
X
Revolved About _____
Name:
Class:
Section:
Date:
Introduction to 3-D Spatial Visualization
Grade:
Page
166

Indicate the axis and the number of degrees (90, 180, 270, or 360) about which the 2-D shape was revolved to obtain the given solid.

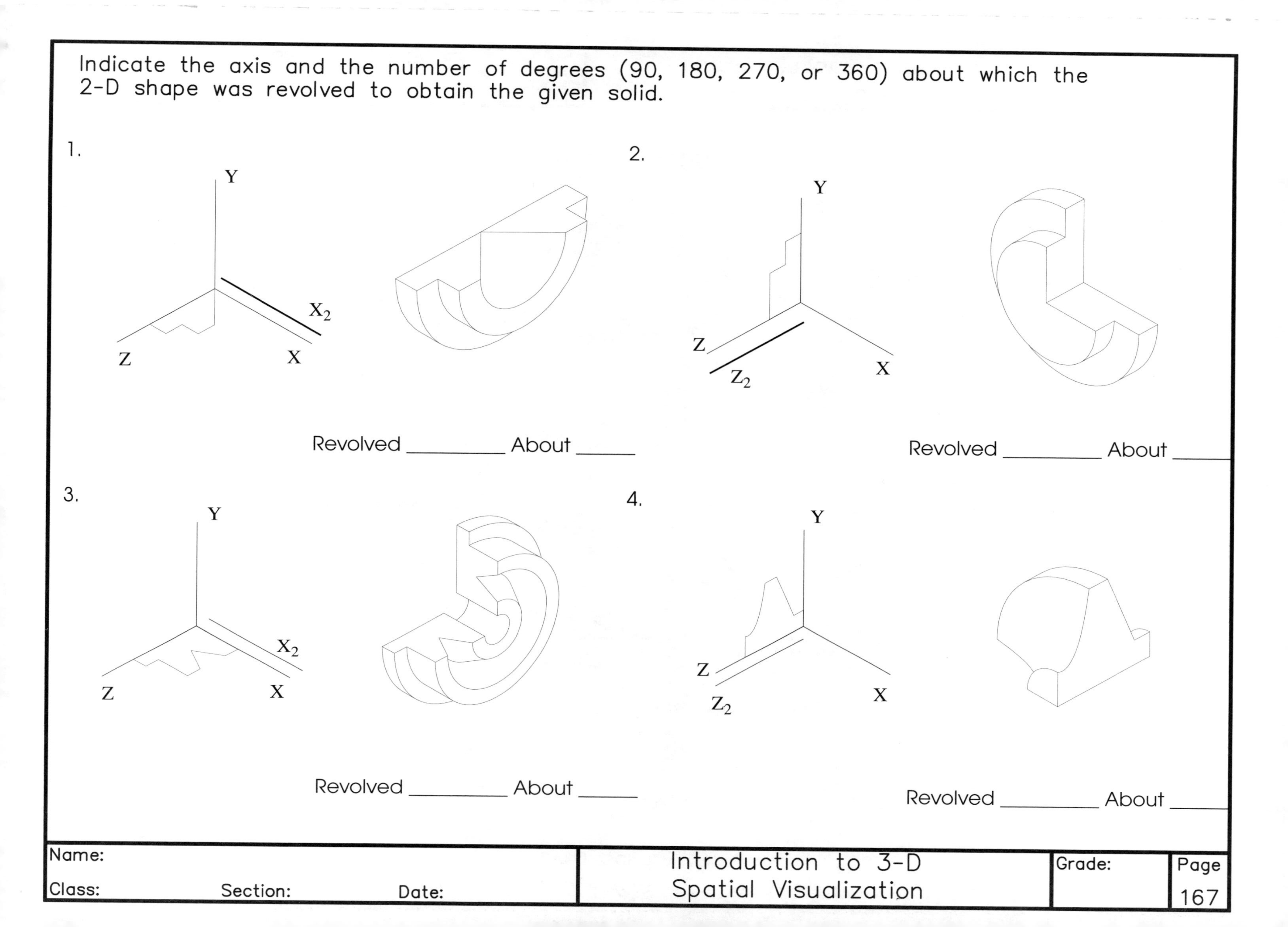


1. Revolved ________ About _____

2. Revolved ________ About _____

3. Revolved ________ About _____

4. Revolved ________ About _____

| Name: | Introduction to 3-D Spatial Visualization | Grade: | Page 167 |
|---|---|---|---|
| Class: Section: Date: | | | |

Indicate the axis and the number of degrees (90, 180, 270, or 360) about which the 2-D shape was revolved to obtain the given solid.

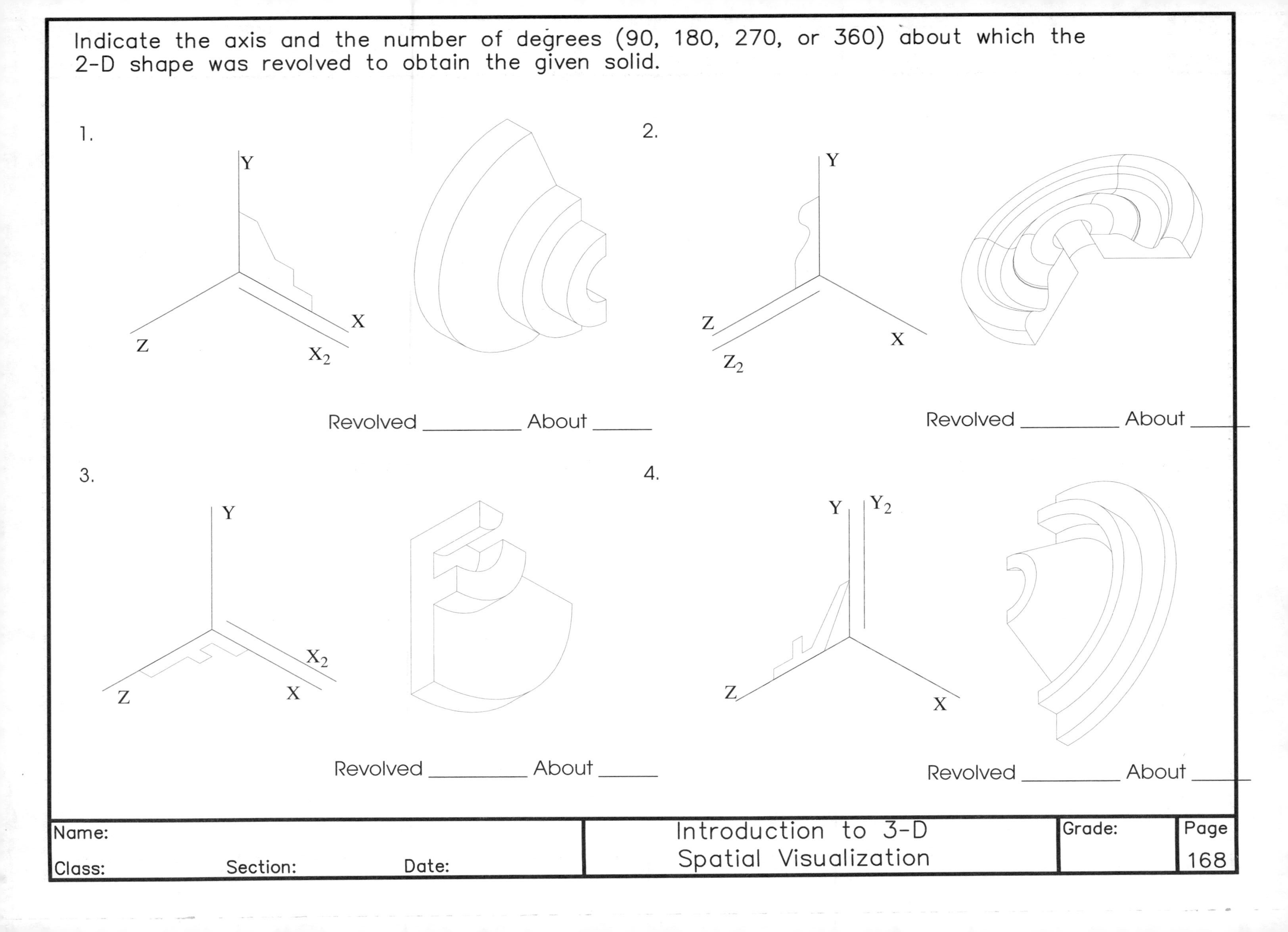


1. Revolved ________ About ______

2. Revolved ________ About ______

3. Revolved ________ About ______

4. Revolved ________ About ______

| Name: | Introduction to 3-D Spatial Visualization | Grade: | Page |
|---|---|---|---|
| Class: Section: Date: | | | 168 |

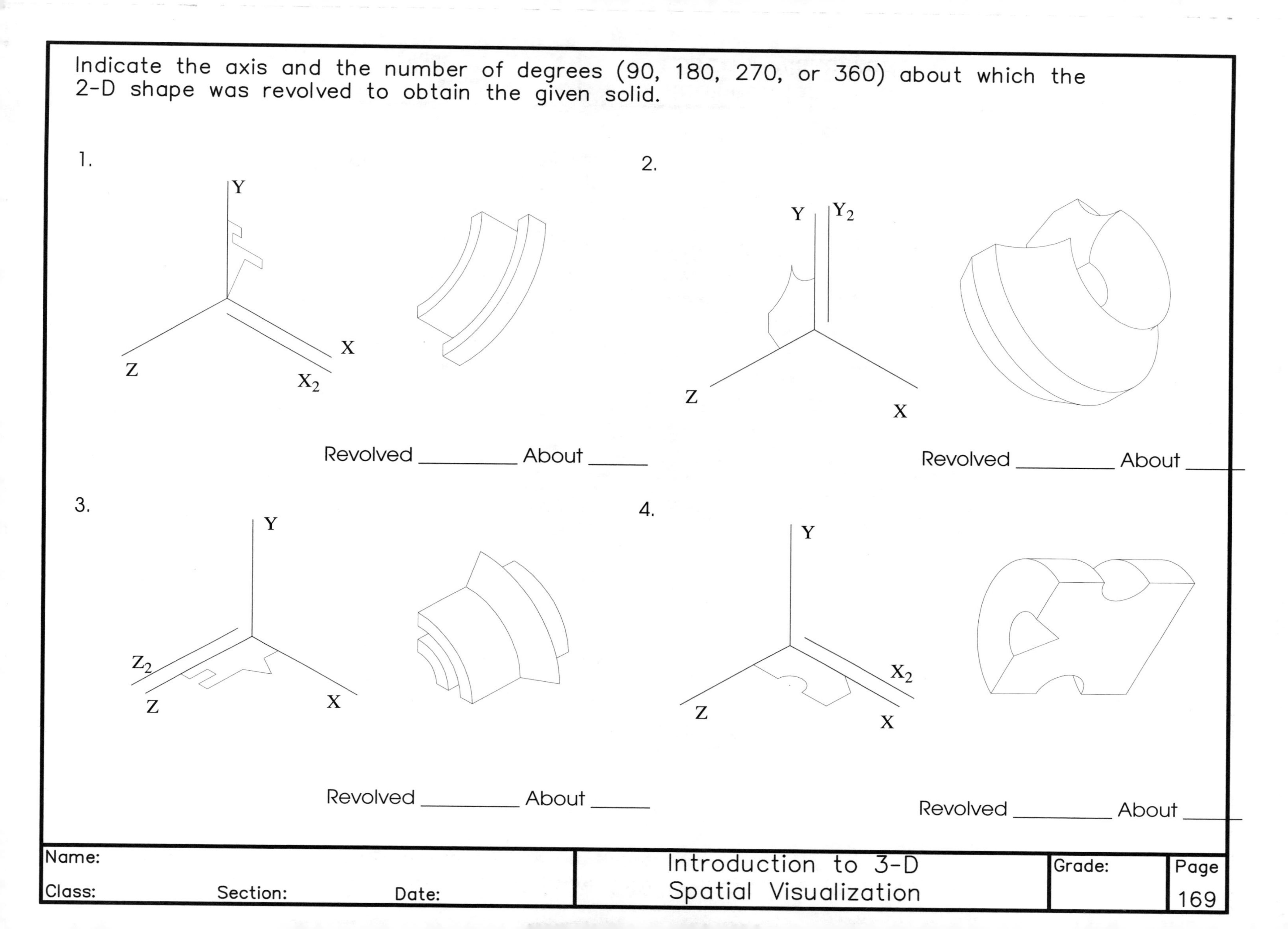
Indicate the axis and the number of degrees (90, 180, 270, or 360) about which the 2-D shape was revolved to obtain the given solid.
1.
Y
Z
X
X2
Revolved ________ About _____
2.
Y
Y2
Z
X
Revolved ________ About _____
3.
Y
Z2
Z
X
Revolved ________ About _____
4.
Y
X2
Z
X
Revolved ________ About _____
Name:
Class:
Section:
Date:
Introduction to 3-D Spatial Visualization
Grade:
Page
169

Indicate the axis and the number of degrees (90, 180, 270, or 360) about which the 2-D shape was revolved to obtain the given solid.

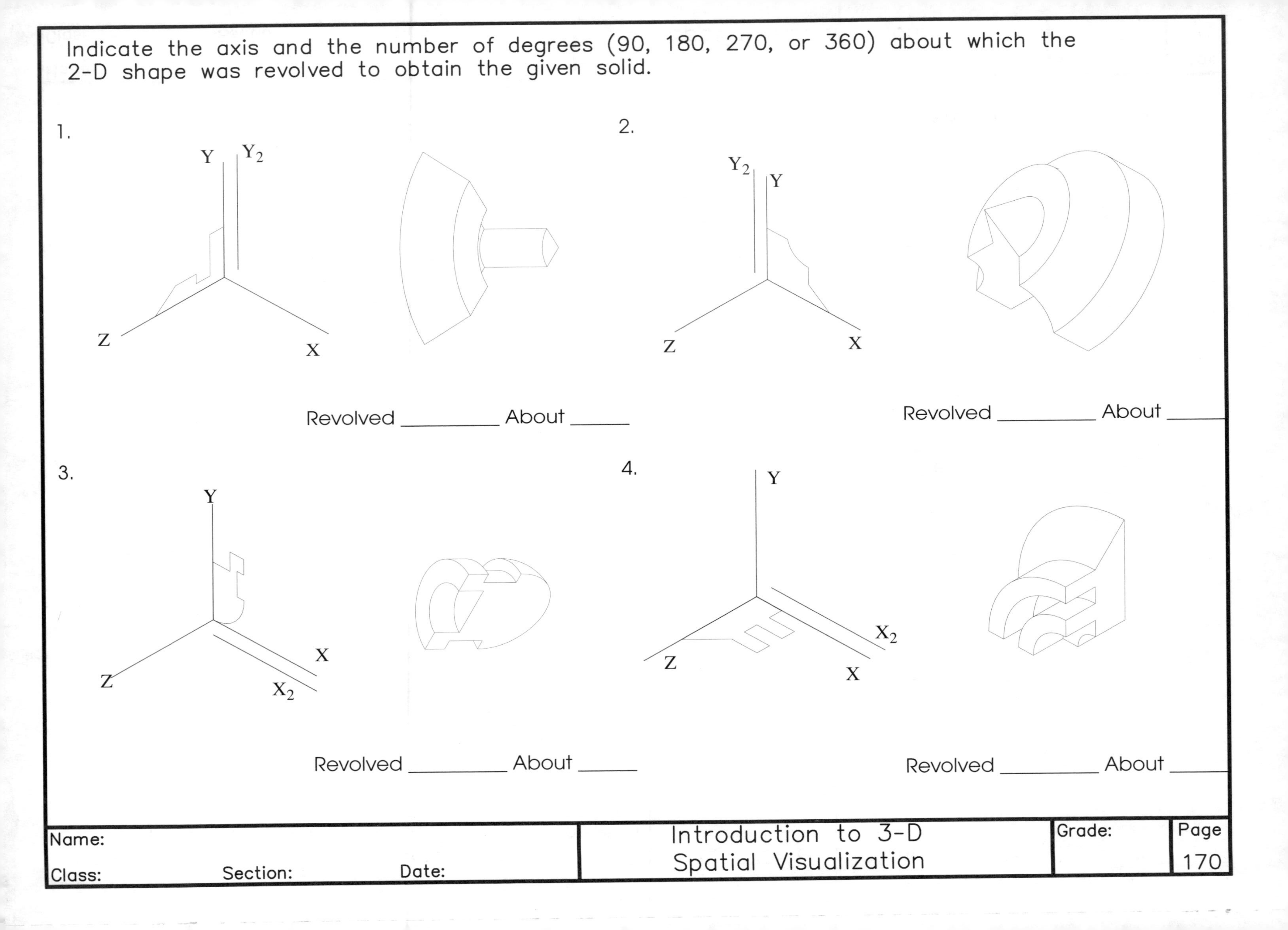


1. Revolved __________ About ______

2. Revolved __________ About ______

3. Revolved __________ About ______

4. Revolved __________ About ______

| Name: | | | Introduction to 3-D Spatial Visualization | Grade: | Page 170 |
|---|---|---|---|---|---|
| Class: | Section: | Date: | | | |

For the objects shown on the left below, circle the letter corresponding to the shape that was revolved to create it.

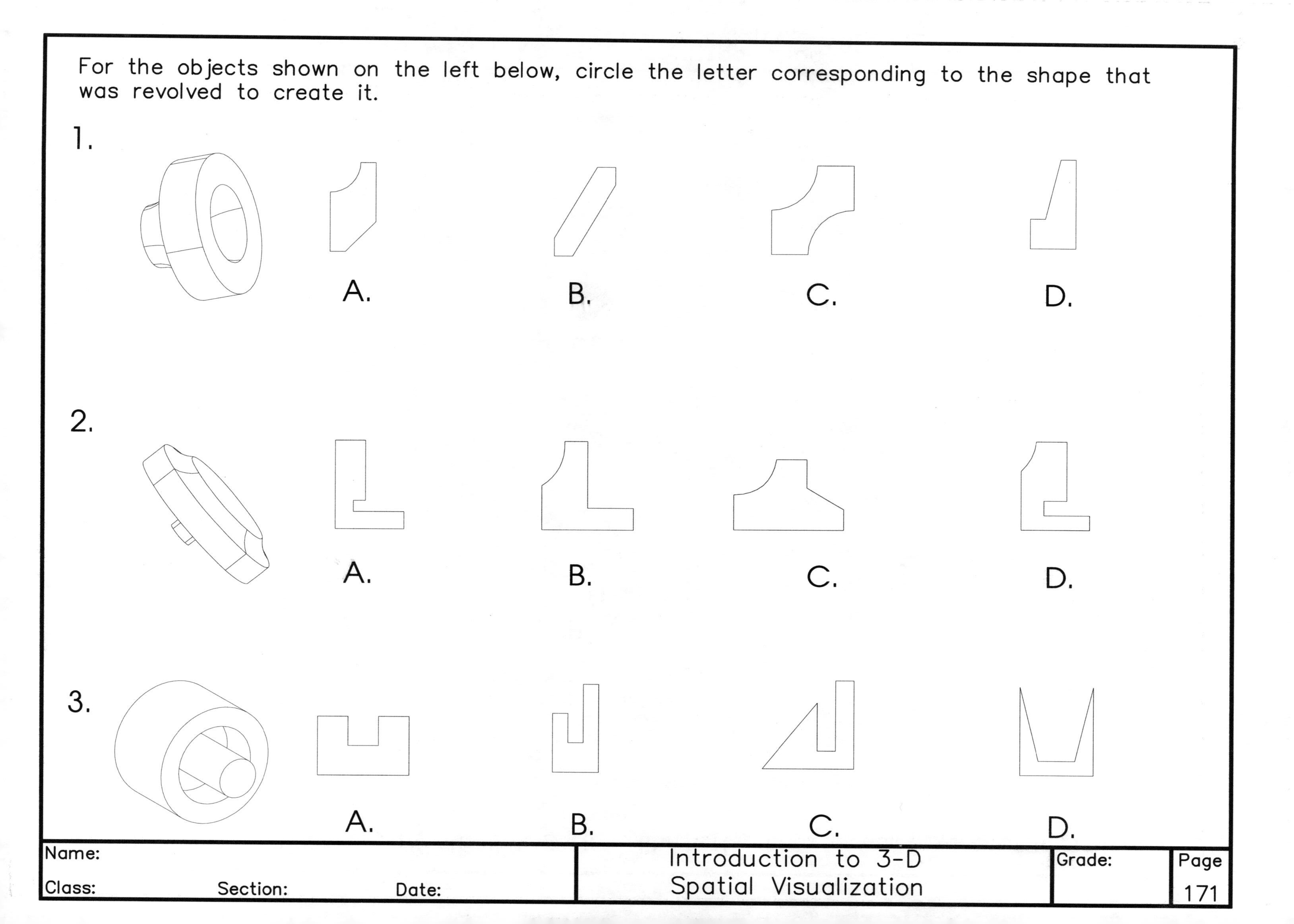

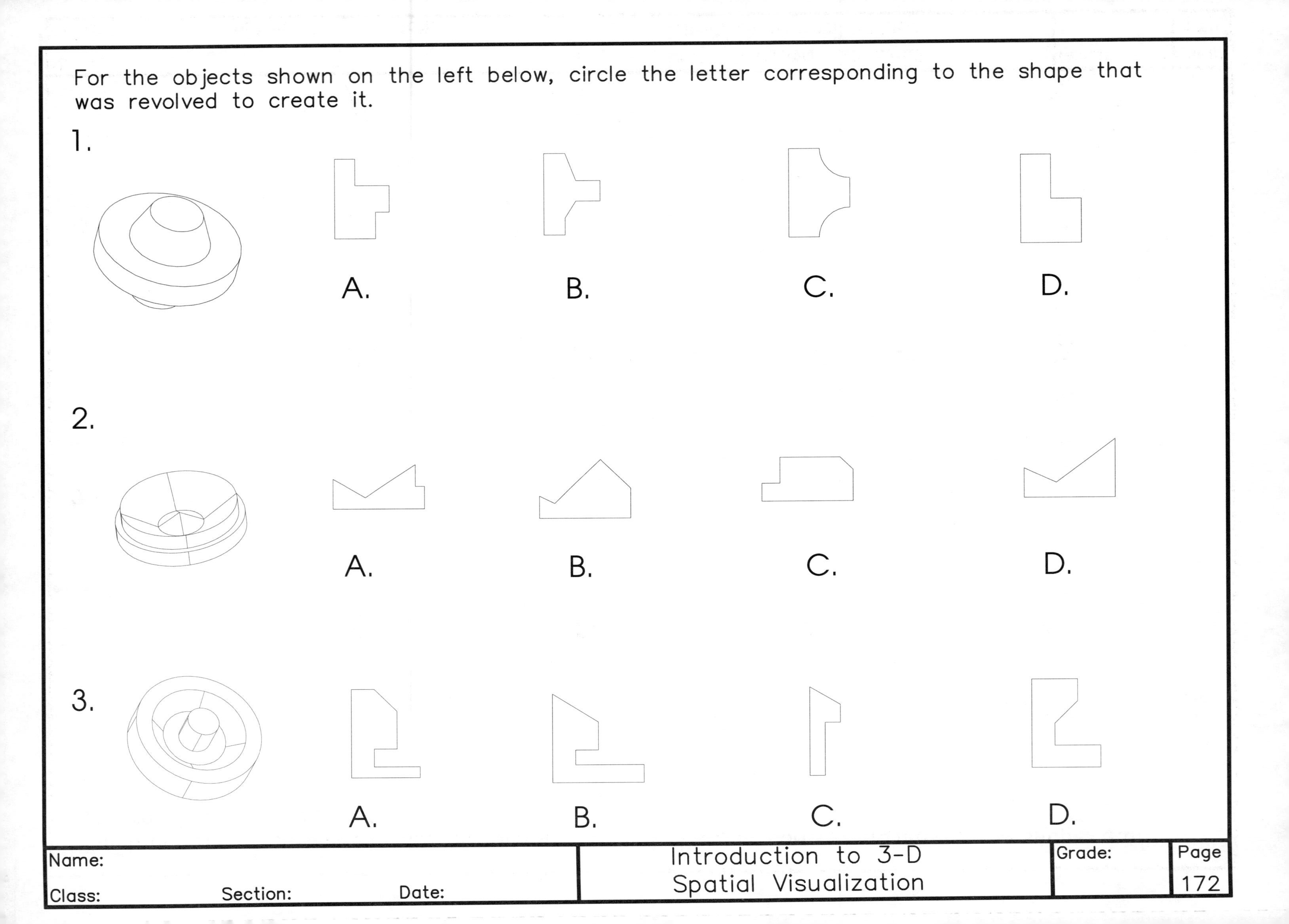

For the objects shown on the left below, circle the letter corresponding to the shape that was revolved to create it.

1.

A. B. C. D.

2.

A. B. C. D.

3.

A. B. C. D.

| Name: | Introduction to 3-D Spatial Visualization | Grade: | Page 172 |
| --- | --- | --- | --- |
| Class: Section: Date: | | | |

For the objects shown on the left below, circle the letter corresponding to the shape that was revolved to create it.

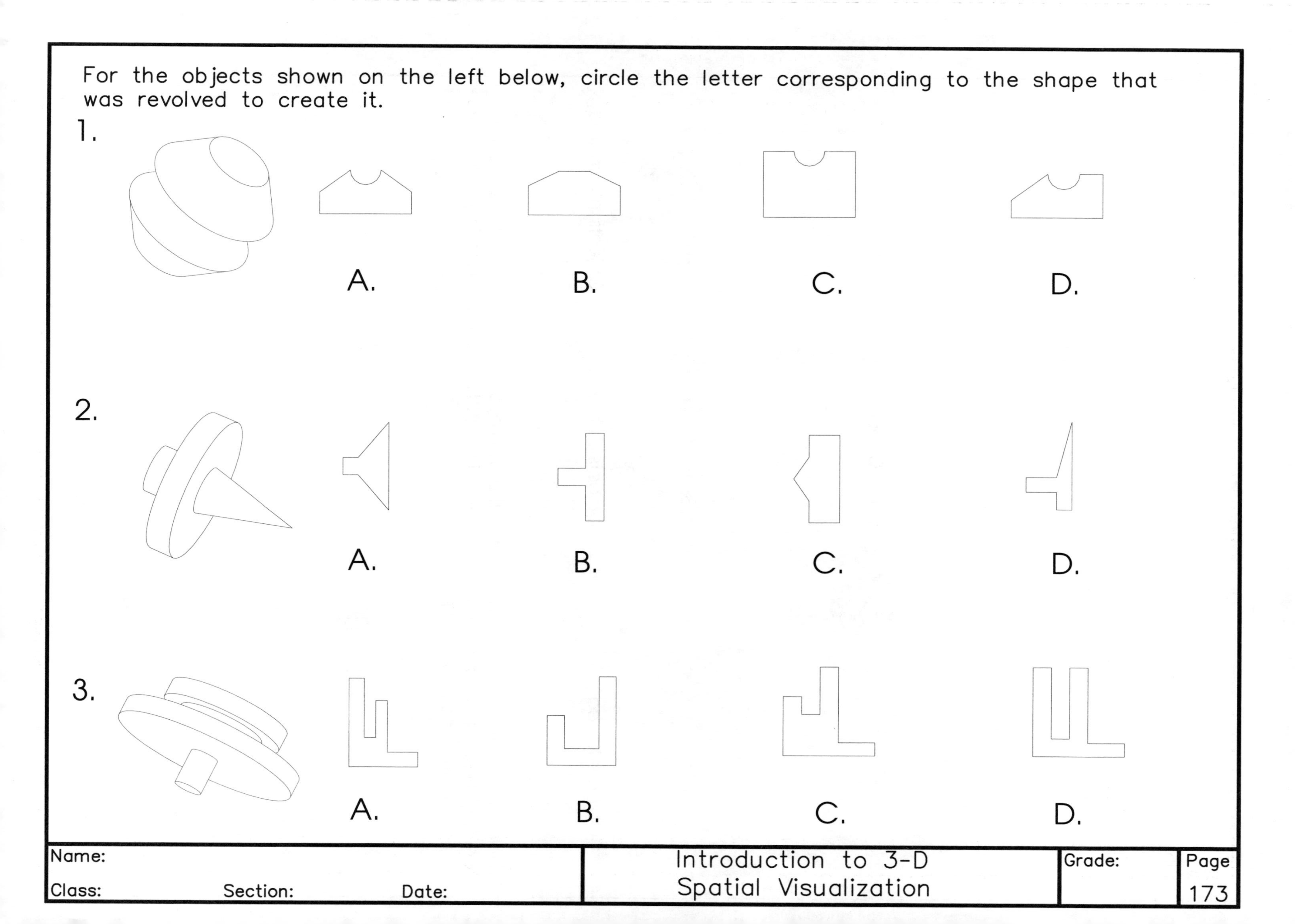

## Combining Solid Objects

Two overlapping objects can be combined to make a new object by joining, cutting, or intersecting.

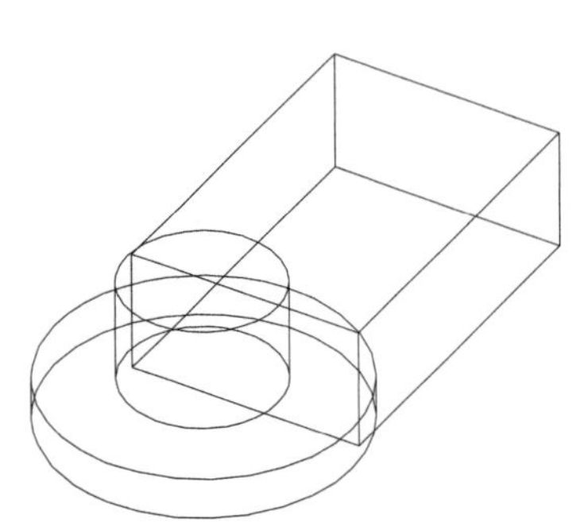

Original Two Objects

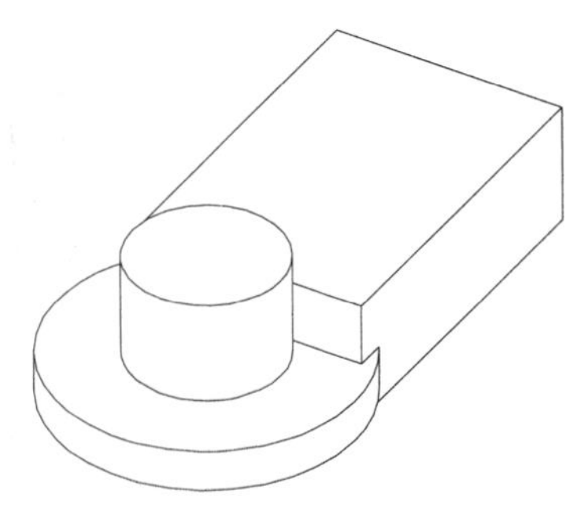

Objects Joined

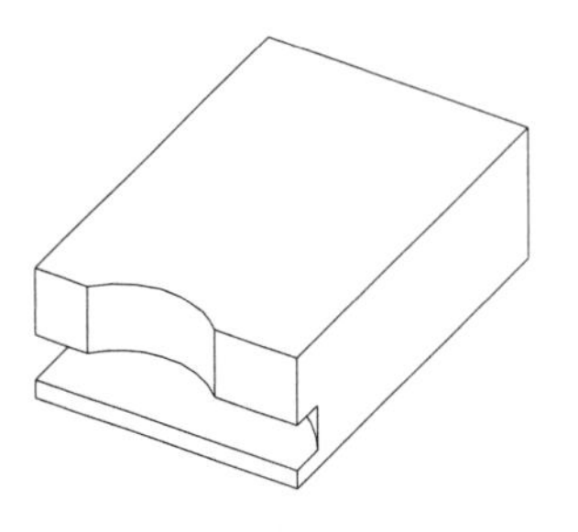

Objects Cut

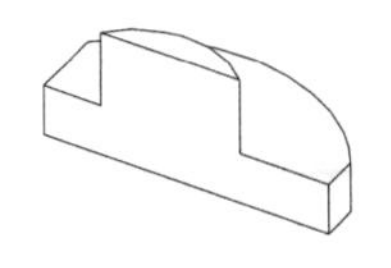

Objects Intersected

The volume of interference is the volume that two overlapping objects have in common. The different combining operations--joining, cutting, and intersecting, use the volume of interference in different ways.

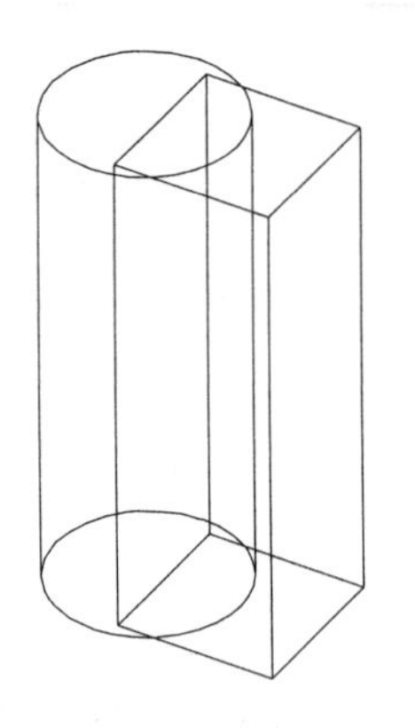

Two Overlapping Objects

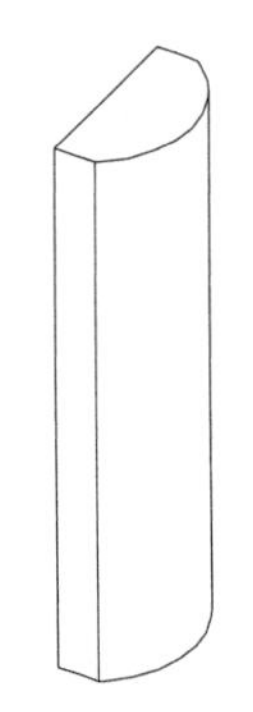

Volume of Interference

When two objects are joined, they become a single object. The volume of interference is absorbed into the resulting object.

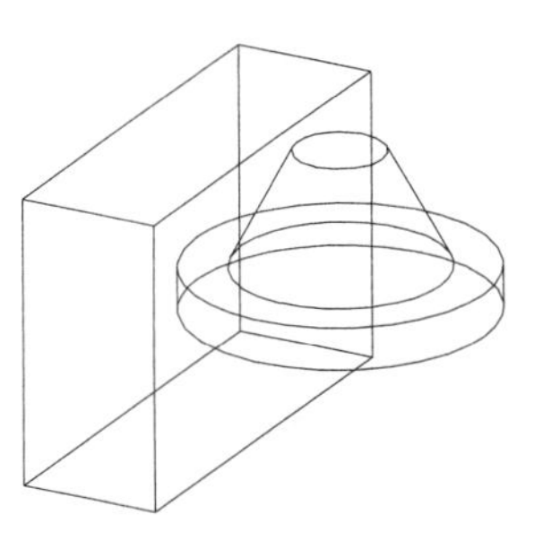

Two Overlapping Objects

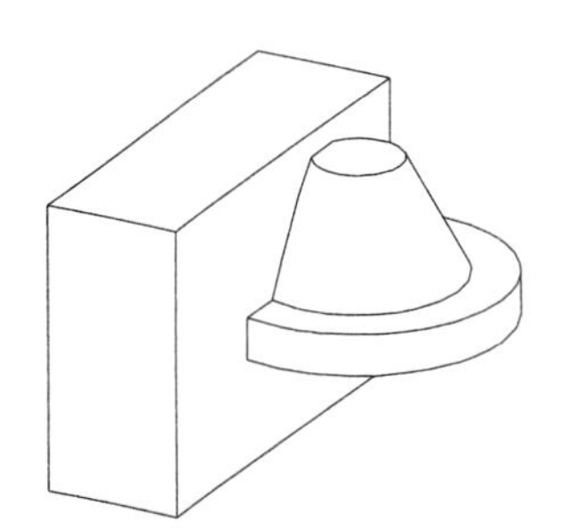

Objects Joined

When two objects are combined by cutting, the volume of interference is removed from the object being cut.

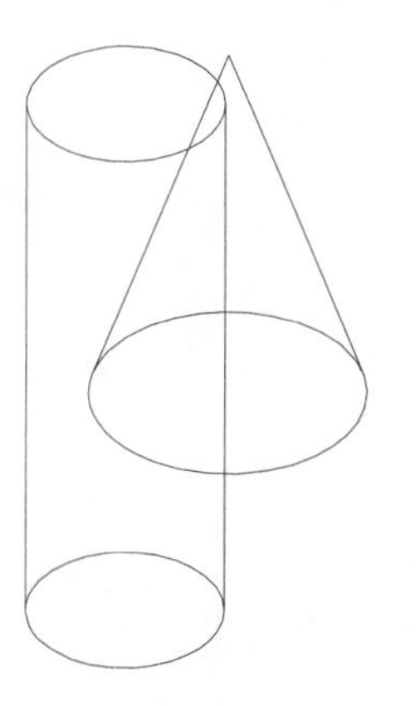

Two Overlapping Objects

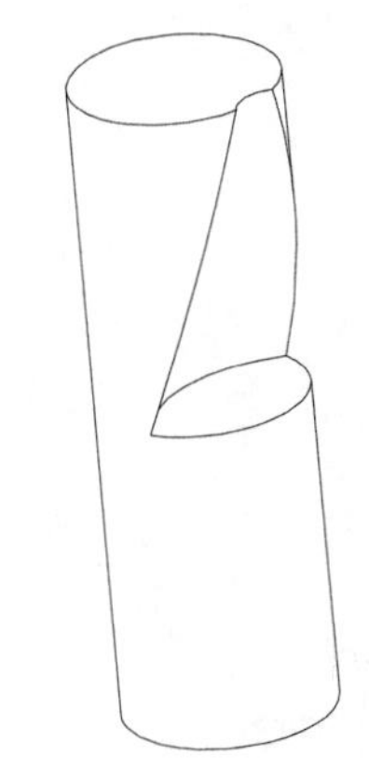

Objects Cut

In a cutting operation, one object acts as a cutting tool on the other object. The final result depends on which object is designated as the cutting tool and which object is the object being cut.

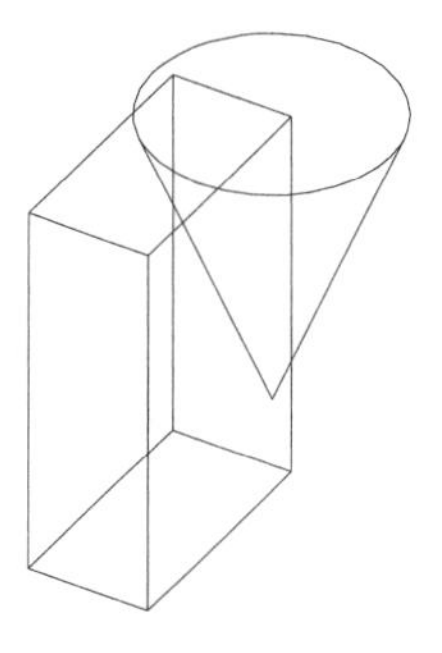

Overlapping Objects

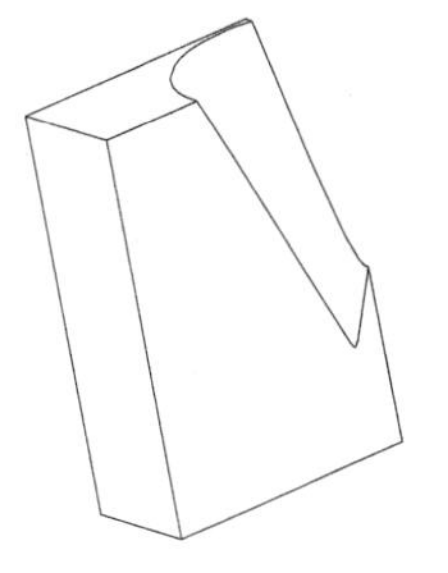

Cone Cuts Block

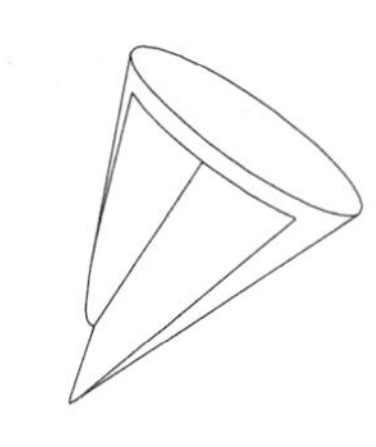

Block Cuts Cone

When an intersect operation is performed, the resulting object consists of the volume that is common to the two original objects (the volume of interference).

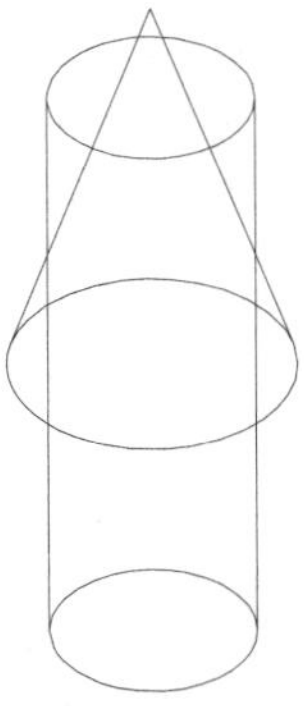

Two Overlapping Objects

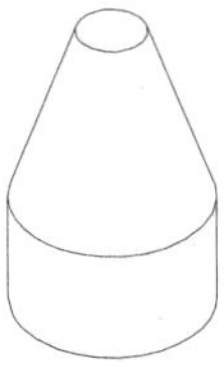

Intersected Objects

More complicated objects can be created by a series of cut/join/intersect operations. In this situation, it is best to look at the final object and think about the building blocks that were part of its creation. The creation of a more complex part is illustrated in the following figures.

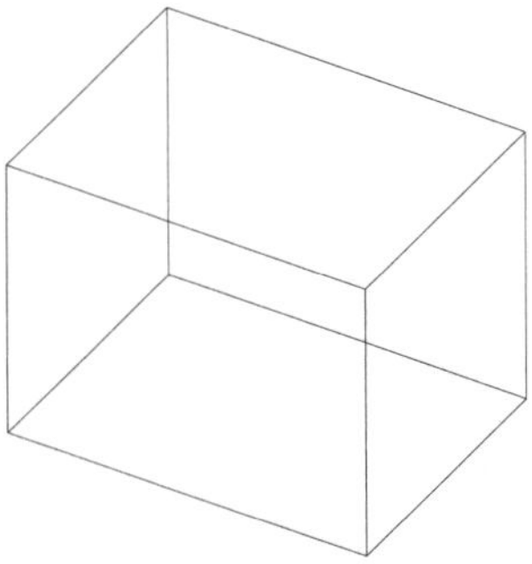

Create Block

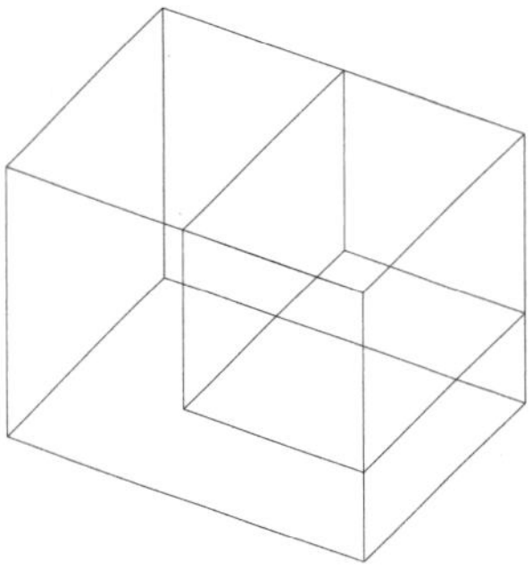

Create Small Block

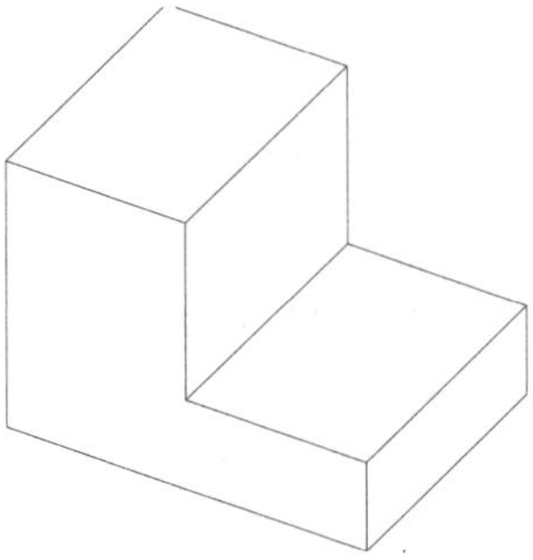

Small Block Cuts Larger Block

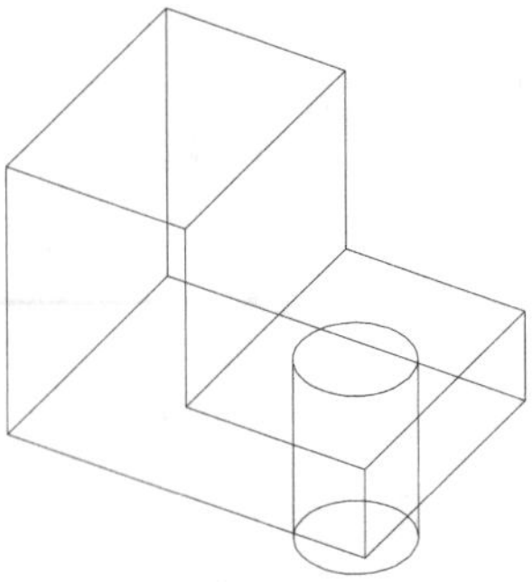

Create Cylinder

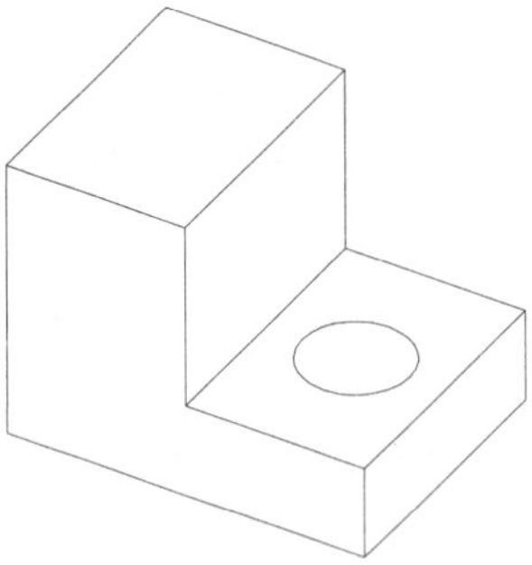

Cylinder Cuts Object

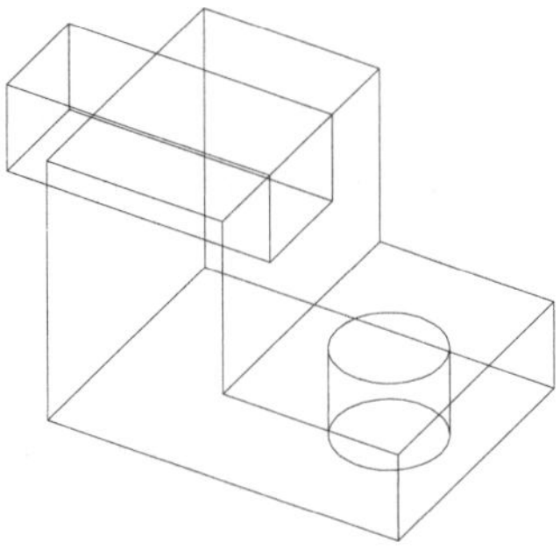

Create Small Block

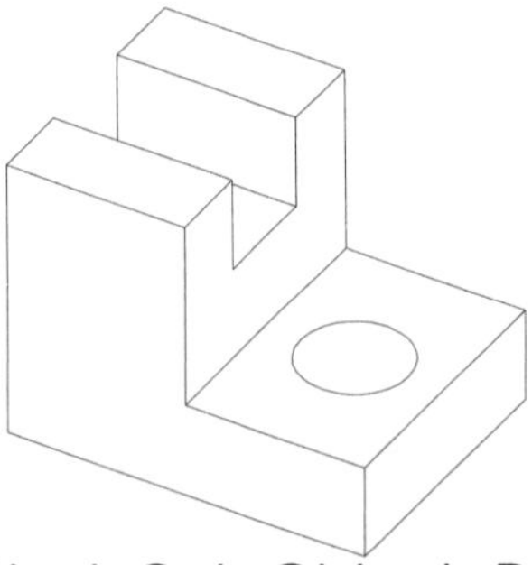

Small Block Cuts Object--Desired Result

Sometimes you will be presented with two overlapping objects and must try to visualize the result of combining the objects. If you think about the edges and surfaces that define the object and then try to imagine which of them will be part of the final object and which part will be removed, you can begin to fill in the shape of the result of the combining operation. In the following figure, two overlapping cylinders are shown. Also shown are the results of a join, an intersect and two cut operations (the larger cylinder cutting the smaller one and vice versa).

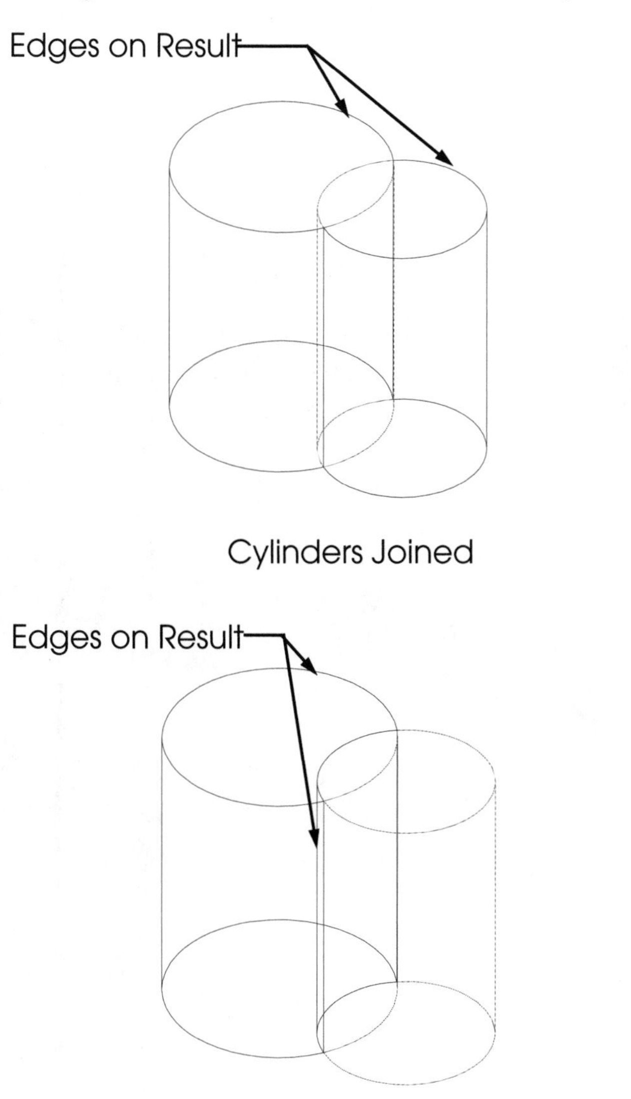


Cylinders Joined

Small Cylinder Cuts Large Cylinder

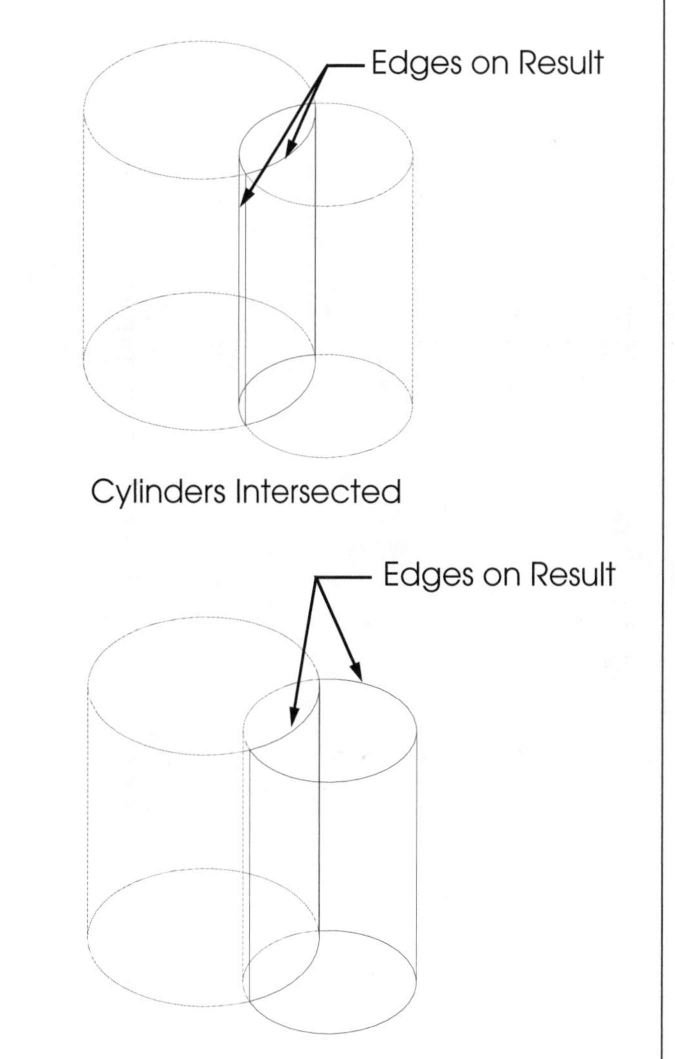


Cylinders Intersected

Large Cylinder Cuts Small Cylinder

The objects shown on the left are to be combined, with the result shown on the right. Circle the appropriate word (cut, join, or intersect) indicating the operation that was performed.

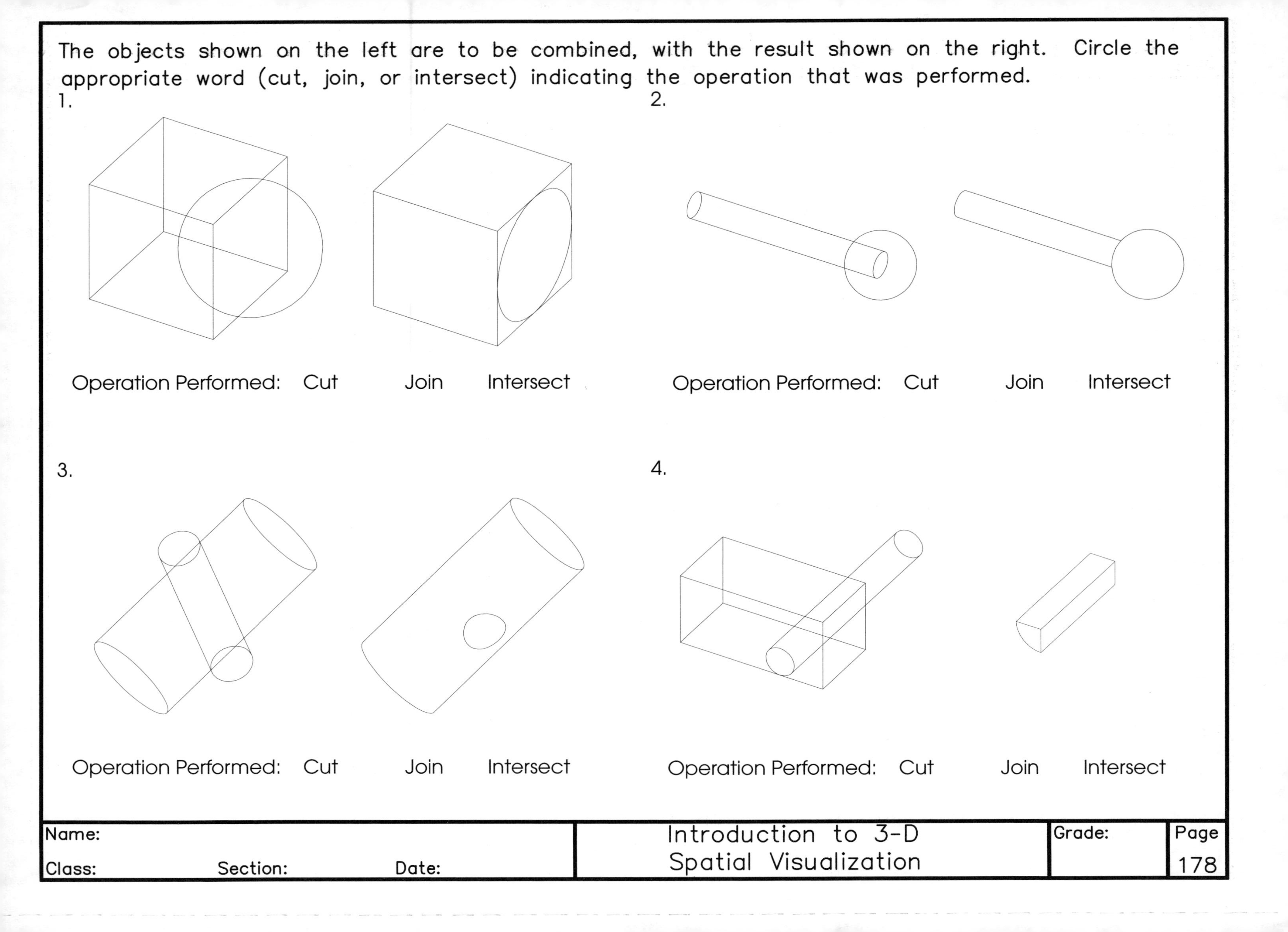

1.

Operation Performed: Cut Join Intersect

2.

Operation Performed: Cut Join Intersect

3.

Operation Performed: Cut Join Intersect

4.

Operation Performed: Cut Join Intersect

| Name: | Introduction to 3-D Spatial Visualization | Grade: | Page |
|---|---|---|---|
| Class: Section: Date: | | | 178 |

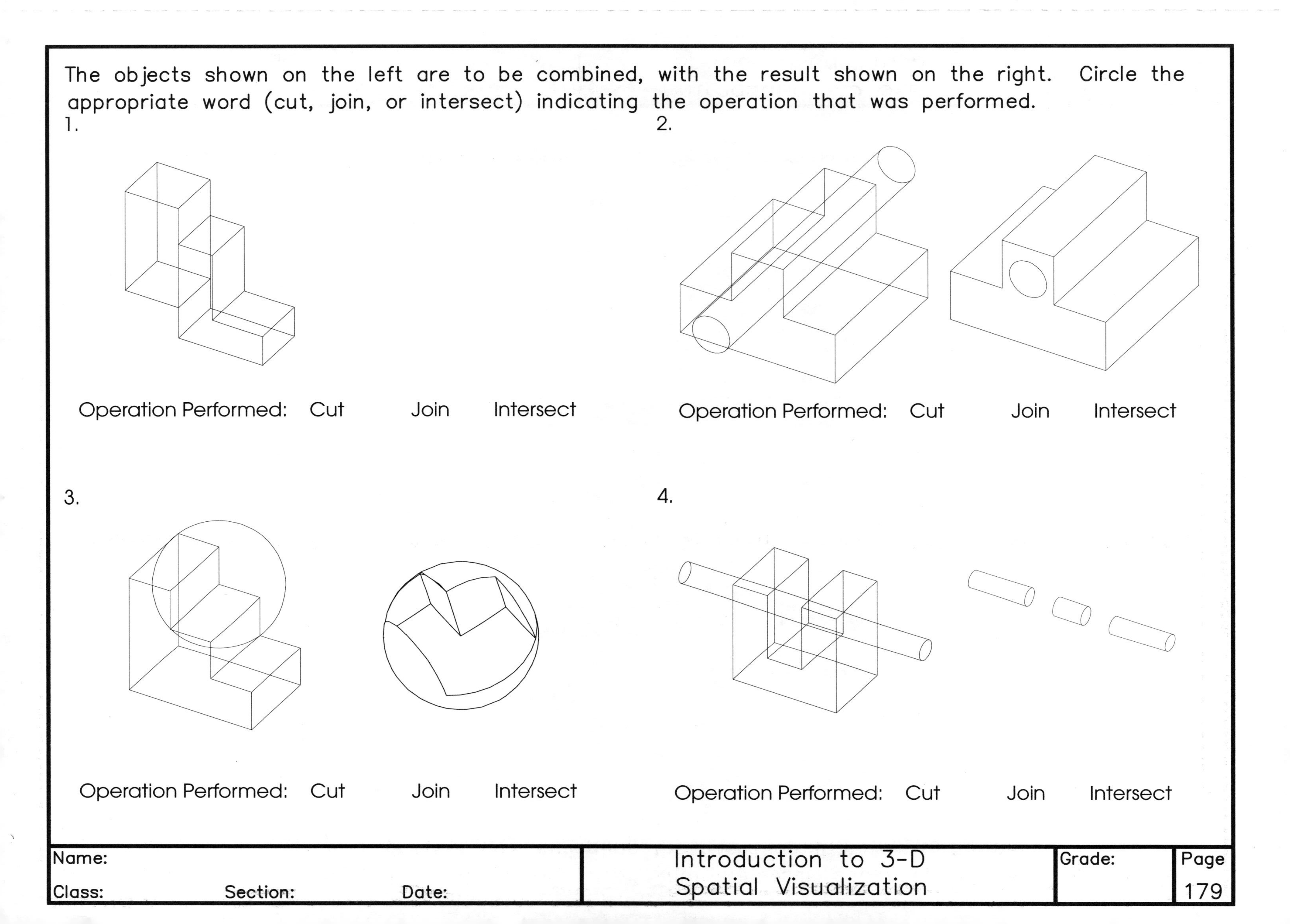

The objects shown on the left are to be combined, with the result shown on the right. Circle the appropriate word (cut, join, or intersect) indicating the operation that was performed.

1.

Operation Performed: Cut Join Intersect

2.

Operation Performed: Cut Join Intersect

3.

Operation Performed: Cut Join Intersect

4.

Operation Performed: Cut Join Intersect

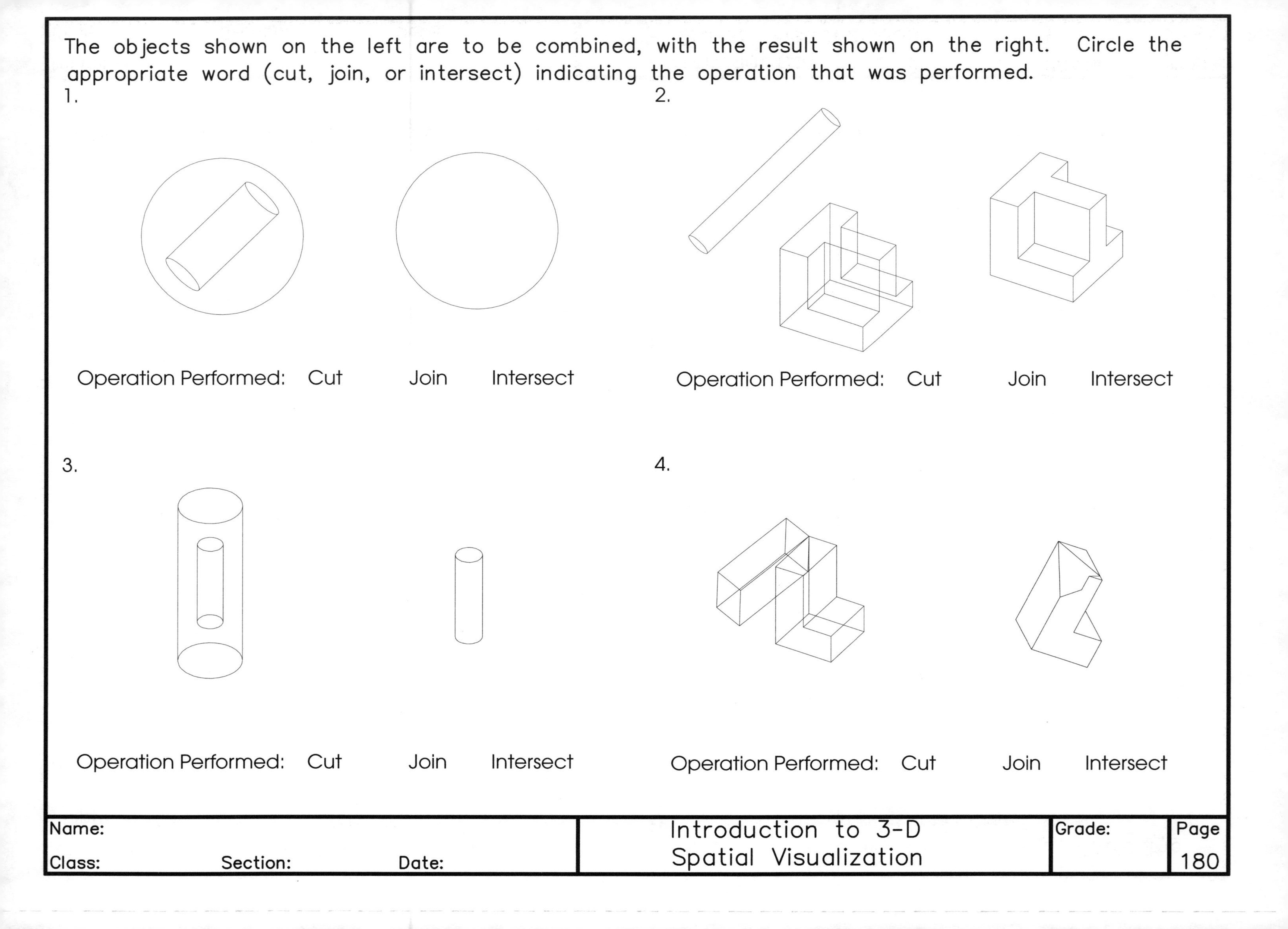

The objects shown on the left are to be combined, with the result shown on the right. Circle the appropriate word (cut, join, or intersect) indicating the operation that was performed.

1.

Operation Performed: Cut Join Intersect

2.

Operation Performed: Cut Join Intersect

3.

Operation Performed: Cut Join Intersect

4.

Operation Performed: Cut Join Intersect

| Name: | Introduction to 3-D | Grade: | Page |
|---|---|---|---|
| Class: Section: Date: | Spatial Visualization | | 180 |

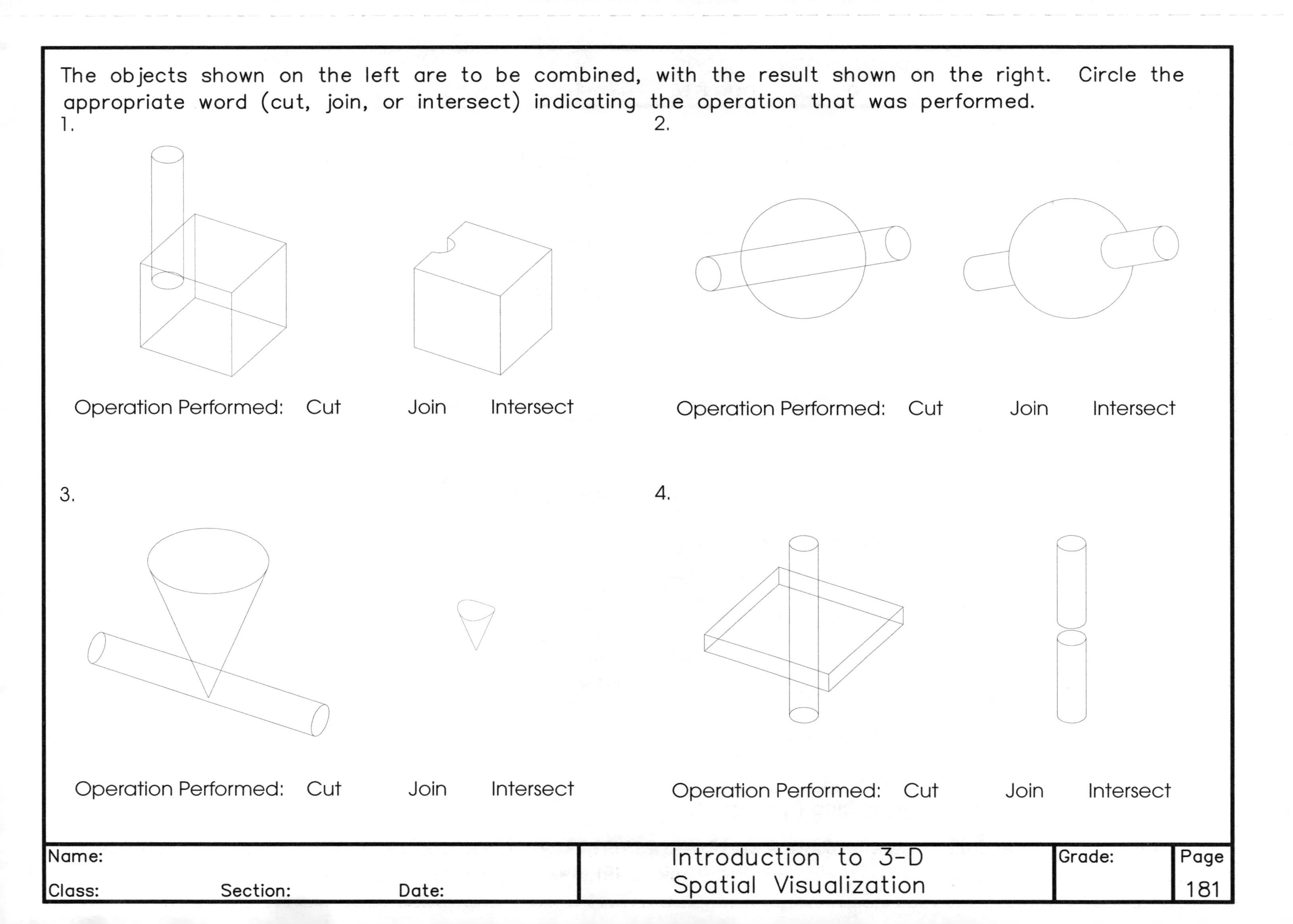

The objects shown on the left are to be combined, with the result shown on the right. Circle the appropriate word (cut, join, or intersect) indicating the operation that was performed.

1.

Operation Performed: Cut Join Intersect

2.

Operation Performed: Cut Join Intersect

3.

Operation Performed: Cut Join Intersect

4.

Operation Performed: Cut Join Intersect

| Name: | Introduction to 3-D Spatial Visualization | Grade: | Page |
|---|---|---|---|
| Class: Section: Date: | | | 181 |

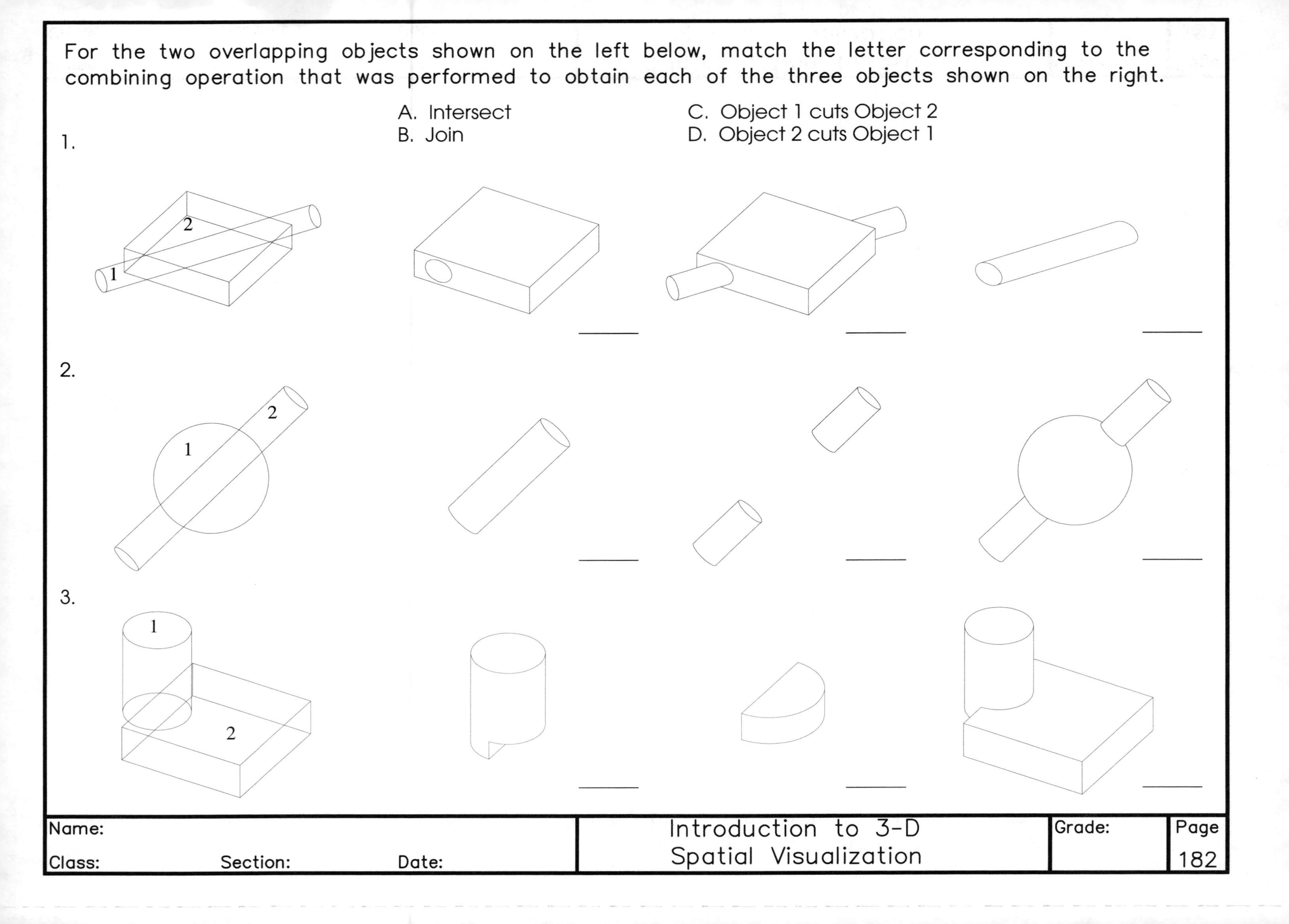

For the two overlapping objects shown on the left below, match the letter corresponding to the combining operation that was performed to obtain each of the three objects shown on the right.

A. Intersect
B. Join
C. Object 1 cuts Object 2
D. Object 2 cuts Object 1

1. 
______ ______ ______

2. 
______ ______ ______

3. 
______ ______ ______

| Name: | Introduction to 3-D | Grade: | Page |
|---|---|---|---|
| Class: Section: Date: | Spatial Visualization | | 182 |

For the two overlapping objects shown on the left below, match the letter corresponding to the combining operation that was performed to obtain each of the three objects shown on the right.

A. Intersect
B. Join
C. Object 1 cuts Object 2
D. Object 2 cuts Object 1

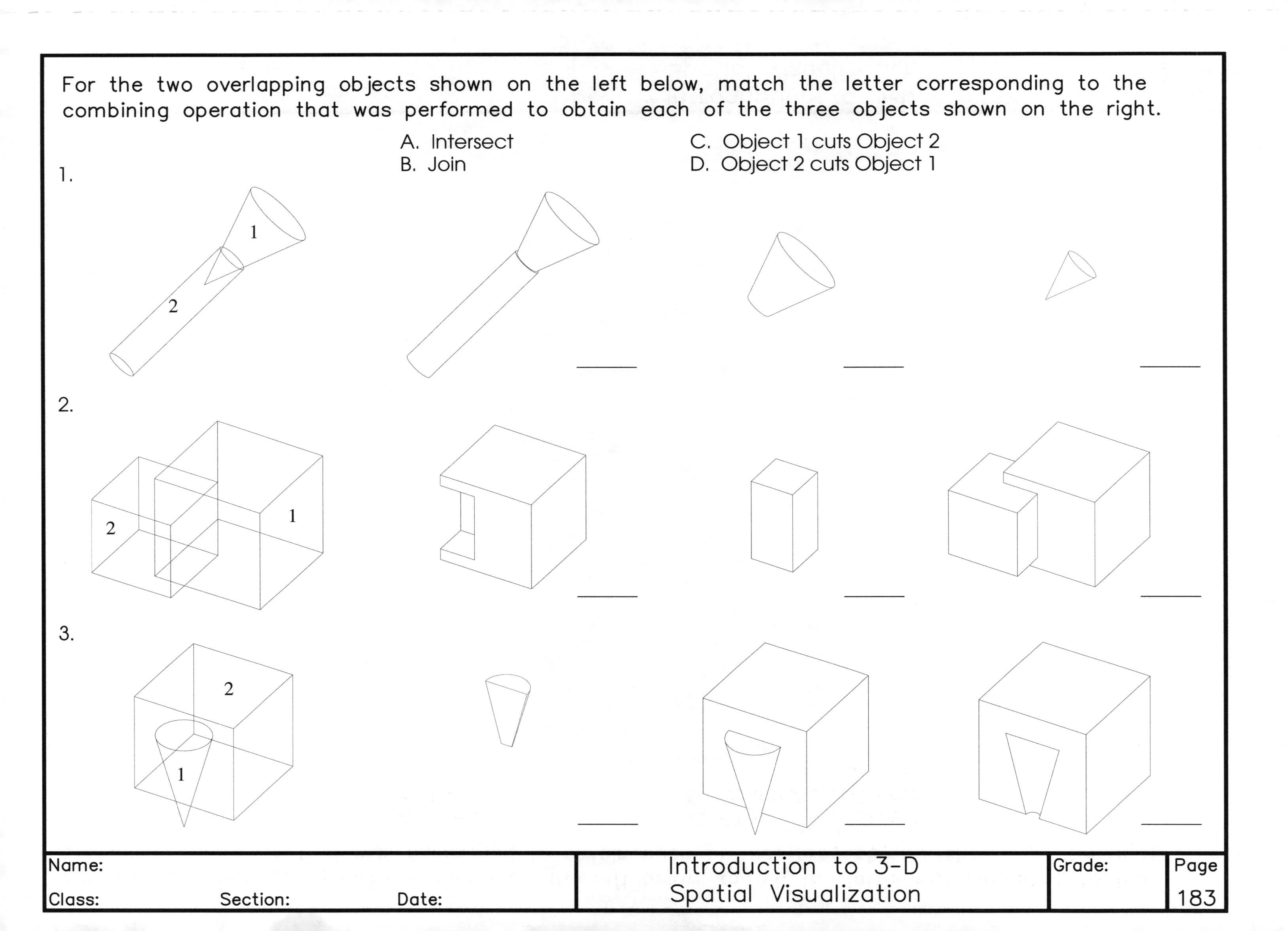


Name:
Class: Section: Date:

Introduction to 3-D Spatial Visualization

Grade:

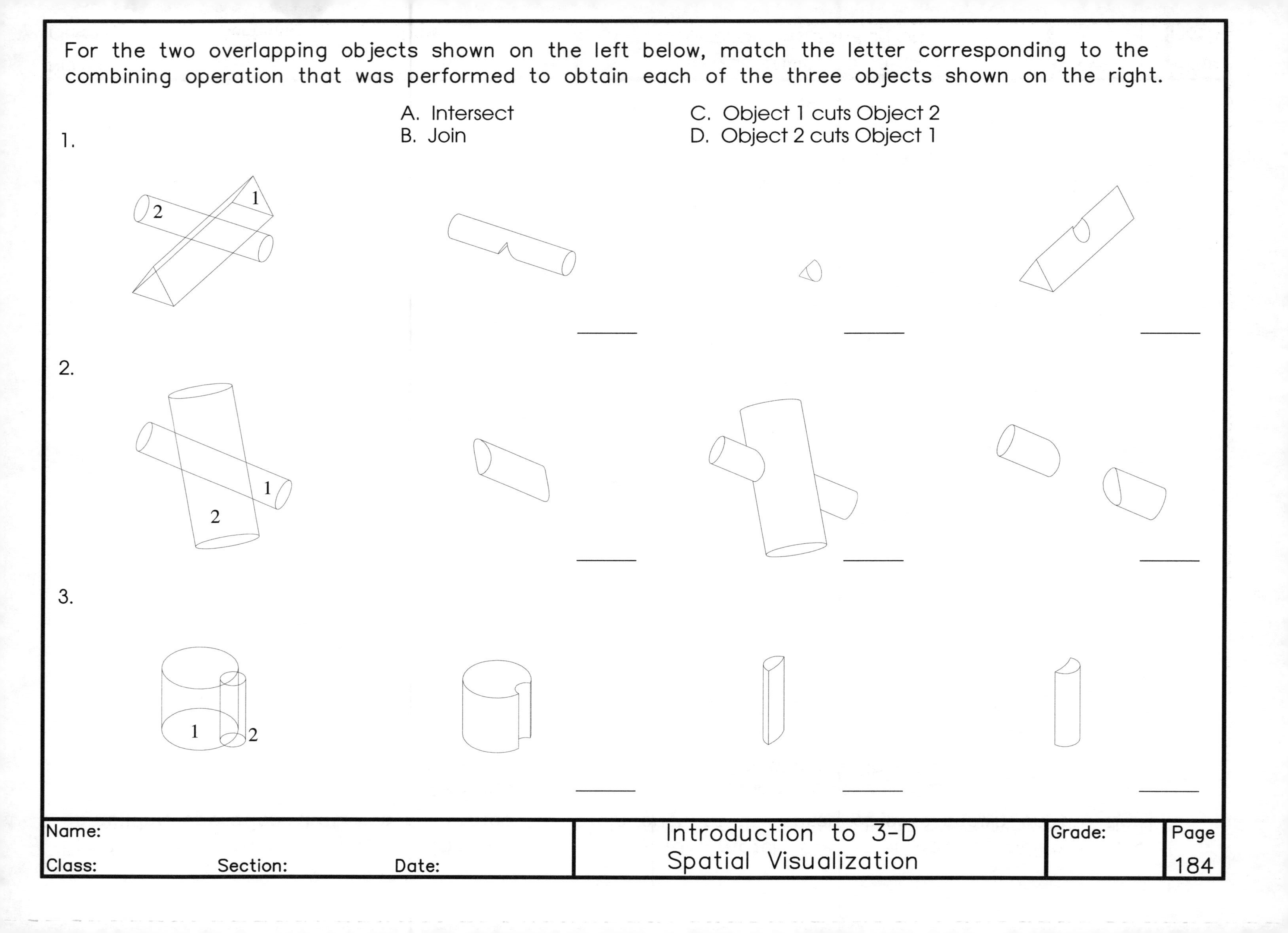

For the two overlapping objects shown on the left below, match the letter corresponding to the combining operation that was performed to obtain each of the three objects shown on the right.

A. Intersect
B. Join
C. Object 1 cuts Object 2
D. Object 2 cuts Object 1

1.



_____ _____ _____

2.



_____ _____ _____

3.



_____ _____ _____

| Name: | Introduction to 3-D | Grade: | Page |
|---|---|---|---|
| Class: Section: Date: | Spatial Visualization | | 184 |

For the two overlapping objects shown on the left below, match the letter corresponding to the combining operation that was performed to obtain each of the three objects shown on the right.

A. Intersect
B. Join

C. Object 1 cuts Object 2
D. Object 2 cuts Object 1

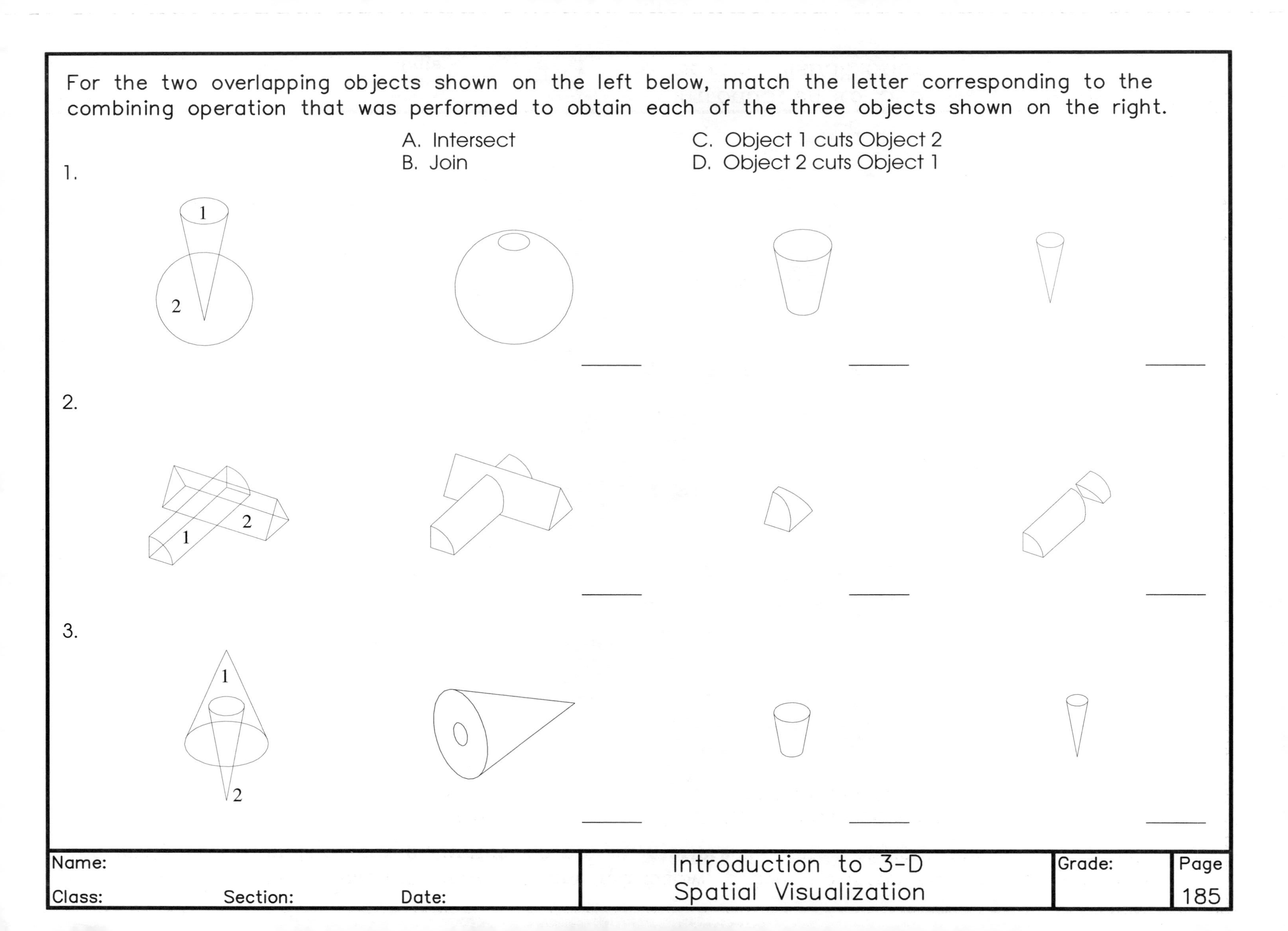


| Name: | Introduction to 3-D | Grade: | Page |
|---|---|---|---|
| Class: Section: Date: | Spatial Visualization | | 185 |

For the two overlapping objects shown on the left below, circle the letter corresponding to the correct volume of interference.

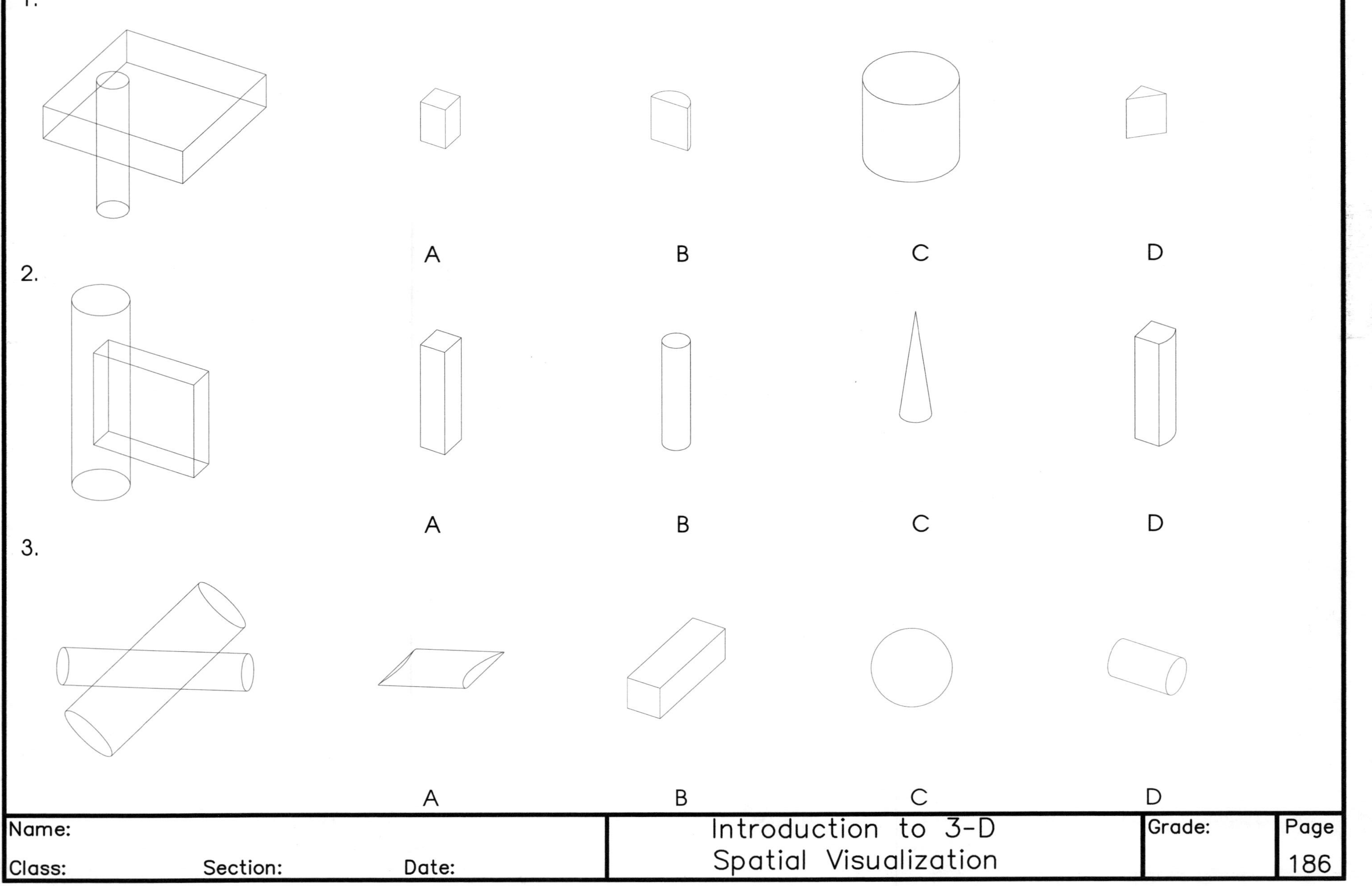


| Name: | Introduction to 3-D | Grade: | Page |
|---|---|---|---|
| Class: Section: Date: | Spatial Visualization | | 186 |

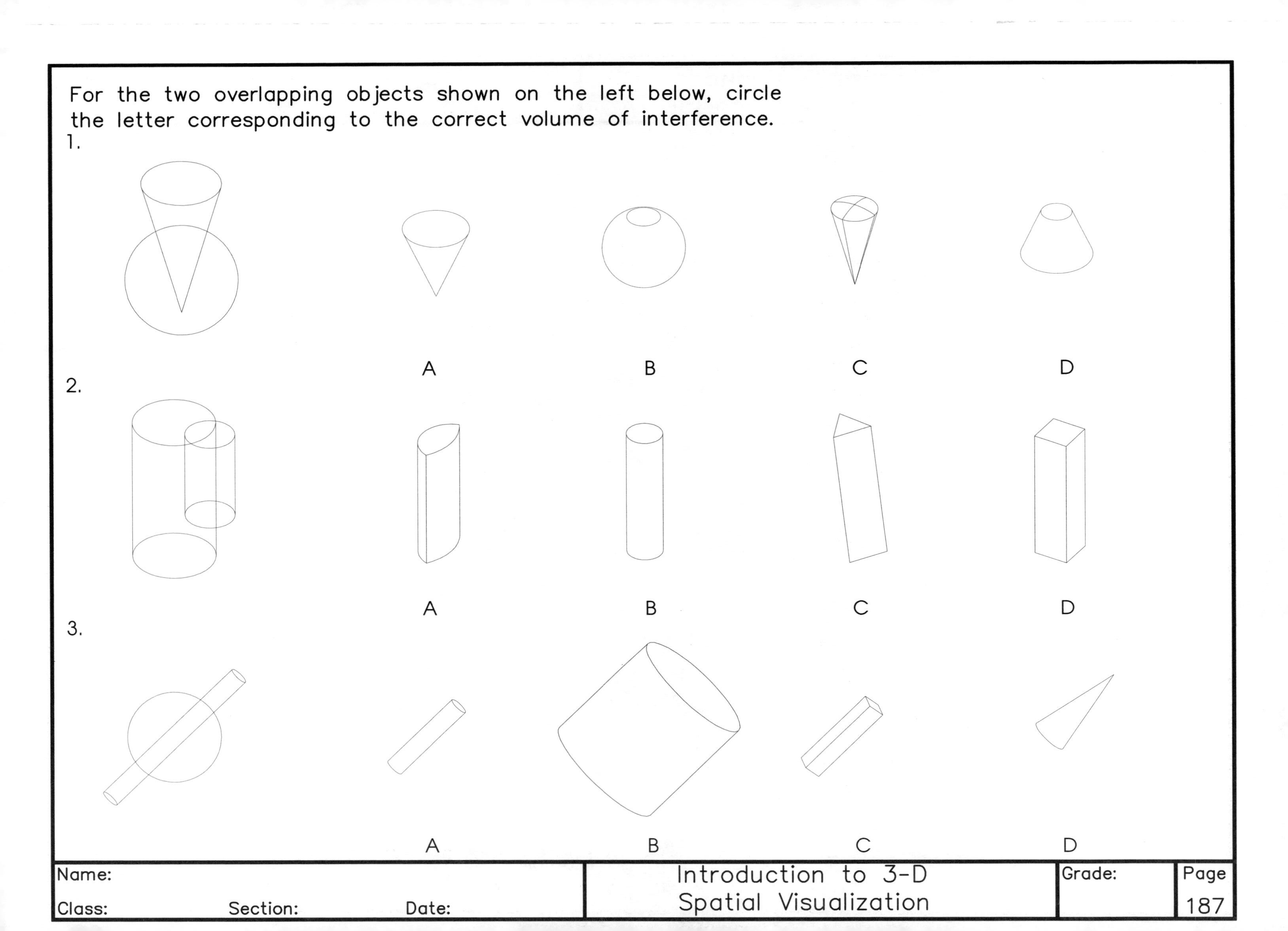
For the two overlapping objects shown on the left below, circle the letter corresponding to the correct volume of interference.
1.
A
B
C
D
2.
A
B
C
D
3.
A
B
C
D
Name:
Class:
Section:
Date:
Introduction to 3-D
Spatial Visualization
Grade:
Page
187

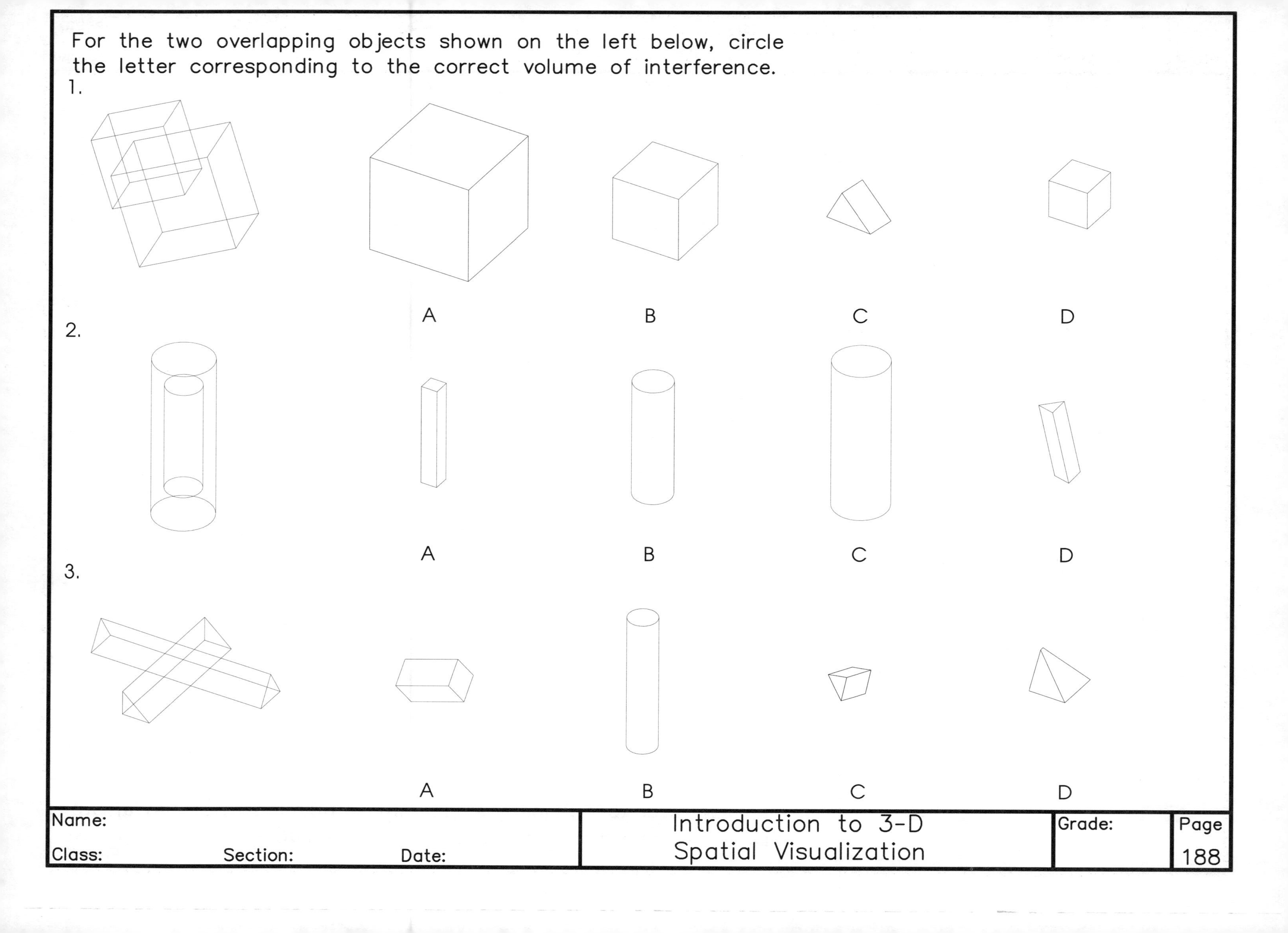
For the two overlapping objects shown on the left below, circle the letter corresponding to the correct volume of interference.
1.
A
B
C
D
2.
A
B
C
D
3.
A
B
C
D
Name:
Class:
Section:
Date:
Introduction to 3-D
Spatial Visualization
Grade:
Page
188

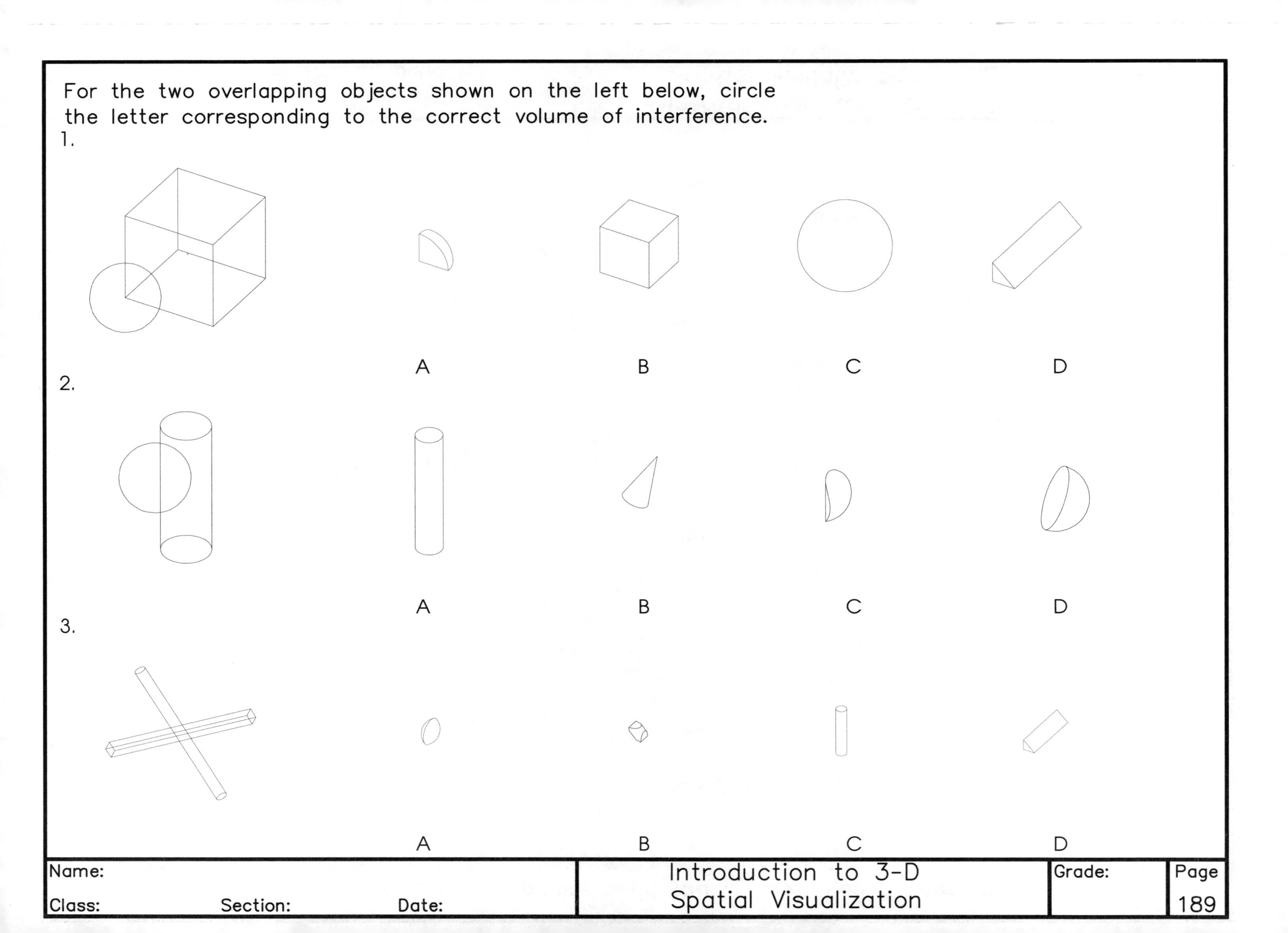
For the two overlapping objects shown on the left below, circle the letter corresponding to the correct volume of interference.
1.
A
B
C
D
2.
A
B
C
D
3.
A
B
C
D
Name:
Class:
Section:
Date:
Introduction to 3-D
Spatial Visualization
Grade:
Page
189

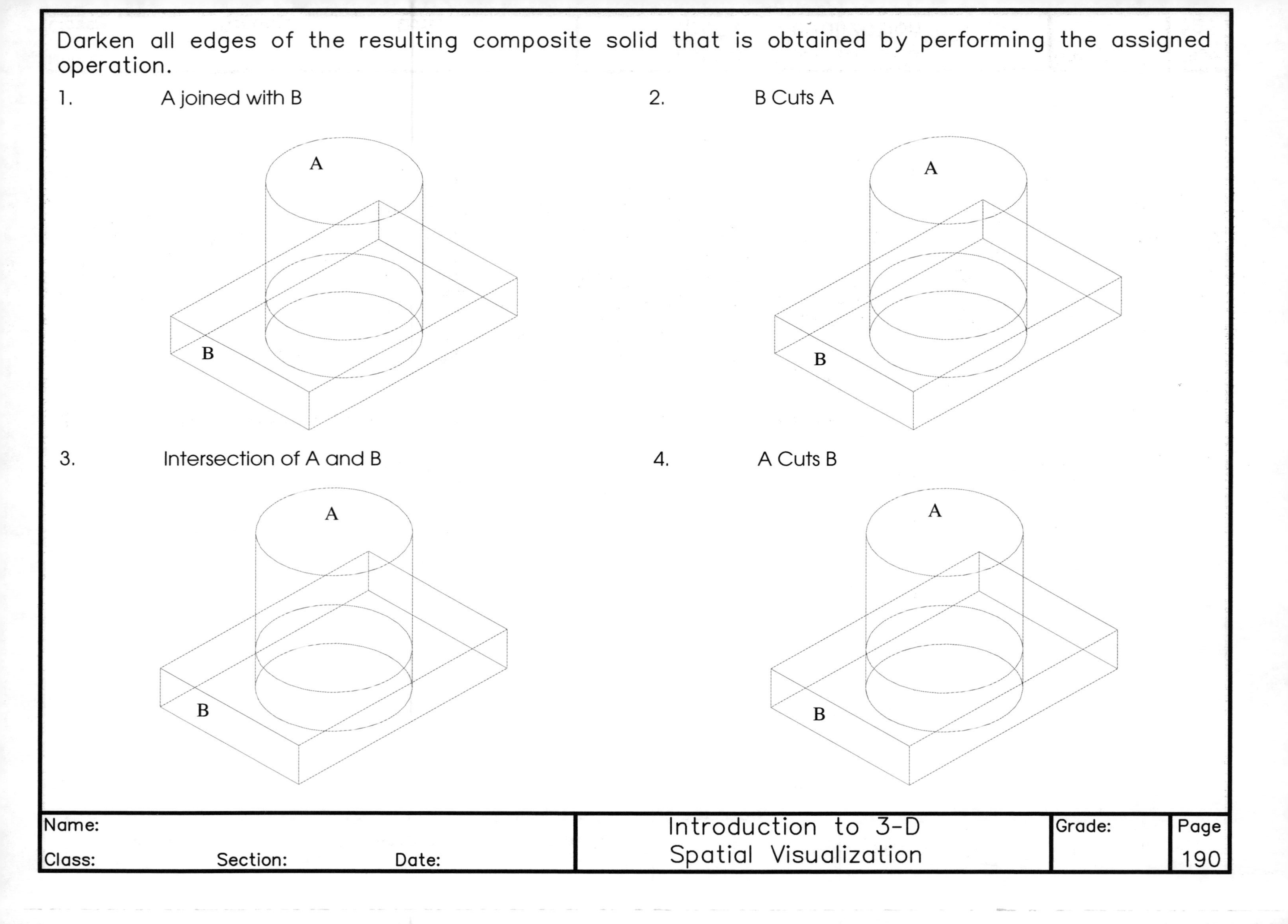

Darken all edges of the resulting composite solid that is obtained by performing the assigned operation.
1. A joined with B
A
B
2. B Cuts A
A
B
3. Intersection of A and B
A
B
4. A Cuts B
A
B
Name:
Class:
Section:
Date:
Introduction to 3-D Spatial Visualization
Grade:
Page
190

Darken all edges of the resulting composite solid that is obtained by performing the asigned operation.

1. A joined with B

2. B Cuts A

3. Intersection of A and B

4. A Cuts B

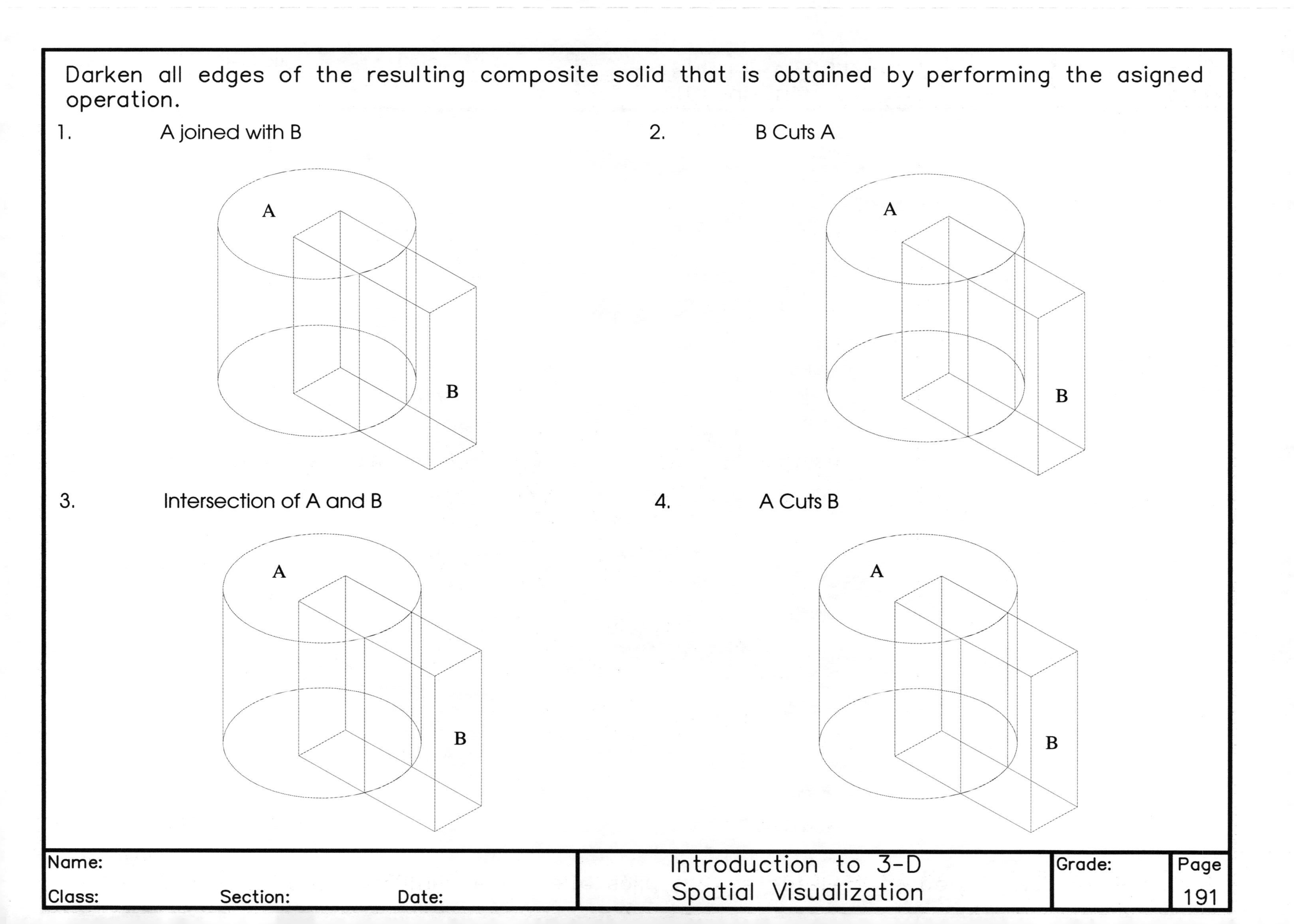


| Name: | | | Introduction to 3-D Spatial Visualization | Grade: | Page |
|---|---|---|---|---|---|
| Class: | Section: | Date: | | | 191 |

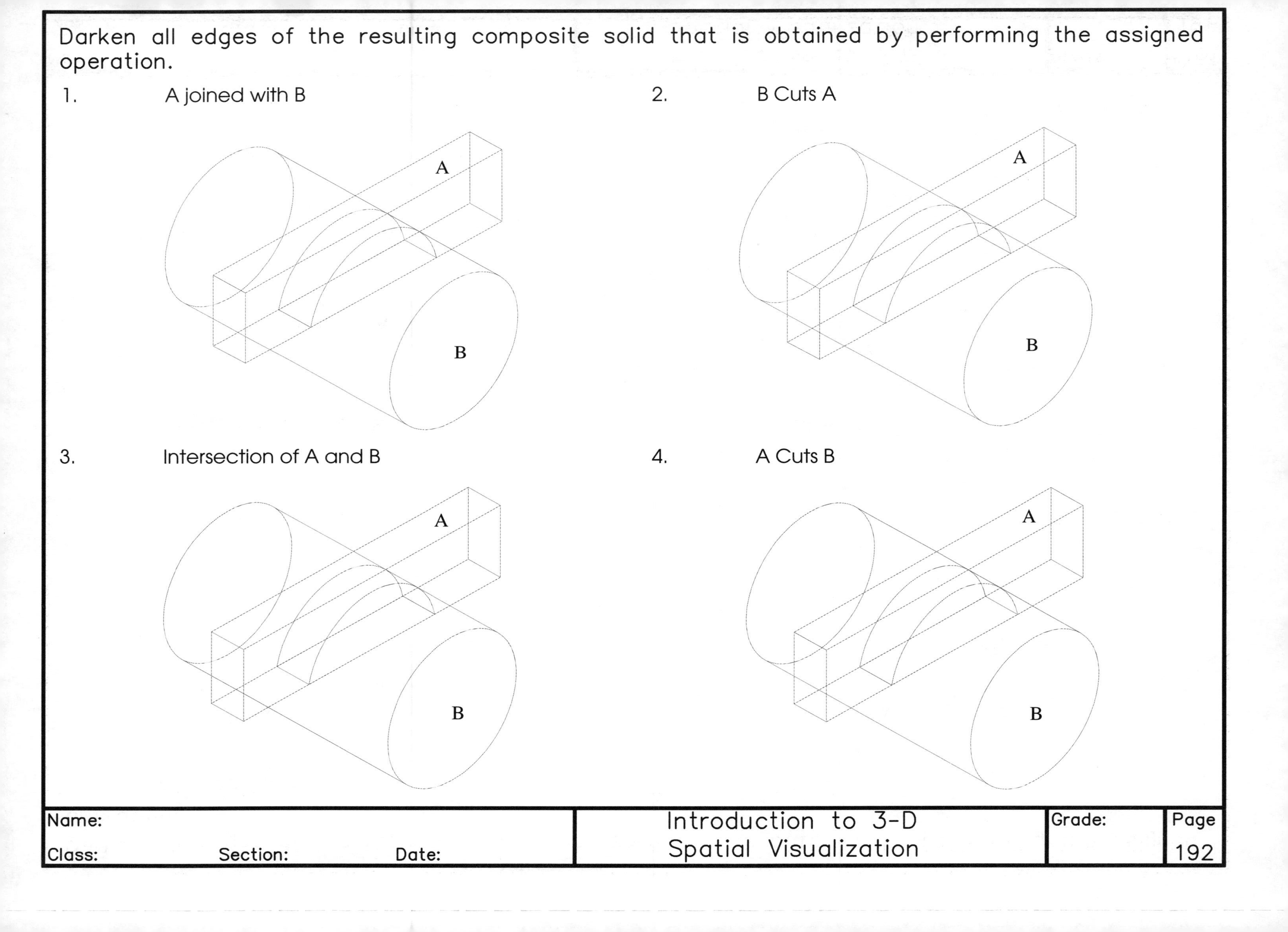

Darken all edges of the resulting composite solid that is obtained by performing the assigned operation.

1. A joined with B



2. B Cuts A



3. Intersection of A and B



4. A Cuts B



| Name:<br>Class: Section: Date: | Introduction to 3-D<br>Spatial Visualization | Grade: | Page<br>192 |
|---|---|---|---|

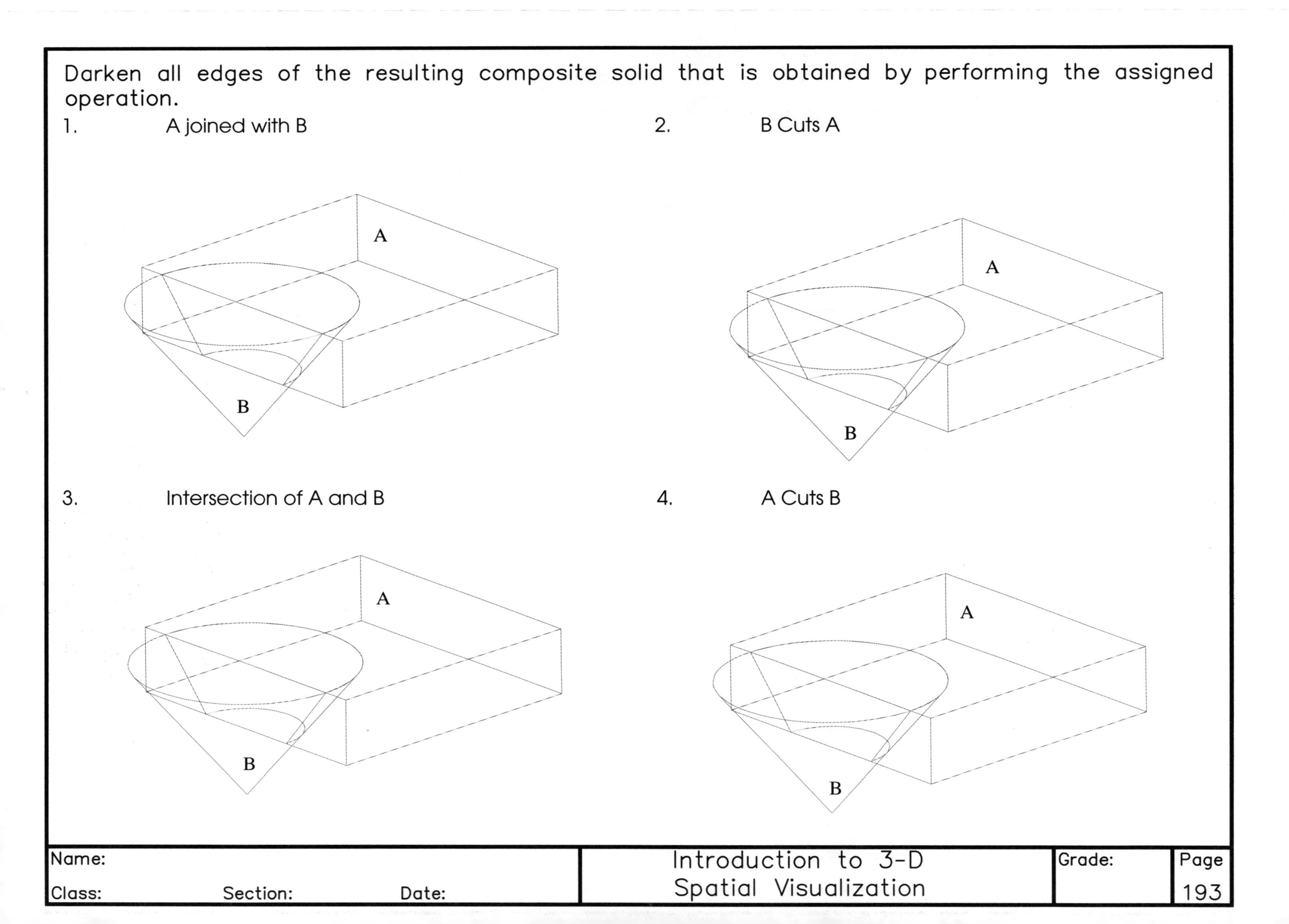
Darken all edges of the resulting composite solid that is obtained by performing the assigned operation.
1. A joined with B
A
B
2. B Cuts A
A
B
3. Intersection of A and B
A
B
4. A Cuts B
A
B
Name:
Class:
Section:
Date:
Introduction to 3-D Spatial Visualization
Grade:
Page
193